VALIDATING CORPORATE COMPUTER SYSTEMS

Good IT Practice for Pharmaceutical Manufacturers

Editor

Guy Wingate

CRC Press is an imprint of the
Taylor & Francis Group, an **informa** business

CRC Press
Taylor & Francis Group
6000 Broken Sound Parkway NW, Suite 300
Boca Raton, FL 33487-2742

First issued in paperback 2019

ISBN-13: 978-1-57491-117-6 (hbk)
ISBN-13: 978-0-367-39856-9 (pbk)

Library of Congress Cataloging-in-Publication Data

Validating corporate computer systems : good IT practice for pharmaceutical manufacturers / editor, Guy Wingate.
p. ; cm.
Includes bibliographical references and index.
ISBN-13: 978-1-5749-1117-6 (hardcover : alk. paper)
ISBN-10: 1-5749-1117-1 (hardcover : alk. paper)
1. Pharmaceutical industry--Automation--Quality control. 2. Pharmaceutical technology--Automation--Quality control. 3. Information technology--Quality control. I. Wingate, Guy.

RS192.V34 2000
615'.19'0285dc21 00-026767

Visit the Informa Web site at
www.informa.com

and the Informa Healthcare Web site at
www.informahealthcare.com

Contents

DISCLAIMER: The information contained in this book is provided in good faith and reflects the personal views of the contributors. These views do not necessarily reflect the perspectives of the contributors' respective employers. No liability can be accepted in any way. The information provided does not constitute legal advice.

For Sarah,
Katherine, and Robert

Preface

Possibly the biggest challenge facing pharmaceutical manufacturers today in terms of computer validation is how to deal with large corporate systems, such as MRP II, LIMS, EDMS and their supporting infrastructure. There has been massive investment by the pharmaceutical industry in these systems. The magnitude of these systems means that the impact of noncompliance in terms of validation could bring a pharmaceutical manufacturer to its knees if a regulator orders a system not to be used. Organizations cannot afford to run stand-by parallel systems, and new systems can take a year or more to institute. Validation must be right the first time. Corrective action in the form of retrospective validation is very expensive, perhaps more than five times as expensive as prospective validation, and rarely meets the high standards of compliance expected by regulators.

Validating Corporate Computer Systems presents practical advice on validating corporate computer systems. I hope readers find the book valuable as a reference. There is a lack of case study material reflecting real-life pharmaceutical industry experience of corporate computer systems. This book begins to address this gap, providing practical validation advice from industry experts. I am indebted to those contributors listed on page xv; biographies and contact addresses are given in Appendix A. Without their willingness to share their hard-won knowledge, this book would not have been possible. I am also grateful to Paul D'Eramo, formerly of the FDA, for his foreword.

Computer validation practitioners seeking case study material for process control systems and analytical laboratory equipment are encouraged to read the companion book from Interpharm, *Validating Automated Manufacturing and Laboratory Applications: Putting Principles into Practice.* This book includes a foreword from the MCA's Principal Inspector of Computer Systems.

Finally, I am indebted again to my wife Sarah for her steadfast patience, love and support during the preparation of this book. Just like the birth of our daughter Katherine during the preparation of my earlier book for Interpharm, *Validating Automated Manufacturing and Laboratory Applications*, the collation of this book occurred over a period which saw the arrival of our son Robert.

Dr. Guy Wingate
Computer Compliance Manager
Glaxo Wellcome

Foreword

Paul N. D'Eramo
Johnson & Johnson
Executive Director
Worldwide Policy and Compliance Management
New Brunswick, New Jersey

During the 1990s, the pharmaceutical industry made a significant investment implementing large corporate computer systems supporting GxP operations. Perhaps most common amongst such applications were Materials Requirements Planning (MRP), Manufacturing Resource Planning (MRP II) systems, and LIMS systems. These systems offer considerable business benefit, establishing common ways of working across a business, efficient inventory management, more responsive sales order management, better supply chain management, more effective management of production information and easier accounting. The return on investment from implementing these systems in some instances repaid their costs in less than a year of operation.

Many of the operations that large corporate computer systems are used to support are covered by Good Clinical Practice (GCP), Good Laboratory Practice (GLP), Good Distribution Practice (GDP) and Good Manufacturing Practice (GMP) regulations. Collectively, these regulations are commonly known as GxP, and they require the validation of computer systems supporting drug manufacture. However, the regulatory requirements for computer validation were developed with process control systems and laboratory systems in mind rather than corporate computer systems. Not surprisingly, the validation of large systems has, for many pharmaceutical manufacturers, been uncertain. Regulators too have also struggled with what they should reasonably expect in terms of validation. Until now, little reference material has been available to professionals implementing corporate computer systems subject to validation in the pharmaceutical industry.

The 21st century will see pharmaceutical manufacturers initiate enterprise IT strategies that can connect corporate computer systems into one overall integrated system. For now, corporate computer systems often exist as islands of automation, perhaps loosely connected together. The challenge is to couple systems tightly and thereby build a single integrated system. Not only will this facilitate more efficient electronic exchange of information across a business, it will also allow a business to move toward a paperless mode of operation. To achieve this, some corporate systems that are not compatible with the strategy might have to be replaced. Equally, new corporate systems might need to be implemented to fill any functionality gaps not covered by existing systems. The high cost of implementing new corporate computer systems means that pharmaceutical manufacturers will need to scope, plan and prioritize new investment in their enterprise IT strategy. With more focus than ever on budgets and the efficient delivery of systems, it is vital that the appropriate level of validation be targeted and that the activity of validation be as efficient as possible.

Validating Corporate Computer Systems: Good IT Practice for Pharmaceutical Manufacturers provides valuable reference material for pharmaceutical manufacturers, suppliers and regulators on Good IT Practice, validation principles for corporate computer systems and case studies on Electronic Document Management Systems (EDMSs), Enterprise Asset Management (EAM) systems, LIMSs and MRP II systems.

The book opens by introducing the concepts behind paperless operations and the practical implications behind realizing such a vision. A stepped transition will normally be required to achieve a paperless operation. It goes on to explain how validation is founded on Good Practice and describes Good IT Practice expectations in the pharmaceutical industry. Such practices are not specific to the pharmaceutical industry. A body of knowledge already exists and is reviewed. Indeed, by collating GCP, GDP, GLP and GMP regulatory requirements and industry guidance on computer validation, it is possible to develop a set of life-cycle practices for the implementation, operation and maintenance of corporate computer systems. Based on recent inspection findings, there are no real differences between the regulatory expectations of the U.S. Food and Drug Administration (FDA) and the UK Medicines Control Agency (MCA). Common sense and documented Good Practice will meet most regulatory requirements. Future inspections of corporate computer systems are likely to become more commonplace, with enforcement more rigorous.

Case studies reflecting practical experiences in the validation EAM systems, EDMSs, LIMSs, MRP II systems, networks and IT infrastructure are presented in next main portion of the book. The well-known V model promoted through the GAMP Guide is adapted in various ways to meet the specific needs of different applications. Checklists are provided that should prove to be useful tools when developing, reviewing or auditing validation documents. Many projects implementing corporate computer systems involve a large number of people, and possibly a centralized project team supporting a number of site implementation teams. Indeed, there may be a separate organization, apart from the manufacturing sites using the corporate systems, responsible for maintaining the systems. These organizational options each have their own benefits and complexities to validation, and these issues are discussed. Planning for inspection readiness is the key to compliance. The end of a project does not mean validation is over. Handover of a project to the support organization must be carefully managed.

The final portion of the book looks at operational compliance, Supplier Audits and the practical implications of 21 CFR Part 11 (the use of electronic records and electronic signatures).

Maintaining the validated status of computer systems is often taken for granted. This topic, however, deserves as much attention as project validation. It is pointless for a pharmaceutical manufacturer to squander its project validation efforts by underresourcing or not establishing a robust Quality Management System (QMS) to support the operational compliance of its systems. Recent industry experience with the millennium bug highlighted the financial impact on a business if it has to replace a corporate computer system within a short time-frame. Regulatory noncompliance may require systems to be replaced, and one thing is almost a certainty—fast-track projects with insufficient planning are usually very inefficient and costly. It is not an unreasonable regulatory and, indeed, shareholder expectation that pharmaceutical companies operate their corporate computer systems in a state of compliance.

Principles for Supplier Audits are well established, but case study material is rarely presented from a supplier's perspective. In the next to last chapter of this book, the world's largest MRP II supplier presents a perspective on audits from customers, and a pharmaceutical manufacturer looks at the audit requirements for hardware platforms. During the late 1990s, SAP AG saw a dramatic increase in the number of Supplier Audits being requested as pharmaceutical manufacturers realized their regulatory obligations to evaluate the quality of critical software products supporting corporate computer systems. To address the issue of numerous customer audit requests, supplier firms are encouraging joint audits by groups of pharmaceutical manufacturers. This approach is currently being explored by the PDA as a possible future standard to increase audit efficiency to auditors and suppliers alike. Of course, the audit requirements for the hardware platforms supporting corporate computer systems are equally important to those concerning software and have their own particular needs.

The concluding chapter of the book discusses probably the most contentious topic currently facing validation in the pharmaceutical industry, that of compliance with 21 CFR Part 11. The GAMP Forum issued draft guidance to solicit feedback at the end of 1999, and the PDA have established a task group to look at providing advice to industry professionals. This last chapter presents current thinking on how to manage the assessments and remedial actions that will undoubtedly be required to bring systems into compliance. The needs of both existing and new computer systems are considered.

This book includes a wealth of knowledge concerning the validation of corporate computer systems contributed by 23 experienced industry professionals from around the world. Those involved in validating, auditing and inspecting such systems will learn much from their lessons and advice. There is currently no comparable source of guidance available, and I highly recommend this book to you.

Contributor List

AUTHOR AND EDITOR

Guy Wingate	Glaxo Wellcome, England

CONTRIBUTORS

Roger Dean	Pfizer, England
Anton Dillinger	SAP, Germany
Paul N. D'Eramo	Johnson & Johnson, United States
Christopher Evans	Eutech, England
Howard Garston Smith	Pfizer, England
Heinrich Hambloch	GITP, Germany
Norman Harris	20CC, England
Kees de Jong	Solvay Pharmaceuticals, The Netherlands
Robert Kampfmann	SAP, Germany
Erna Koelman	Solvay Pharmaceuticals, The Netherlands
Richard Mitchell	Pfizer, England
Gert Moelgaard	Novo Nordisk, Denmark
Kees Piket	Solvay Pharmaceuticals, The Netherlands
Brian Ravenscroft	AstraZeneca, England
Chris Reid	Integrity Solutions Limited, England
Tony Richards	AstraZeneca, England
Owen Salvage	CSR, Australia
David Selby	Selby-Hope International, England
Nicola Signorile	HMR (Gruppo Lepetit), Italy
Melanie Snelham	Eutech, England
David Stokes	Motherwell Information Systems, England
Torsten Wichmann	SAP, Germany
Peter Wilks	Glaxo Wellcome, England
Guy Wingate	Glaxo Wellcome, England

OTHER INTERPHARM PRESS PUBLICATIONS BY THE AUTHORS

Howard T. Garston Smith, *Software Quality Assurance: A Guide for Developers and Auditors,* 1997.

Teri Stokes, Ronald C. Branning, Kenneth G. Chapman, Heinrich Hambloch, and Anthony Trill, *Good Computer Validation Practices: Common Sense Implementation,* 1994.

Guy Wingate, *Validating Automated Manufacturing and Laboratory Applications: Putting Principles into Practice,* 1997.

1

Paperless and Peopleless Plants: Trends in the Application of Computer Systems

Gert Moelgaard
Novo Nordisk Engineering A/S
Bagsvaerd, Denmark

Paperless operations have been heavily promoted in recent years within the pharmaceutical industry. However, only a few companies have been successful in the transition toward paperless operations. The reasons for this are manifold. There are significant factors: little of the promising technology is mature enough to facilitate a low-risk, large-scale, corporate roll-out, and the regulatory requirements of the pharmaceutical industry have raised questions on "how much needs to be done" to ensure the compliance of a paperless pharmaceutical operation. There are few published, real-life case studies that report clear financial benefits from implementing corporate/enterprise computer systems.

Until recently, paperless manufacturing operations were held back in the pharmaceutical industry because the U.S. Food and Drug Administration (FDA) did not accept electronic signatures. This obstacle was formally removed when the U.S. GMPs (Good Manufacturing Practices) were enhanced to accept electronic records and signatures under specified conditions by

21 CFR Part 11 on 20 August 1997. Some uncertainties in the interpretation of certain specific issues in the rule, however, continue to discourage the pharmaceutical industry from embracing the use of electronic signatures.

When the industry started discussing the application of paperless systems and electronic signatures more than 10 years ago, the issue of electronic signatures seemed very specific to the pharmaceutical industry and a few other critical businesses. In the meantime, electronic transaction has evolved into a mainstream issue as the rest of society is rapidly transforming itself toward electronic commerce and other paperless business processes. Now key issues from the pharmaceutical discussion of electronic signatures and records have become mainstream society challenges, for example, trustworthy electronic signatures, security of electronic transactions, risk and benefits of encryption technology and so on. International agreements on electronic transaction standards will be needed and may make some of the long-discussed pharmaceutical security issues almost obsolete as they become part of general regulation throughout the industrialized world. What used to be an area with very specific pharmaceutical industry needs now has turned into common issues for mainstream business of the future.

As part of a highly regulated industry, pharmaceutical companies have high requirements on documentation in all important areas of their operations. All companies have realized the potential of paperless operation, not only improved control and potential cost savings but also the rest of the industrial society is rapidly evolving toward more paperless operation.

This in turn implies that technology is developing very fast, not least due to the huge infusion through Web technology making computer systems throughout the world easily accessible through fairly uniform operator interfaces and information access methods. It also means that the rest of society is turning toward paperless thinking as part of everybody's daily life. Children grow up with computers and start using them for games and education early on. Education is being changed to prepare young people and older, more experienced people to work with information technology (IT). Most people in highly industrialized countries have access to computers not only at work but also at home as a natural way of doing things. With access to the Internet, people start to expect more and more information electronically rather than on paper because electronic media have up-to-date information for everything that can reasonably be put into a computer system. As more information becomes available on the Internet, electronic access is evolving from desktop programs, E-mail and simple Web-access browsers to on-line radio broadcast, TV broadcast, voice mail, video telephones, and so on. This in turn will lead to a much broader application of personal computers (PCs) in many areas of our daily life.

Thus "paperless operation" is far beyond the point of being a special focus area for the pharmaceutical industry, as it has become one of the most intensively discussed focus areas in industry generally. We are moving more and more into a digital world.

THE "HIDDEN FACTORY"

Said with some cynicism, a typical pharmaceutical operation produces both products and paper. The paper production is the "hidden factory" which some people estimate to be as costly as the "real factory". Documentation of an operation must be an integral part of the operation itself. This is, of course, a huge simplification, but it illustrates a very important point. The physical input and output from an operation not only includes raw materials and finished products but also documentation. Inside the operation, the "hidden production" of paper have many similarities to the "real production" of products (Figure 1.1).

Documentation in the hidden factory has its own production process. The raw material of paper is prepared into documents and data entry forms as production plans, master production and control records, batch records, raw material sheets, Standard Operating Procedures (SOPs) and so on, which in turn are enhanced with more and more manual data entry and signing at each step of activity with approvals. The paper flows from desk to desk as collections of related documents. There is a huge "add ingredient" or "assembly" process of data to the physical medium, which at each step of activities has to be at the right place at the right time—just as the physical products.

Thus, the "hidden product" of paper also has its own logistics. One needs to define "raw material" storage, dispatching, logistics flow paths and "intermediate" storage with some buffer capacity, where the papers are out of

Figure 1.1. Two Parallel Production Streams

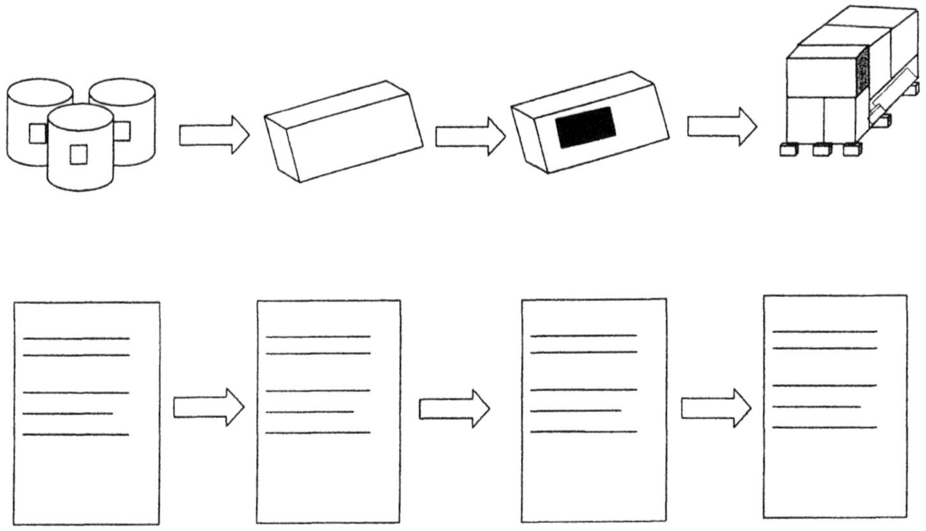

activity and waiting for the next process. If documentation is temporarily removed from its flow stream, then it will cause delays for the next operation. This part of the document's logistics is not planned. The workflow and handling procedures of documentation usually get much less attention than the "real" product. Yet inefficient document management can incur significant costs in the stock of in-process papers waiting for approvals, transportation, archiving and so on.

Quality management for documentation should follow the same principles as quality management for manufacturing the actual drug product. As documents are prepared and copies made, they must be verified. At each production step, a person must find the right sheets, record data and sign; critical data requires another person to be present to verify and countersign the entry. Each person responsible for an activity may look back to the previous data entries and then carry on with his or her own. Finally, quality responsible people will be looking through at both the quality of the production that the documents represent and the quality of the documents themselves. Thus, the paper product has its own quality criteria, sometimes less consciously treated than on the real product.

This hidden production with its execution, logistics and quality management is what the paperless operation replaces. Therefore, the hidden factory needs to be exposed—both to set the goals for the paperless operation, which should be based on a streamlined version of the existing hidden factory, and to set the benchmarking goals that are important for evaluating the business case of a paperless operation on both the cost and benefit side.

THE SITUATION WITH PAPER

On top of the actual cost of running the hidden factory, there are some serious disadvantages of paper media, which up until recently have been accepted without question because there were no alternatives.

The paper-based operation is based on a split between execution and approval of operational steps in the pharmaceutical environment. Paper is an unintelligent medium. The real execution steps are separated from the approval steps that cause both a delay in time and a supervision of the production status, which is far from real time and can lead to a bottleneck in the production process.

Workflow is not transparent. As paper documents exist only as a physical media during the production process from which data can be read only by physically handling the actual papers, only one person at a time can get insight into the content of the information. Additionally, the information work has to be conducted in a strictly serial manner, in which each operation awaits the paper to enter the area of that particular operation. This in turn requires the information to go through a stream of waiting points for each processing step; as the information processes are an integrated part of the production progress, these waiting points implies similar delays on the overall processing lead time.

Paper-based information cannot be easily reused. Although papers may be copied, they can be used only as the specific type of information they contain. In contrast, electronic information may be copied from one information type to another immediately. For example, a series of batch records may be converted immediately into a trend study if they are contained in a database; if they are on paper, a separate work activity is required to establish overview trend evaluations. The consequence of this is very poor information management, in which there are only few ongoing overview activities on the information, which in turn is a huge waste of the potential compared to what could be deducted from the many observations.

To summarize, paper is an excellent short-term choice because

- it is flexible,
- people feel comfortable with papers,
- it is low tech (a simple medium that can easily be afforded),
- paper handling procedures are well known and
- archiving is almost a non-issue as it is a well-known standard to anyone.

However, in a longer perspective, paper is bad because

- it is unintelligent,
- it is difficult to make significant improvements,
- there are no technological advantages,
- there need to be too many manual procedures,
- there is no easy retrieval method if going across archiving indexes and
- recorded data cannot be reused for other purposes such as data analysis without significant manual intervention.

WHAT IS "PAPERLESS"?

"Paperless" is possibly one of the most oversold technology concepts currently being promoted. There are no signs indicating that we are turning into a truly paperless way of life. Paper consumption throughout the pharmaceutical industry, as in other industries, has continued to increase. We are still consuming more and more paper with the more computer systems we apply. Companies have thought of themselves as implementing the paperless office for many years, but the more computers they put in, the more papers these systems have tended to create. With new computer systems facilitating access to printers, people have tended to take paper copies by default without thinking if they really need a print, perhaps as a backup or just because they feel more comfortable with this traditional medium. It may not remain so forever, but this is definitely the current situation.

Thus, though it may seem that the term *paperless* is a natural and obvious concept that has become a natural part of our highly computerized life in today's industrial society, this is really not the case. In order to describe paperless operations in the context of a corporate/enterprise computer system, it is important to be specific on what is meant with the term *paperless*.

There is an important distinction between whether "paperless" means *no paper* or *less paper.* The distinction is important when setting up goals for paperless facilities, whether they are changes in an existing operation or implementing new green-field operations. The concept of paperless meaning no paper is a total concept, whereas less paper can be implemented step-by-step by eliminating islands of paper-based operation. When companies decide to implement a paperless operation in a certain area, they typically mean to reduce the amount of paper by eliminating some (or most) of the paper-based systems that were previously used; they mean "less paper" rather than "no paper". It is an important distinction because there is a huge difference in complexity between a situation where much of the traditional, paper-based activities are retained in certain areas and a situation where everybody is expected to manage all information by means of electronic media.

We will introduce a similar distinction between a *paperless operation* as a total plant concept and a *paperless activity* as individual step changes toward the concept of a paperless operation. Both are highly relevant in the real world, but there may be quite significant differences in what can be achieved by a paperless operation and a number of paperless activities. A paperless operation is established through a "big bang" project that may be established in more than one step toward a cautiously planned paperless goal. A set of paperless activities may add up to a paperless operation, but typically an operation with paperless activities will remain a hybrid until a major change project turns it into a paperless operation. In most companies today, the concept of a paperless operation is achieved through a few projects for paperless activities in selected areas. So far there are only few examples of truly paperless operations.

"Paperless" can be obtained by different means, not all involving computers, but all important when reshaping an operation. Paperless can involve the following:

- Omit redundant paper-based activities, for example, identify certain steps and activities which are maintained for historical reasons that are no longer valid. Paperless projects conducted as full-scale Business Process Reengineering (BPR) projects typically identify many such activities that have continued to exist due to a lack of organizational transparency where a certain report or chart is kept in one department because it is believed to be important for another department.

- Replace paper with a different medium, e.g., computers, which is what people typically have in mind when talking paperless. The immediate example is simply to transfer procedures and documents in a paper-based operation into screens and reports in a computer system, but it can also be the replacement of some procedures and instructions by drawings, video clips and other electronic media which replace the paper.
- Replace paper by doing something else, for example, identify that a certain activity creating papers can be replaced by a different activity. This can be a manufacturing process step which is changed from manual in-process control based on control charts to integrated measurement with full control of the actual quality parameters and reject of defects. In many cases, it is more effective to entirely automate an activity rather than to support the manual operation with electronic controls. Similar examples are found in many areas of automation, not only in traditional production processes but also in laboratories, materials management functions and warehouses.

All three aspects are worthy of consideration when moving toward paperless operations, especially since paperless projects are in danger of focusing too heavily on technological opportunities to the detriment of the original intent of going paperless. The purpose should not only be to replace paper-based ways of working by computers but to create a streamlined, competitive and manageable environment that enables and motivates people to cooperate across physical and organizational boundaries with a common goal of achieving competitive, well-functioning business processes.

PAPERLESS AND PEOPLELESS?

A controversial area over at least the last 20 years on the topic of implementing computer applications has been the replacement of manual jobs with computers. What are the impacts of paperless operations from an employment perspective? And to be more specific: Do paperless operations imply peopleless manufacturing plants?

The answer is typically both yes and no. Paperless manufacturing plants will not be totally peopleless plants, but most of them will have lower manning levels than traditional plants. The transition toward paperless operations should be seen as part of a larger transition throughout the industrialized world toward more streamlined supply chains with more customer focus, less lead-time and more efficient operation. This in turn allows companies to apply computer applications as part of a larger changeover in the way they conducts business.

Management philosophy and operating requirements are changing concurrently with the progressive introduction of computer applications, each influencing the others. Industry trends toward lean manufacturing, team-based production organizations and application of Kaizen-inspired principles of operator empowerment on the plant floor are encouraging a broader and more process performance approach to manufacturing facilitated by computer applications. Paperless systems enable real-time overview of processes and product status that integrates the quality focus into the direct control of products.

Roughly speaking, there are two ways of applying the benefits of paperless operations:

1. Conduct "business as usual" but replace some of the paper (e.g., records, documents and procedures) that people use during their daily work. This will enable people to work paper-like on computers with little impact on the company's business processes. Less paper should mean fewer people to run the business but only marginally in this case. There may also be less data entry and hence fewer data errors.
2. Use the transition toward paperless systems to facilitate operator empowerment. Operators are allowed to take a broader responsibility and to use electronic media as an enabler to supervise performance and quality of their production. This enables in-line control of quality and deviations. Multiskilled, team-based working may also be introduced with the aim that broader job roles will make work more challenging and interesting. Significant lower manning levels are usually expected. All this must be achieved with due consideration of current GMP regulatory requirements.

Today's typical pharmaceutical business has many people working with paper-based information throughout its supply chain. The operations of the future may involve fewer people, but, on the other hand, a more streamlined and dynamic operation, for which the paperless operation concept can be an important enabler, will probably be a necessary condition for business survival.

However, the consequences on the technical side of the transition to paperless operations should not be underestimated. Paperless systems are normally much more complicated to operate and maintain from the technical side than traditional equipment. Thus, an operation must expect to have more staffing of competent technical people to support the operation and maintenance of the systems and to have clearly defined maintenance agreements with their suppliers.

The concept of paperless operations is not just an issue of replacing manual information operations with automatic, electronic operations; it is a challenge of transformation toward a vastly changed organizational setup where automation no longer just replaces human labor but also provides a much more information-focused production organization.

"PAPERLESS" DRIVERS

What makes companies go for paperless operations? Paper is one of the most important cornerstones in the pharmaceutical industry today for several reasons.

Documentation is a critical and strongly emphasized aspect of pharmaceutical regulation throughout the world. It is clearly described as a requirement in most areas of GMPs, Good Laboratory Practices (GLPs) and so on, all requiring the pharmaceutical company to document its practices, qualifications, results and so on, as a key part of regulatory compliance. Today, "document" means establishing paper documents and paper-based routines for each of these activities. Paper is and has been important in most industries. The pharmaceutical business is one of the most paper-intensive industries by nature due to its quality focus and regulated requirements.

Paper-based documents are very easy to work with and are a natural part of our daily life. It should not be forgotten, however, that insufficient operator literacy and attention to detail can pose a significant problem which may compromise the implementation of paper-based activities. It requires training to enable people to strictly follow written procedures and to document their observations and activities on paper. Special training is required to document according to GMP requirements when dealing with corrections, signatures and so on, but the basic skills are normally in place. Such training is not very difficult to implement.

Although it seems straightforward to establish a paper-based system, most organizations have realized that it is both expensive and very resource consuming to maintain a proper and foolproof, large-scale paper-based systems. The initial step of putting things on paper is easy, but setting up a system that ensures that paperwork is sent to all relevant persons, outdated documents are withdrawn, critical papers are being archived and so on is both an exhaustive task to establish and very resource consuming to operate. More and more companies throughout industrialized society are realizing this—if not before implementation, then at the point where they establish well-controlled, paper-based systems as a logical consequence of the GMP regulations as well as of quality systems based on, for example, the ISO 9000 standards.

Paper is also a slow and primitive medium. Paper as technology has undergone a huge development since its origin as a medium for handwritten information. Since Gutenberg established printing technology around 1440, replacing the tedious work of copying through writing with the imprinting technology and thus changed paper from being an information conservation medium to a true mass medium, the technology has undergone only few major changes (e.g., the typewriter, the copying machine and the telefax). There have been other technologies such as punch cards which were transitory and are no longer widely used. Paper is a physical medium and in a sense very similar to the product of a manufacturing company: Information must be handled and managed as physical items just as the main product. Compared to electronic information technology, the physical attributes of paper technology can be a burden.

While all this and more can be said about paper's insufficiencies, it must be realized that paper is not entirely avoidable. Most paper can be replaced; in fact, almost all paper can be avoided, but there will still be some left. We cannot live without paper, and even so-called paperless operations have many printers. To manage paperless operations in the real world means managing a hybrid system of paper and electronic media. It is still worthwhile to call it paperless but to remember that some paper always sneaks in.

On the other hand, the concept of paperless operations contains some promising advantages. Some of the most important attributes driving paperless operations are cost based, but there are also some other important advantages.

Properly implemented paperless systems can reduce deviations during operation. Paper is a dumb medium that receives everything people record on it, but computer media have the potential of guiding or controlling operations as well as tracking deviations and errors directly during the operation. In properly implemented paperless systems, this has lead to clear deviation reductions for people, processes, materials, equipment and lead time.

Due to the nature of information technology, paperless operations can obtain significant savings on the cycle time, thus improving the utilization of the capital invested. This is not just by implementing computer systems to replace paper-based operations but also where automation and computers are applied to integrate business processes such as planning, manufacturing, distribution, reporting and so on without any requirement for paper. This in turn enables shorter lead times, inventory savings both in operation and in stock, operation transparency and other advantages—all included in the concept of "lean manufacturing".

"PAPERLESS" CONTROLS

When dealing with paperless operations, there are several issues that are self-evident in the paper-based world but must be dealt with specifically to ensure the control of documents in the electronic world. When implementing a paperless operation, these issues must be carefully evaluated and resolved, preferably with a standard solution across an operation or whole company. It is relatively easy to modify paper-based systems compared to electronic systems. In electronic systems, requirements on process flows must be carefully considered and strategic decisions made. Some of these important decision areas are listed below.

Document Life Cycle Activities

Documents have a life cycle from creation to destruction. In the paper world, we take for granted that each physical instance of a document is created, distributed, archived and destroyed. In the paperless world, it is also important to decide what steps the document life cycle consists of, i.e., what stages a document can be in—"created", "reviewed", "approved" and "archived" as an example (see Figure 1.2).

Roles and Responsibilities

As mentioned in several of these focus areas, there is a need for specifying not just individuals but also their roles and responsibilities. This will be based on the representation of the organizational positions, but there is much more to

Figure 1.2. Example of Document Life Cycle

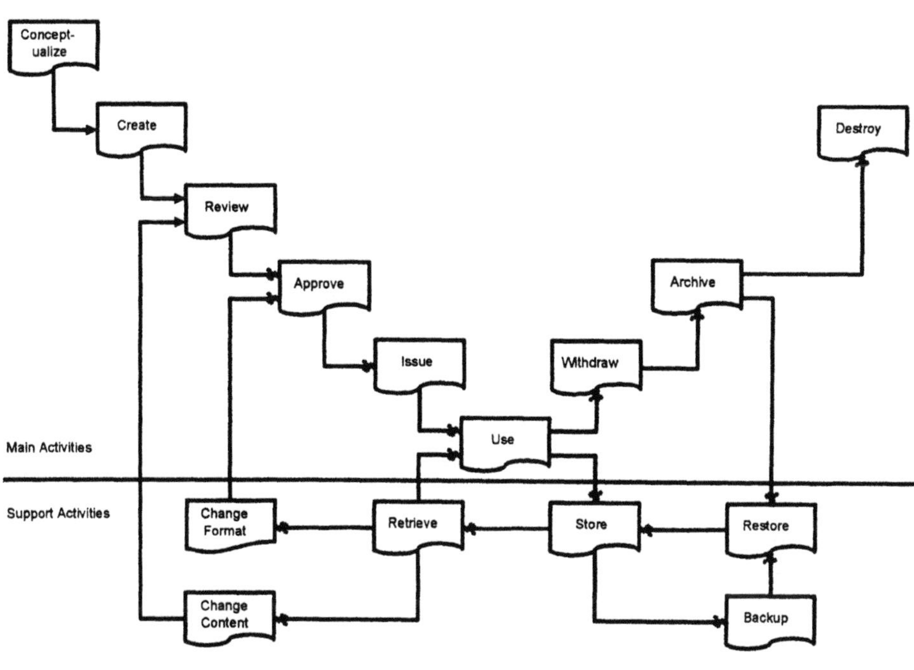

it because most people have different roles on different document types and even on individual documents. For example, a person may be the "author" on one work instruction, "reviewer" on another and "approver" of yet another document. This is another area where planning is most important but often difficult, but it is an important cornerstone for a successful implementation of a paperless system.

Document Types and Categories

Types of documents may be "procedures", "instructions", "interoffice memos", "master production records", "batch records" and so on; there are many document types in a modern organization. The types of documents need to be described and categorized into specifically agreed categories. There is a balance between using a few general categories and using specific terms that people understand in their daily work.

Document Status Categories

A document will have a certain status during its life cycle, which will change with different activities. For example "draft", "approved", "released", "withdrawn" and so on must be defined if they are not already standardized in a paper-based system.

File Formats and Portability

A separate and complex technical issue is the standardization of file formats. The configuration management issues described above and the electronic storage of documents over a long time with certainty will make previous file formats obsolete. There are two important aspects:

1. Electronic documents must be viewable and printable with all their attributes throughout their retention period and available in a native version that enables update, reuse and so on. The first part may be addressed by using special file formats, so-called portable formats, that exist as both international standards (such as Standard Generalized Markup Language [SGML] format) controlled by the International Organization for Standardization (ISO) or the XML (Extensible Markup Language) format controlled by the World Wide Web Consortium) and proprietary standards (such as the Portable Document Format [PDF] owned by Adobe).

2. Software types, versions and file formats can change rapidly and significantly over time. One software vendor may guarantee compatibility between different versions of its software but only as long as the vendor exists and often only to a limited degree. In real life, the preservation of an electronic document during software updates can

require significant effort, e.g., maintaining tables or graphics in a document after a software update. The impact of this issue is set to increase as more complicated data types can be embedded into documents, thus making them dependent on not only the document creation software but also the software used for the embedded document objects.

Document Supporting Activities

In addition to the life cycle stages, supporting activities for a document must be defined, i.e., what can happen to a document at all stages of its life cycle, for example, storage, retrieval, backup, format changes and so on. This may not be inherent in a paper-based system but may need to be specified when implementing a paperless system.

Document Distribution and Recipients

The issue of managing document distribution is very different in a paperless operation where the transportation of documents no longer involves physical activities. As transportation and distribution are one of the key areas where the electronic world is superior to the physical paper world, one may believe that distribution is a minor issue. Everybody familiar with the World Wide Web knows that distribution of a new Web page is just a question of putting it on the Web server and making it accessible. Here distribution is no longer an issue. It is not quite so easy when implementing a paperless operation; distribution methods and ensuring recipients use controlled copies must be carefully planned. An important issue which clearly differs from the physical world is that individuals and groups will need to be named not just by the name of the individual persons but by the role they have in the organization of the distribution setup for the paperless operation. There is a need for more general and abstract planning, which complicates implementation of a paperless system compared to the paper-based system.

Configuration Management

Everybody using computers has experienced the problems involved in just printing a document from somebody else, from a previous version of a program or on a different printer. Though solutions are available which guarantee printable file formats across computers, programs and printers, the configuration of hardware and software must be carefully managed in a paperless operation. There will be a need for configuration management of computer systems to ensure that the applied hardware and software for not only the individual computers but also the network, printers, file formats and so on are kept in control. Computer systems also change—that is a basic fact of life. The challenge is to control how changes are managed in the paperless operation at all times. Documents live for a long time (much longer than software versions). Thus there is a need for migration plans to deal with the transitions of software versions. This challenge is much more evident with computer systems than with paper documents, where it is merely a question of using durable printing and writing technology and storing the documents under appropriate physical conditions.

Updates and Notification

In a paper-based operation, documents are typically distributed by what could be called "push technology": They are sent ("pushed") actively to their recipients. Though some documents may be distributed on a request basis ("pull"), these are few in normal organizations. Most new or updated documents are physically sent to people. For controlled documents, where the recipients are to replace documents in binders, the real-life experience of the organization's alignment may be frightening. In the paperless world, updates may be much simpler and much better controlled. However, where electronic documents are distributed both by push and pull, for example, sent by email or copied from a file server or a Web page, the methods for updating and notifying affected users will need to be specified. We must avoid users continuing to use out-of-date document versions when they potentially might not be aware that an update exists.

Workflow

The actual path that various documents must follow in the organization may or may not be fixed in a paperless operation, but it must be decided whether workflow automation is within the scope and, if so, how it should be implemented. Again, it is different from the paper-based world in that such definitions will not just be for specific persons or specific documents but on the roles of the persons and on the categories of documents. Possible workflow paths for the various document types and document phases will also need to be decided. Paperless operation may be achieved without tying electronic documents and their management to a workflow. When doing this, it is important to retain flexibility.

Practicalities

In today's paperless operation, it may be necessary to keep two versions of each document—one in the portable file format (which cannot be edited) and one in the native format of the creating program (which can be edited). Future technologies may change this as the issues become important in more and more areas of society and as such technologies as Hyper Text Markup

Language (HTML) and XML on the World Wide Web provide more standardized solutions for both portability and object embedding.

Many of the issues mentioned in this section put requirements on the attributes used to characterize the individual documents. In the paper-based world, it is comparable to the content of letterheads, page headers and footers and so on that are standardized for each document and across documents. Thus, it must be decided what the characteristics of each document type will be, including, for example, document title, author, approvers, dates, status, version, language and so on.

There are a number of system implementation decisions that need to be made. The computer system project will involve system requirements, project organization, the selection of suppliers, software, hardware and system network configuration, user access points and user access rights, systems integration, acceptance testing, validation and so on. These aspects of system development are discussed later in this book. Ironically, such project activities typically build up a huge pile of paper, even for projects in paperless facilities. Hopefully this situation will change.

"PAPERLESS" IN THE GxP ENVIRONMENT

The importance of documentation is obvious in the pharmaceutical industry, as there are both regulatory reasons and good business practice requirements for documentation in almost all areas of the pharmaceutical business.

The requirements of the pharmaceutical product and its effects must be documented from discovery through clinical testing to final production. Both regulatory requirements—the amount of documentation this involves—and the time-to-market pressure on the whole pharmaceutical industry have made most pharmaceutical companies very aware of the potential benefits of electronic document management and paperless operations.

GxP regulations on production and quality control involve a lot of documentary requirements. The whole life cycle of the product is impacted from raw materials receiving, storage, release, planning, production, process control, cleaning, testing, approvals, release, shipping and so on, which is to be controlled by written procedures and carefully documented during execution. This in turn involves both material management from receiving through every production step and the execution of each production process controlled by master production and control records and documented in batch records and other production data.

There are documentary requirements for a whole range of supporting activities across the pharmaceutical operation, including organizational charts, job descriptions, skills and training, operating procedures, quality management, equipment calibration, maintenance, qualification and validation activities and so on. For many of these supporting activities, there is a potential benefit in having a paperless operation across departments and sites, as they are typically controlled by similar means across the company.

Though many of these areas in practice have been done electronically for several years, using traditional office automation tools, e.g., word processors and file servers, the rush towards putting them into an integrated environment is fairly new. Most pharmaceutical companies have started implementing Enterprise Resource Planning (ERP) systems, Electronic Document Management Systems (EDMS), clinical trials systems, Laboratory Information Management Systems (LIMS) and so on, but few pharmaceutical companies have established a successful integration between these corporate systems, thus taking the step into paperless operation.

The emergence of Web technology is a great enabler. It has given various electronic documents a uniform environment and user interface. Intranet systems will lead the way in simplifying all areas of a paperless operation as part of a transfer into Web-based systems, providing a uniform accessibility to relevant, corporate-wide, real-time information throughout the paperless operation of the future corporation.

The immediate benefits of paperless activities are not only in document management and databases but also in automatic identification of materials, equipment and personnel, automated materials handling, automated production execution and so on—a whole range of enabling systems and technologies for the paperless operation. Thus "paperless" is not just a question of implementing the electronic management of data and documents but also electronic identification methods (for example, the selection of bar coding standards), control system strategies, equipment supplier cooperation and so on in each of the production areas throughout the supply chain of pharmaceutical products. This includes traditional automation and new areas of control such as electronic inspection and in-line printing of packaging materials (an unavoidable paper part of even a paperless operation). There are significant opportunities to simplify the large number of variants on international packaging materials, formats and content and their reconciliation checks. A truly integrated environment should reap the rewards of reduced lead times to develop a drug product coupled with more efficient manufacturing facilitated through the clearly defined and transparent document storage structure, real-time access from all participating persons to all relevant information at any time and structured data management.

The paperless operation does not limit itself to the borders of a company. Communication with customers, suppliers, partners, service providers and so on is being equally reshaped, especially through the application of Internet technology and the initiatives on electronic commerce, including, for example, Electronic Data Interchange (EDI) and Extensible Markup Language (XML). More and more companies are giving each other access to feed and pull defined business information, and many are seeing the advantages of close communication with their customers and suppliers.

GOING PAPERLESS

Acknowledging the advantages of paperless rather than paper-based operation, it should be realized that the benefit of a paperless operation can only be truly exploited through top-down planning, similar to the concept of BPR. An operation going paperless needs to map what the workflow is, what papers are created and how they interact. Like a BPR project, this needs to be addressed by a cross-functional effort mapping the present situation, the envisioned future and the steps to be followed to get there.

One successful way to address it is through a series of cross-functional workshops. One simple tool which has proved highly efficient to achieve this is the concept of so-called "paper-wall" sessions. By establishing a "paper wall" of very large paper sheets, the work group establishes an overview of the present situation by putting the existing paper templates on the "paper wall" and describes how the individual documents progress through the various operational steps (see Figure 1.3). Contrary to data architecture–oriented approaches, this approach is easily understood by most people and enables all participants in a workshop, whether or not they are familiar with data analysis, to participate actively in the mapping.

During paper-wall sessions, the existing work and document flow can be changed to take advantage of options made available through a paperless operation. Although individual processes may be optimized, the main outcome

Figure 1.3. Paper Wall Session

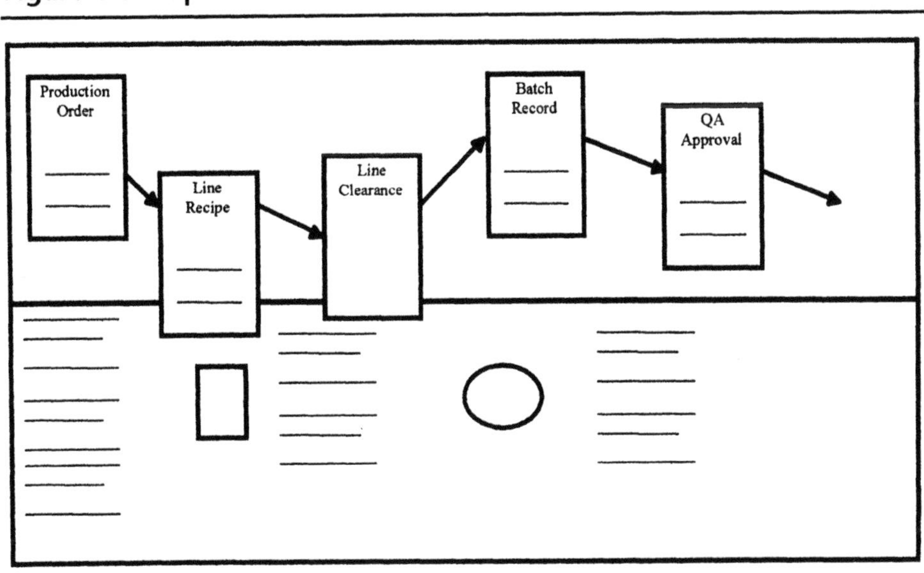

is that the overall flow can be greatly streamlined when the key properties of paperless information are taken into account. With creative thinking turned into planning the top-down planning, may be combined with the bottom-up experience from the daily operation.

Paper walling is only one method among many available, but it cannot stand alone. The operation of an existing as well as a streamlined operation may be illustrated in many ways, inspired by formalized information and logistic architecture methods. The main objective of ensuring a lean and streamlined operation will have to be translated into system requirements and in turn a system design that enables the actual system to be built. A key success criterion is how well the outcome in the computer system design reflects a user-friendly and streamlined flow of operation and information in the paperless facility.

GOING PAPERLESS WITH THE COMPUTERS OF TODAY

The change from paper-based operations to paperless operations may have more advantages than are immediately apparent. Computers provide an interactive medium that provides a fundamentally different way of working than the passive medium of paper. They can respond back on data entries, guide the operator and ensure dynamic updates with up-to-date information throughout the company.

Furthermore, computers have now become a much more "rich" medium than paper ever was. Within the last 10 years of development of the Graphical User Interface (GUI), the technology has evolved from a mere "what you see is what you get" (WYSIWYG) concept into an intelligent and multimedia-based environment. Today's computer systems can replace papers in the streamlined operation and incorporate a much more functional and intuitive way of describing process activities and outputs. A rapid development of standard systems interfaces, including Web technology, has enabled systems to contain more automatic updates and much more intuitive information access methods than the paper-based medium ever could.

Thus planning for paperless operations today poses the challenge of empowering operators with a much better insight to the process, the required operational sequence and the related information than would ever be possible in the paper-based world. Computers can make use of information resources and description methods other than the traditional text to bypass many of the previously mentioned problems of instruction through written procedures. Modern computer systems not only streamline the information flow but also improve the understanding and quality of operations to be executed. Trend displays may provide an overview that no paper-based system comes close to, and embedded video clips may describe operations in a way that even the best written procedure could never achieve. This not only applies to daily operations but also to more rare tasks such as maintenance operations on equipment that previously required long training and very specialized personnel.

In this vision of paperless operations, the main challenge may not be the computer system technology but the operators' ability to understand and manage the information available in a way that makes sense for the business. This requires very conscious management decisions on the scope of "going paperless". Although the rich medium of modern computing contains many useful options, the goals and especially the business drivers for a project and the daily use of the project deliverable are key. Many of the options presented by modern technology seem very useful. It must be remembered, however, that when aiming for paperless information management, 80 percent of the mainstream advantages may be easily exploited, but the last 20 percent of what it takes to become truly paperless may be very expensive.

LEVELS OF A PAPERLESS OPERATION

Paperless operations may be handled in many different ways. These differ by the level of ambitions of the system and by the intelligence put into the computer system. The decisions on how far to go may be helped by a simple and pragmatic categorization of four main levels of the paperless operation.

Passive "Paper on Glass"

The simplest approach to paperless operations is to replace papers by the electronic medium of computing, putting existing procedures and forms into a computer system and making them accessible to the operators. This approach exploits only a small fraction of the potential of computer-based processing but nevertheless holds many of the quick win advantages of getting rid of the paper medium. The term *passive paper on glass* can be used to describe systems where existing SOPs and operational documents such as the master production and control documents are directly transferred into electronic documents by which the operation is being executed. This typically implies only a few changes to the workflow but holds the advantage of being fairly easy to implement; it may easily give some of the advantages of paperless operations because it makes documents readily accessible for more than one person at a time.

Active "Paper on Glass"

The difference between active and passive paper on glass is that not only is the paper-based information transferred into the electronic medium but some of the important properties of information, such as calculations or quality limits, are enforced actively within each document. This approach will capture some of the most common errors of data entry, giving immediate feedback on compliance to quality requirements that cannot be ensured through the paper-based execution. It also implies verification of the data that can be used, for example, to determine that the operator identifies the correct materials, equipment or persons, thus enabling a level of ongoing compliance that cannot be achieved without computers.

Active Control

Active control implies that the "paper on glass" documents are no longer static. Data may be collected automatically so that only a few of the data entries are manually entered, most of them are entered through automatic data capture, possibly with a border value check and some kind of compliance enforcement. This requires a certain level of integration with physical data equipment, such as control systems, instruments, bar code readers and so on, which may prevent common errors from manual data entry and additionally provide enforcement of basic quality requirements. Active control requires the paperless system to be a true application that actively assists the user's data entry with automatic data collection and possibly interfaces to other systems.

Proactive Control

The highest level of a paperless operation not only captures and controls data entry but also provides the operator with the means for ongoing automatic supervision, an overview of the present state, and predicting future states or events if no intervention is made. This level may include statistical tools or integration with simulation models to give a proactive approach to the control of the relevant operation. This level goes far beyond what can be achieved on paper and holds the potential of supplementing the experienced operator's judgments with more proactive information that can truly improve the operation on an ongoing basis.

PAPERLESS ENABLERS

Depending on the scope of a paperless project, it is important to consider both the system strategy and the system architecture. Often, there is only a limited choice of relevant systems that meet a pharmaceutical company's IT strategies and preferred platform. Furthermore, there will often be some existing systems in place that do not support the scope of replacing paper used in daily operations. Paperless operations have in the past been used as a sales promotion feature for specific products that purport to enable paperless operation. As configurable standard systems have gained acceptance in many areas of pharmaceutical operations, the scope of the paperless operation has increasingly become an issue of selecting specific paperless system enablers as part of an overall strategy rather than relying on a specific supplier to provide a paperless system.

As no single system alone can fulfill the requirements for a paperless operation, it is necessary to consider a set of enabling technologies. Two key areas need to be evaluated, namely the *system's degree of openness* (describing the method and the degree of complexity involved in integrating data to and from the paperless enabling technology) and the *system's operational platform.*

The degree of openness of a system depends mainly on its basic technological platform and its specified interfaces. Without going deeply into the technical issues involved, it is important to realize that in the industrial environment, there are a few basic technology platforms that enable a significantly more straightforward implementation than was possible only five years ago. However, most of these more recent technologies are closely tied into specific system platforms. Previously, one would attempt to select systems that were open across operational platforms, but over the last few years, it has become obvious that some system platforms, most notably from Microsoft, provide much more system integration facilities than others.

System Integration Platforms

Only few years ago, system interfaces required special programs developed in a language such as Structured Query Language (SQL), C or C++ to link applications directly to each other or through interfaces of shared files or database tables.

In recent years, technologies based on proprietary standards set by Microsoft in the Windows environment have gained common acceptance throughout most system enablers. Most systems in the industrial environment are now offered on a Windows platform (especially based on the Windows NT/Windows 2000, etc., technology), and many provide a client/server-based approach which makes it fairly easy to integrate systems. Although these technologies do require a certain degree of customization, data access and application integration do provide a set of interfaces for newer generations of software, which often includes an object philosophy that breaks the integration challenge into manageable pieces.

Competing technologies based on recognized open (i.e., nonproprietary) standards are available but have not gained the same market popularity as the Microsoft technology. This situation may be changing as more open technologies have emerged with the popularity of the World Wide Web and system tools such as Sun's Java technology. As these various technologies evolve and gain popularity within key enabling technologies, the system integration challenges may become less of an issue. Recent shifts toward data-oriented technologies such as the XML standard, which enable uniform user access of

Web browsers and an information-aware data format that allows Web information to be processed by a computer without manual data intervention.

It would go beyond the scope of this book to discuss the technical issues of the enabling system platforms, but it is important to bear in mind that the integration challenge of most enabling technologies will largely depend on the system platform on which they operate. Some platforms will facilitate more integration than others.

Process Control Systems

To date, process control systems like Programmable Logic Controllers (PLCs) have only enabled paperless operation in combination with SCADA (Supervisory Control and Data Acquisition) systems or as part of a DCS (Distributed Control System), which enable measurement and control actions to be recorded and used as part of batch documentation. Process control systems have the advantage that they focus on real-time data as a necessary part of both control and supervision. The real-time focus is very useful for implementing both active and proactive control when combined with, for example, statistical tools or predictive algorithms.

In the past, process control systems have been based on proprietary computer platforms, acting as "islands of information" from which production reports were printed out and stored as part of the critical production information. This situation is rapidly changing as most process control systems now operate on open standard platforms that are much easier to integrate. Recent development in control communication protocol standards has made such system integration even easier. Nevertheless, many process control systems currently used have been in operation for many years, leaving companies with the challenge of interfacing these proprietary systems in order to release the benefits of paperless operation.

Major suppliers of process control systems now offer batch packages as key paperless enablers supporting recipe management, production execution and production information management in one integrated environment. Though it still requires some effort to integrate the process control systems with other paperless enablers, the integration challenge is significantly less than it was a few years ago, partly due to a recent standardization of batch handling principles (ISA S88–1 standard on Batch Control, also approved as IEC 61512–1 by the International Electrotechnical Commission). This in turn has enabled some suppliers to offer separate "component" batch software packages that can be easily "bolted on" to a core system.

Another important area of process control systems is the ability to handle trend curves as part of production documentation. Recent software packages (called Historians) provide such capability over an extremely long time span and with batch production facilities which enable paperless handling of batch data that should be documented by graphical trend curves. Some suppliers offers separate software packages for such historical data storage and access which can interface to most types of process control systems and thus provide another recent enabler for paperless production.

Enhanced process control systems functionality is not limited to batch packages and historical data packages. Over the last few years, many process control system suppliers have expanded their product range to cover much broader functionality, such as materials tracking, scheduling, batch record management and so on to create a broad set of paperless enablers. Many of these enhanced functions are offered with standard interfaces to other paperless enablers such as MRP II (Manufacturing Resource Planning) systems, LIMS and even EDMS. With recent generations of process control systems also offering Web-based operator interfaces, the potential for paperless operation is becoming even more important. There remains as a key success criterion the supplier's awareness of the special pharmaceutical requirements for GMP validation—electronic records and electronic signatures.

Manufacturing Execution Systems

Manufacturing Execution Systems (MES) have been subject to considerable hype on their potential for paperless production, yet these systems have proven to be some of the most difficult to use in the pharmaceutical industry. So far, only a few suppliers have been able to deliver their customer's expectations beyond initial pilot installations, and many MES implementations hardly go further than the "active paper on glass" functionality described earlier.

The MES concept describes an area of functionality rather than a specific type of system, namely the area in a classical system hierarchy between high level planning systems (MRP II systems or ERP systems) and shop floor control (which includes both manual and automated process control systems). MES systems support manufacturing processes by providing planning, execution and reporting functions.

Common MES functions in supplier's offerings include the following:

- Scheduling
- Materials receiving and management
- Resource allocation
- Quality management
- Process management
- Recipe management
- Batch reporting
- Batch tracking
- Product history tracking
- Dispatching

- Maintenance management
- Document control
- Trend analysis
- Performance analysis
- Deviation handling
- Data acquisition
- Labor management
- System interfaces (e.g., to SCADA/DCS/PLC, MRP II/ERP, LIMS etc.)

As the term *MES* has become widely accepted, suppliers have started describing themselves as MES suppliers and their systems as MES systems. MES applications are typically factory or site specific. MES systems can roughly be divided into three types:

1. Custom-built MES systems
2. MES enabling software tool sets
3. Customizable standard MES systems

Custom-Built MES Systems

Custom-built systems are developed for specific customers from scratch or from a previous system for another customer. Typically, these are either in-house developments or developed by a software company that does customer specific programming rather than configurable off-the-shelf software products. They are often based on a relational database and a high-level client/server programming tool. Some of the systems on the market are generalized to a certain degree to fit not only the initial customer's needs but also other customers with closely similar needs. Most of these software companies only operate on a local or national level, and these solutions are only feasible for a company or a production site in an individual country, not for a global strategy in an international corporation. Many of these systems also have significant overlap with the functionality of the process control systems and MRP II/ERP systems.

MES Enabling Software Tool Sets

MES enabling tool sets are systems sold as flexible tools for building MES applications. These are off-the-shelf software systems but are primarily software tools that require extensive configuration and/or customization by a project compared to "standard" MES systems. System integrators (which may also be the system vendor) will do the customer specific programming and set up the MES system for use. Some software tool sets contain templates for specific operations, such as dispensing, mixing, tableting and so on, from other customers that can facilitate project start-up. Pharmaceutical companies basing their MES strategy on one of these MES enablers implement a basic standard setup which they use in their rollout to plants of similar types, for example, tableting plants. By doing so, a pharmaceutical company can achieve a fairly standardized MES system and similar working procedures and production methods across similar plants.

Customizable Standard MES Systems

Customizable systems are largely preconfigured but with some options open for customization. They are mostly based on initial custom-built systems but are generalized into commercial off-the-shelf software that takes only little customization. A significant amount of functionality is preconfigured functionality, often suitable for pharmaceutical business processes, which makes it fairly easy to establish an initial setup. The basic system often contains the skeleton for typical production flow, both life cycle (i.e., raw material receiving, dispensing, granulation, tableting etc.) *and* supporting activities (i.e., material release, in-process control, approval, deviation handling etc.). With such functions in the basic system, they do not need as much preconfiguration as the MES enabling software tool sets. The challenge is that this category of MES is fairly new and has demonstrated only a few mature products to date.

MRP II or ERP Systems

MRP II or ERP systems provide functionality for not only production planning and manufacturing but also ordering, receiving, warehouses, shipping and so on. These systems support supply chain management and may include related functions such as maintenance, calibration control and human resource information.

MRP II/ERP systems enable paperless operation within a facility and with supply chain partners (suppliers and customers). System integration standards such as EDI bring the possibility of transferring business information automatically between systems in a very secure manner, which not only replaces paper but also makes many manual operations obsolete. In combination with Web interfaces on the Internet to enable customers and suppliers to enter data directly into the business systems, the whole issue of electronic commerce with its huge implication on business processes within a company becomes an area with many important implications for paperless operation.

There is a clear overlap between the functions of MRP II/ERP systems and MES systems. Indeed, many MRP II/ERP functions are moving toward the plant floor, bridging the gap between business and production systems. Some modern process control systems are starting to offer interfaces "upwards" to MRP II/ERP (e.g., enhanced batch management functionality). Meanwhile, major MRP II/ERP system suppliers are including MES features in their product range, typically as separate modules that are closely integrated with the rest of the system. Some also have specified interfaces to third-party software certification. MRP II/ERP systems that have "standard" interfaces to other important paperless enablers such as process control systems, LIMS and EDMS provide a significant backbone for integrated pharmaceutical operations.

Caution is required, however, when linking MRP II/ERP systems to legacy systems, as some older systems are poorly suited for integration or for dealing with electronic records and signatures. Many of these legacy systems were originally implemented to support financial aspects of supply chain management, and such systems may only with significant effort be turned into truly paperless enabling systems. It may be cheaper and easier to initiate a wide ranging replacement program for legacy systems rather than reengineer them.

Laboratory Information Management Systems

LIMS are an important paperless enabler in the laboratory area which contains information regarding the quality and product characteristics of raw materials as well as intermediate and finished drug products. Users inside and outside the laboratory may typically access LIMS information. LIMS resemble in scope MES, as they support the operation directly in the daily work perspective and are prepared for direct interfacing with most common laboratory equipment and instruments. However, LIMS are a mature and well-defined category of systems, unlike the present situation with MES.

LIMS are a key enabler in pharmaceutical operations for two main reasons: (1) They enable paperless operation in the laboratory environment, thus handling procedures and specifications for laboratory analysis as well as laboratory analysis records. (2) They provide results and status information for materials, which enables systems in other environments such as production to close the quality loop by providing critical feedback on the results from suppliers and the pharmaceutical company's own production processes.

As LIMS typically contain both the analysis values and the associated procedures, they hold much potential for integration with other paperless enablers to provide a streamlined paperless operation. Integration with other systems may include, for example, preassignment of analysis when a new batch is started on the plant floor, material status information for MRP II or ERP systems, electronic certificates of analysis for batch documentation, raw material characteristics for dispensing operation and others. It should be noted, however, that integrating LIMS with other paperless enablers outside the laboratory environment is a complex task.

Electronic Document Management Systems

EDMS are a key component for most paperless operations as a means to store and control the use of electronic documents. As most data in the manufacturing environment are documents that can be stored as files, EDMS have a key functionality in the ability to attach much more document attributes to files than "raw" file servers can do on their own. EDMS features include document administration, work flow, review facilities (e.g., viewing and redlining of all major file types), Web access to documents and so on. EDMS provide interfaces to many desktop applications and enable a fairly seamless integration.

Developing EDMS applications requires both strong user involvement and technical knowledge. Many key document attributes and business process characteristics need to be formalized and set up in the document management system. As with LIMS, when combining EDMS with other paperless enablers, the integration task may be significant, for example, an EDMS integrated with batch records created by a MES or a process control system to provide paperless production documents.

Examples of EDMS are well known, but some companies have implemented similar functionality based on general groupware platforms such as Lotus Notes®. This may provide a simple start-up for a simple document management project, but it lacks many of the special characteristics of genuine EDMS and may prove a complicated solution in a longer-term strategy toward paperless operation.

Automatic Identification Technology

When implementing paperless operations, there is a separate area of enabling technology which falls slightly outside the scope of other paperless enablers, namely Automatic Identification Technology, such as bar codes, radio-frequency tags (RF tags), magnetic cards, smart cards and so on. These are not traditional computer systems but separate physical systems.

When the encoding standards of, for example, bar codes are combined with data definition standards such as ANSI/FACT or UCC/EAN-128, which enables a unique identification of the type of data read in a bar code (e.g., a batch number or an expiry date), the level of quality in the paperless operation can be significantly higher than with manual data entry. However, to gain full advantage of these technologies, pharmaceutical companies need to define internal standards, which enables, for example, bar codes to be printed and read throughout the company and extensively apply the technology to maximize the benefit. This is a nontrivial task which is very important in order to make these technologies true enablers for a paperless operation.

Other Paperless Enablers

There are many more enablers for paperless operation that may be considered. Systems for maintenance, calibration management, warehouses, transportation, test equipment and Computer-Aided Design (CAD) are just a few of the other systems of relevance within the scope of paperless operation. Some of these enablers can be through custom developments and hence be difficult to integrate for the same reasons as described earlier.

There is often a dilemma of how far to go toward integrating a paperless environment, and the cost/benefit consideration will leave many "islands of information" not connected electronically. This is inevitable because much of the transformation to a paperless operation in real life is a transformation from individual computer systems fed with data from papers in turn producing their own reports on paper. Many of these islands of information provide significant business benefits without being integrated into the scope of the paperless operation. There is a need for pragmatically planned phasing within a longer-term strategy. Some systems may yield greater benefit if they are integrated into the paperless operation at a later stage.

PAPERLESS CHALLENGES

There are some important challenges to paperless operations that can pose significant threats to the implementation of systems. The vision and the feel of urgency due to expected business benefits may lead one to underestimate the severity of the challenges involved. Several key questions have to be very carefully addressed.

Is the Scope of the Paperless System Clear?

Two of the most critical steps are scope definition and later scope control. They may seem simple initially, but in real life implementation, the purpose of paperless is typically difficult to define clearly, especially as many issues that seem obvious in a traditional, paper-based business process need to be explicitly defined when established in paperless systems. Scope control as well as the project management of resources, budget and time are pitfalls where many such implementations have fallen short.

Is There a Realistic Relationship Between the Vision and the Actual Systems Selected?

There is always a gap between the vision and the reality of what a system can do, especially as things that seem simple may come across what a system is designed to do. Though the vision part often seems straightforward and system solutions may be demonstrated in a way that seem readily applicable, the actual system's true capability and the ease with which it may be tailored to specific needs may be very different from what one expects. Unrealistic functional requirements that are not in the system yet but are promised in the next version may suddenly bring an implementation project from being at the "leading edge" to the painful position of the "bleeding edge", when it turns out that the assumptions made were not fully secured.

Is There a Clear Indication of the True Benefits and the Costs Involved?

Cost versus benefits is always a key issue for system implementations, but as paperless systems often deal with benefits that are very difficult to quantify and typically have significant intangible benefits, the real values involved are quite a challenge to deal with. Also, some of the key advantages related to error prevention and daily business process deviation may be very difficult to estimate, as they deal with issues that few managers and employees are willing to give true and realistic figures on.

Is the Organization Well Prepared and the Training Needs Clearly Understood?

Although operator training is always a key issue in system implementation, the challenge of changing the information flow and getting people to leave their well-known papers on which everyone can write whatever is needed, and even attach yellow stickers to add further information or questions(!), is quite significant. Some parts of what the organization use to do in less formalized ways will have to be dealt with, and some of it will not make it into the paperless system.

Is There a True Management Commitment to Going Paperless?

Many paperless projects run into problems, either on the technical or the organizational side. The system may end up being not fully implemented and possibly combining the worst of both worlds, with inconsistencies between the paper-based systems and the electronic system—Beware of the hybrid system turning into a low breed. Also, the system may end up being fully implemented but not truly used because the real, implemented system is not as simple and intuitive to operate as it was expected. For several reasons, organizations tend to be conservative, and this is especially true if the daily benefits are less obvious than anticipated. A strong sponsorship on the management level as well as a clear commitment to deal with technical and organizational challenges in a responsible manner are keys to success when inevitable crises occur during the project.

Is There a Clear Distinction Between the Long-Term Strategy and the Implementation Steps?

Unfortunately, it is very easy to be unrealistic in estimating how much can be achieved in one step. There needs to be a long-term plan or vision and a set of clearly obtainable operational steps which demonstrates success and progress toward the long-term goals. Some of the quick wins are important to take, but obviously the long-term implementation will also have to go through more tedious progress steps. Therefore, a realistic balance between the long-term goals and the individual steps is a very important key to success.

Is There a Clear Understanding of the Regulatory Challenges?

The regulatory challenges are at least as important as the technical and organizational challenges. Although all regulatory bodies agree on the potentially higher level of control and documentation in well-applied computer systems for regulated purposes, there are some clear obstacles that need to be addressed for successful implementation in the pharmaceutical industry. As most areas of potential benefits are dealing with clearly GMP critical information and procedures, system implementation as well as quality management needs to be carefully planned and executed to demonstrate and maintain regulatory compliance, i.e., validated. This must be addressed not only during implementation but also ongoing during the operation and maintenance of the systems throughout their life cycle.

In addition, the implications of the FDA regulation on electronic records and electronic signatures (21 CFR 11) need to be carefully understood. The regulation not only affects computer systems and their suppliers and integrators but also the organization in which the system is implemented. These requirements must be met in both implementation and the ongoing operation and maintenance of computer systems. 21 CFR 11 also affects systems that are taken out of daily operation (decommissioned), as they would normally contain important GxP electronic records that need to be preserved. Both the validation approach and 21 CFR 11 compliance is thoroughly described in later chapters of this book.

REAL-LIFE EXPERIENCE

There are many examples of highly automated pharmaceutical facilities aiming for paperless operation. Some of the most important experiences from a highly automated aseptic facility in Denmark give an indication of some key benefits as well as challenges when moving toward paperless production. The facility includes formulation, filling, inspection, assembly and packaging with automatic warehouse and material handling. The plant went live on 1 January 1998. It is mainly based on configurable standard software as enablers for a certain degree of paperless production and integrated into a fairly seamless overall system (Figure 1.4).

The main system components include an interface system to the corporate ERP system, a central MES, an EDMS, a Material Control System, a set of cell-oriented process control systems based on SCADA, PLCs and a few

Figure 1.4. Plant Process Overview

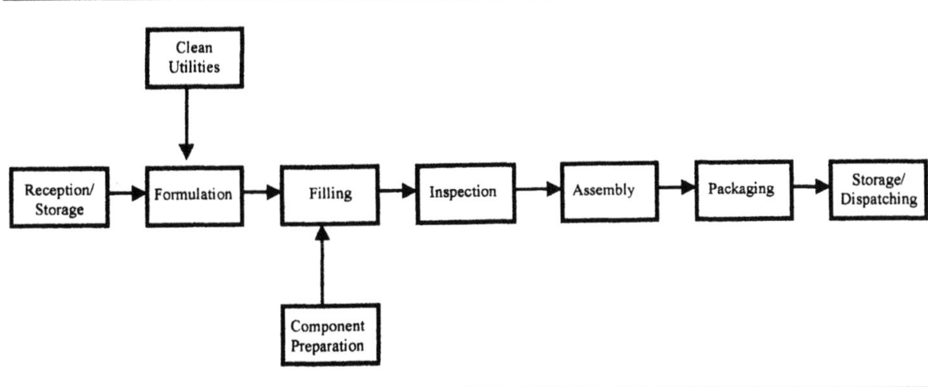

custom (bespoke) programs for some additional functions. A schematic of the facility's computer systems within the Computer Integrated Manufacturing (CIM) hierarchy is given in Figure 1.5.

The aim of the plant was to achieve state-of-the-art automation, productivity and quality control and to ensure a lean and competent workplace, where activities by personnel are either directly adding value or handling exceptions in the production process. The information systems used were to measure and ensure that quality and productivity goals were fulfilled at all times during production by increasing flexibility and decreasing change over time during production as well as preventing errors and defects. The systems employed needed to keep track of all operations and support a team-based organization by facilitating flexibility and human interaction.

It became a key success criterion for the project that all computer systems and utilities were in place at the point in time where equipment was delivered on-site. The aim, which was successfully achieved, was to ensure that the plant would go into production with fully functional and validated systems from the beginning in order to avoid the tedious job of implementing the paperless concepts after production had started.

The overall project goals were achieved. The experiences from the first months of production from the plant are summarized below.

Obviously, such a facility becomes very dependent on its computer systems. The overall run-in time takes longer than a traditional factory. The work flow has become less flexible than in manual operation as a consequence of a well-defined work structure controlled by the systems. This in turn enables a very clear exception handling and a highly traceable operation with data and many detailed records to support analysis and follow up on all significant

Figure 1.5. Functional Reference Architecture

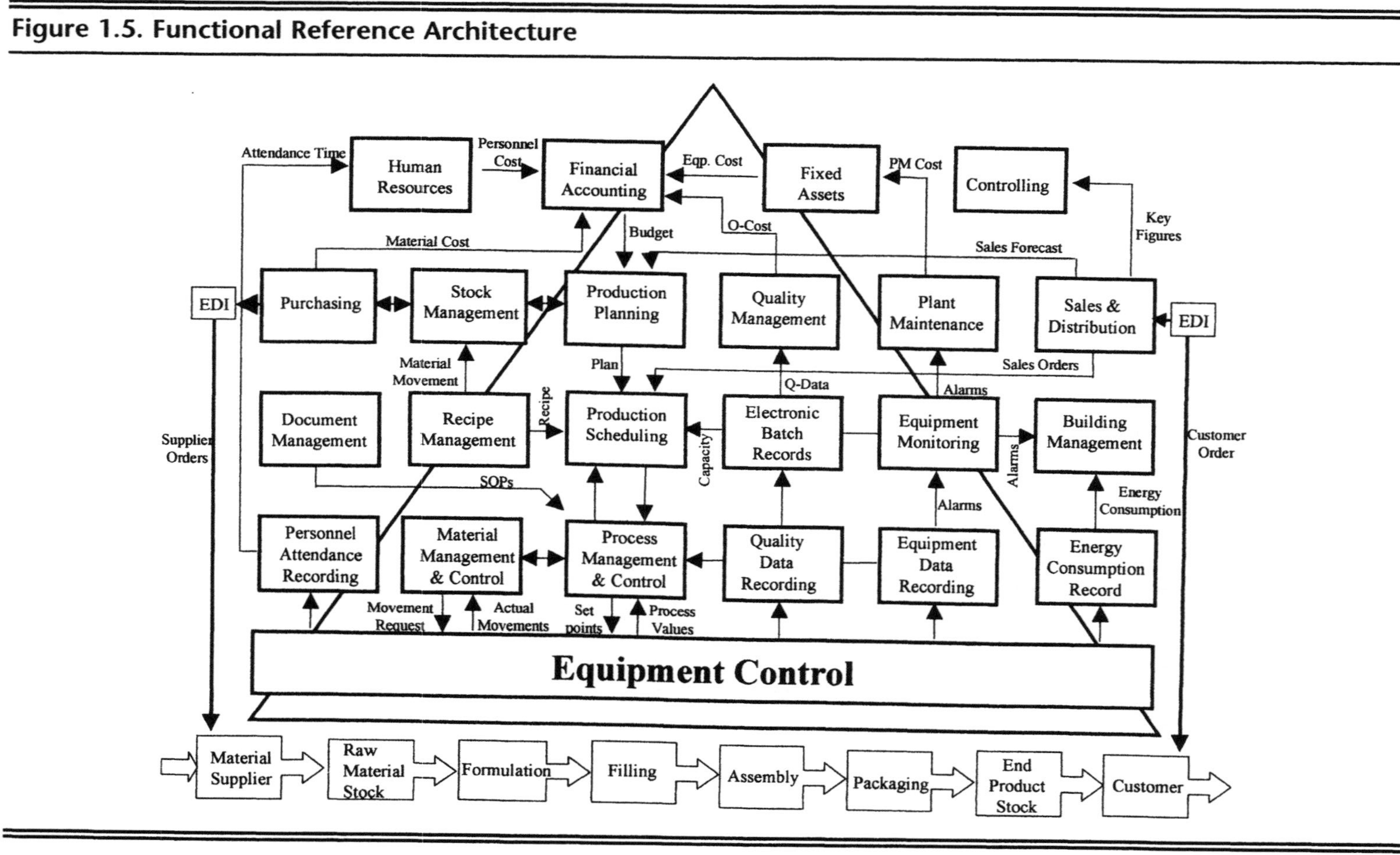

events. Batch records have become highly informative, and quality management is clearly more transparent than possible in similar traditional plants.

The most critical part of the computer systems became the interfaces between the various software packages. Operational experience indicated how important access to skilled computer professionals helped on a daily basis the production teams achieve their goals.

It was proven to be very important that the systems were designed to be clearly selective on the amount of information presented to the users. In hindsight, it would have been all too easy to overwhelm users with information. The selection of important display information could have gone even further than it did in the cited example here, as detailed information is only usually needed on an exception basis or for regular trend analysis.

Manual data entry should be kept to a minimum. As much of the information as possible was to be captured through either integrated data collection or Automatic Identification Technology. Workflow optimization was a key area and taken very seriously during system design. It has also proven important that the system architecture be kept simple and clean. This was not as easy as it may have seemed for an overall system based on standard software packages from different suppliers, because the systems inevitably had overlap in functionality. There were some difficult decisions to make on the architectural level as well as in the customization. Some areas of the overall functionality may not seem logical in some situations. It was very important that error messages were clearly informative and identified the system/function producing the erroneous condition.

Some of the standard systems were less mature than expected based on the suppliers' promises. This had to be addressed by enhanced customization efforts to meet some of the requirements in the system implementation. Furthermore, it was more difficult than expected to ensure required procedural controls across different systems, which sometimes led to additional and more complicated operations on individual systems due to the overall integration.

The integrated systems facilitated a quality management philosophy based on quality circles, with direct feedback to production management on the outcome of a process, the quality status and the immediate handling of exceptions or deviations.

Overall, the facility required fewer operational staff than similar traditional plants, and the integrated computer systems clearly enabled operator empowerment. Operator training was conducted both in classroom and on the shop floor, with the best results coming from hands-on training.

Despite the overall advantages, it has proven very difficult to make those outside the new facility understand how important the advantages are. The production philosophy is very different, and it must be understood as a whole rather than by a few of the important aspects in order to achieve the benefits. A shift toward more integrated and paperless operation also implies a major shift in the production paradigm and technology as a new way of conducting the business.

CONCLUSION

Progress toward paperless operations is a stepwise process. Some pharmaceutical companies claim to be paperless, but what they have achieved is a "first generation" implementation. This evolutionary rather than revolutionary development has been true for most other major technology shifts in recent history. For example, the shift from the horse-drawn carriage to automobiles (see Figure 1.6) led to a first generation of cars that resembled the carriage much more than they resemble the streamlined cars we see today. The information flow of a first generation paperless facility can be clearly more efficient and transparent. The potential of further exploitation of computer technology will become evident. This will enable an even more streamlined way of doing business. This in turn will influence the GxP regulations and move some of these newly acquirable benefits, especially those associated with quality control, into basic expectations in the new millennium.

On the outset of implementing paperless operations, however, either top-down or bottom-up, it is very important to master the learning curve. It takes time for people to understand and shift paradigms. Integrated computer systems do imply a significant business risk; unless carefully managed with clearly defined and realistic goals, it will fail. With that in mind, the preparation for paperless operations should not just be a conversion but a thorough re-engineering of the business processes involved. Computer systems have provided electronic media with a much higher potential than the "dumb" paper medium will ever achieve. Although the transfer to paperless operations is a long process, it is important to identify the quick win's that will help the organization understand the concepts and the benefits *and* the long-term goals to which the individual steps lead.

A high level of paperless operation is not easy to implement, and this makes it very important to have clear step decisions on "how much is enough—at least for now", because achieving the ultimate paperless goal is going to be a long journey.

Figure 1.6. First Generation of a New Technology Is Never Achieving the Full Potential

BIBLIOGRAPHY

1. FDA. 1999. 21 CFR Part 11 Electronic Records; Electronic Signatures; Final Rule. Washington, D.C.: Government Printing Office.

2. GAMP Forum. 1998. *Supplier's Guide for Validation of Automated Systems in Pharmaceutical Manufacture*, Version 3.0. Available from the International Society for Pharmaceutical Engineering.

3. ISA. 1995. *ISA-S88.01–1995: Batch Control Part 1: Models and Terminology.*

2

Developing an Information System Strategy

Chris Reid
Integrity Solutions Limited
Middlesborough, United Kingdom

Norman Harris
20CC Limited
Croydon, United Kingdom

Brian Ravenscroft
AstraZeneca
Loughborough, United Kingdom

UNDERSTANDING THE BUSINESS

The first objective for any team developing an information systems (IS) strategy for a pharmaceutical organization must be to define the business operation and the impact of Good Practice regulations on those operations. The definition of the business operation may already be well established and documented, or there may be a need to embark on a complex Business Process Reengineering (BPR) exercise in order to establish the drivers, purpose and processes of the business before considering an IS strategy.

For the purposes of this chapter, we assume that the business operation is defined and documented and that the organizational culture is largely in place to support the business.

The business model described in Figure 2.1 shows the high-level structure of a pharmaceutical business organization. Company policies are set within an environment of the company mission—the purpose for existence, e.g., to research and develop pharmaceutical products, and external influences including legislation and Good Practice regulation. As this model

Figure 2.1. Business Model

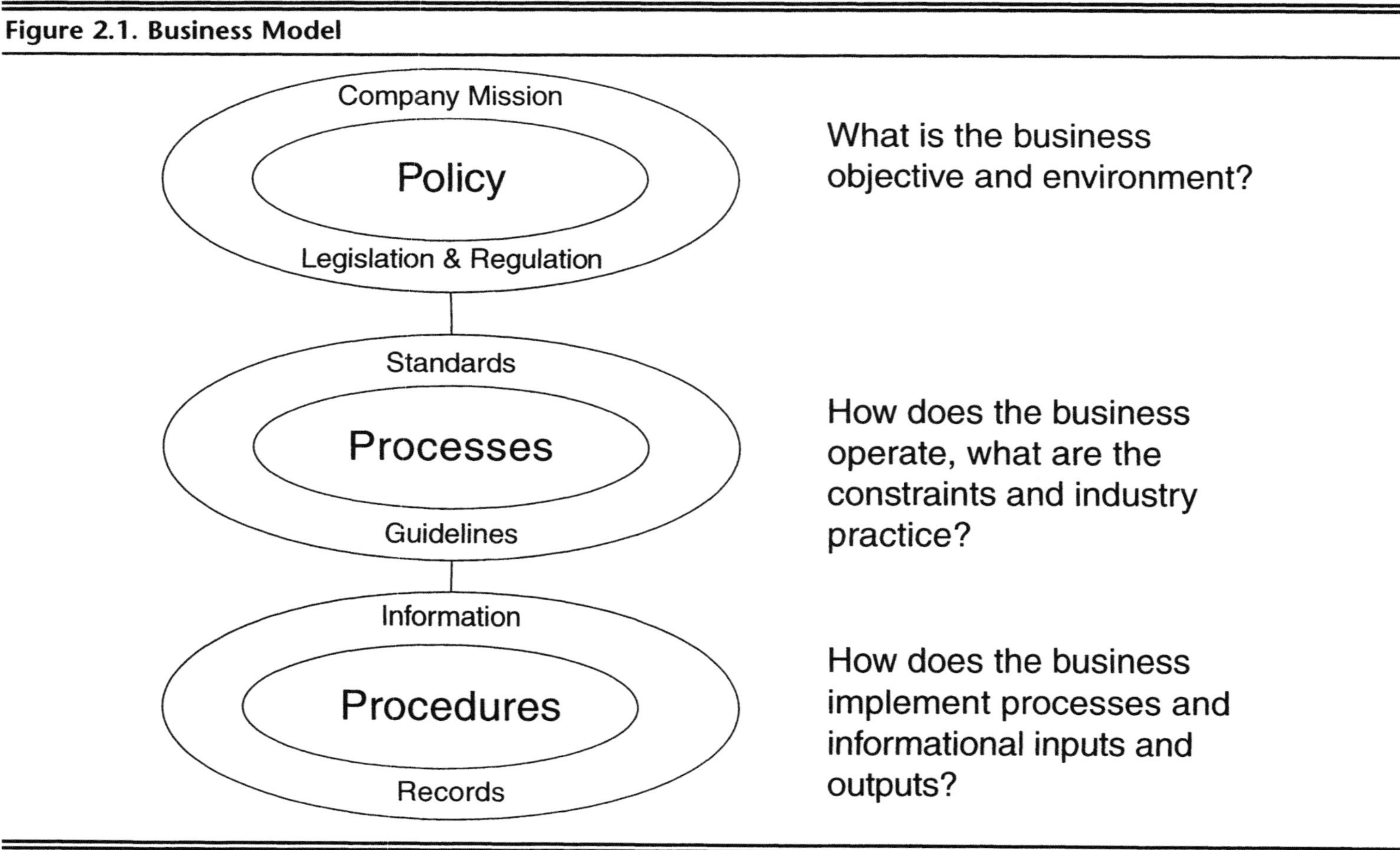

evolves to reflect departmental structures, it is clear that internal influences such as relationships between different sites and internal operating businesses will also be applied. The model demonstrates that all layers of the business have informational inputs and outputs that are essential to the effectiveness and efficiency of the operation.

IMPORTANCE OF INFORMATION TO PHARMACEUTICAL BUSINESS MANAGEMENT

Modern information technology is rapidly expanding the volume, types and, above all, quality of information that can be brought to bear effectively on the needs of a pharmaceutical business. To an organization that has a clear view of its information needs, this can be the basis of crucial, strategic and tactical, business advantages and ultimately a positive impact on profitability. If, however, a pharmaceutical organization does not have such a vision, it can be either a costly lost opportunity or the source of either endemic information overload or information starvation; it is arguable which may be worse. The net effect obscures the way forward for the management and supervision of the organization, leading to decision-making errors at all levels.

There are several basic prerequisites of an effective information strategy within a pharmaceutical organization:

- Clear business goals, derived from an articulated corporate mission statement or departmental remit
- Known, stable business processes
- Known impact of Good Practice regulations on business processes and information strategy
- Personnel who are informed and trained in these processes and the common, keyboard-based tools of information systems
- A coherent IT (information technology) infrastructure

This chapter deals principally with the systems to support the first four prerequisites and will concentrate on IT aspects and the crucial foundation of computer support in the organization.

Infrastructure

A distinction should immediately be made between information technology and information systems. The IT infrastructure comprises the cables, hubs, switches and servers on which the systems run. It is also necessary to add to this the basic set of desktop tools, e.g., Microsoft Office Professional®. There can be cases where the business application drives major step changes in infrastructure direction. These instances, almost by definition, will be very large, strategic, high risk and infrequent opportunities.

Whilst the basic hardware and desktop software strategy is a "given" in most major organizations, this is not invariably the case unless conscious effort has been made. Without such a strategy, simple tools like E-mail may not work and certainly will fail to meet their potential.

IS refers to the software packages and databases that provide the functionality to implement the business processes. These include such packages as Enterprise Resource Planning (ERP), Manufacturing Execution Systems (MES), Electronic Document Management Systems (EDMS) and Maintenance Management Systems.

Management of the corporate and increasingly global infrastructure, due to mergers and international growth within the pharmaceutical sector, cannot function to the satisfaction of the organization unless management can overview the information needs and development plans of the whole organization. All developers must accept the basic tenets of the infrastructure policy and contribute to its enhancement. The result is the easy flow of information to and from any part of the organization. A by-product is the ability of any individual to sit at any personal computer (PC) connected to the infrastructure and do his or her work efficiently, whether that PC is at home or at a site 10,000 miles away. A classic case in modern business is the EDMS, which cannot function unless desktop facilities are consistent and of a high standard. Such a development will require a degree of cross-departmental consistency in how information is portrayed.

The development of the pharmaceutical business information strategy will propose enhancements in scale and extensions of the scope of the infrastructure that is in place. So an essential element in producing a business-related information strategy is to ensure that the IT foundation is in place and constructed satisfactorily.

Information

Information permeates throughout the pharmaceutical business in what tends to be a random, parochial manner, unless it is analyzed and structured as a deliberate action of corporate and departmental management.

Many ways of analyzing information flows have been used in the past. Modern theory, which arguably combines most of the best features of its predecessors, focuses on business processes. The mere process of recording the key processes can be exhausting but results in real benefits as well as conflict as overlaps and gaps become self-evident. The resolution of these overlaps and gaps has to be carried out by a coherent management team committed to the same goals and able to subordinate their departmental interests, since activities will be created as well as abandoned.

The definition of a business process produces the need to report information through a document, database or any other format. These "documents" are windows onto the information resources of the organization and should be structured into

- procedures,
- asset information and
- real-time operational information.

Leading edge analytical practitioners can argue against documents and focus on "information" and "objects". It is common ground that the "document" is the most user-friendly medium to carry information. How information is placed onto the document is the focus of information- or object-driven methods. If adopted, the latter is often one of these infrequent instances, where it is liable to revolutionize corporate computing philosophies as well as all aspects of the organization's business activity.

The developed process is conveyed through Process Route Maps (PRMs) (Figure 2.2). The PRM defines the fundamental tasks and interfaces of the process. Where further decomposition of a process activity is required, detailed Standard Operating Procedures (SOPs) and information to support these SOPs are developed.

Procedures are controlled instructions for implementing specific business processes. Clearly they are an information resource in themselves and have a life cycle as a document, usually textual, but capable with an advanced IT infrastructure of incorporating audio and video clips to aid their explanation and operational use.

Real-time information management is inextricably linked to real-time control systems. These are beyond the scope of most conventional information strategy developments. However, much of their information is for record purposes and should be stored and reported from within the main information resource.

In the pharmaceutical area, the need for validated systems is of a paramount importance. Normal business drivers still apply, and systems must be robust, providing timely information which is relevant, accurate and safe to use.

MAPPING THE INFORMATION STRATEGY TO BUSINESS NEEDS

The development of an information strategy is an interesting challenge that must be clear in its objective. These objectives largely fall into two camps: those developed in order to exploit current, available technology (technology pull) and those developed to provide the best vehicle for improving business efficiency and effectiveness (business push) (see Figure 2.3). In practice, both approaches provide a valuable contribution to business improvement; however, the balance between business push and technology pull must be achieved in order to maximize the benefit of the IS strategy.

In reality, business pull must be the primary factor in the development of the information strategy whilst recognizing that deviation from

Figure 2.2. Process Route Map

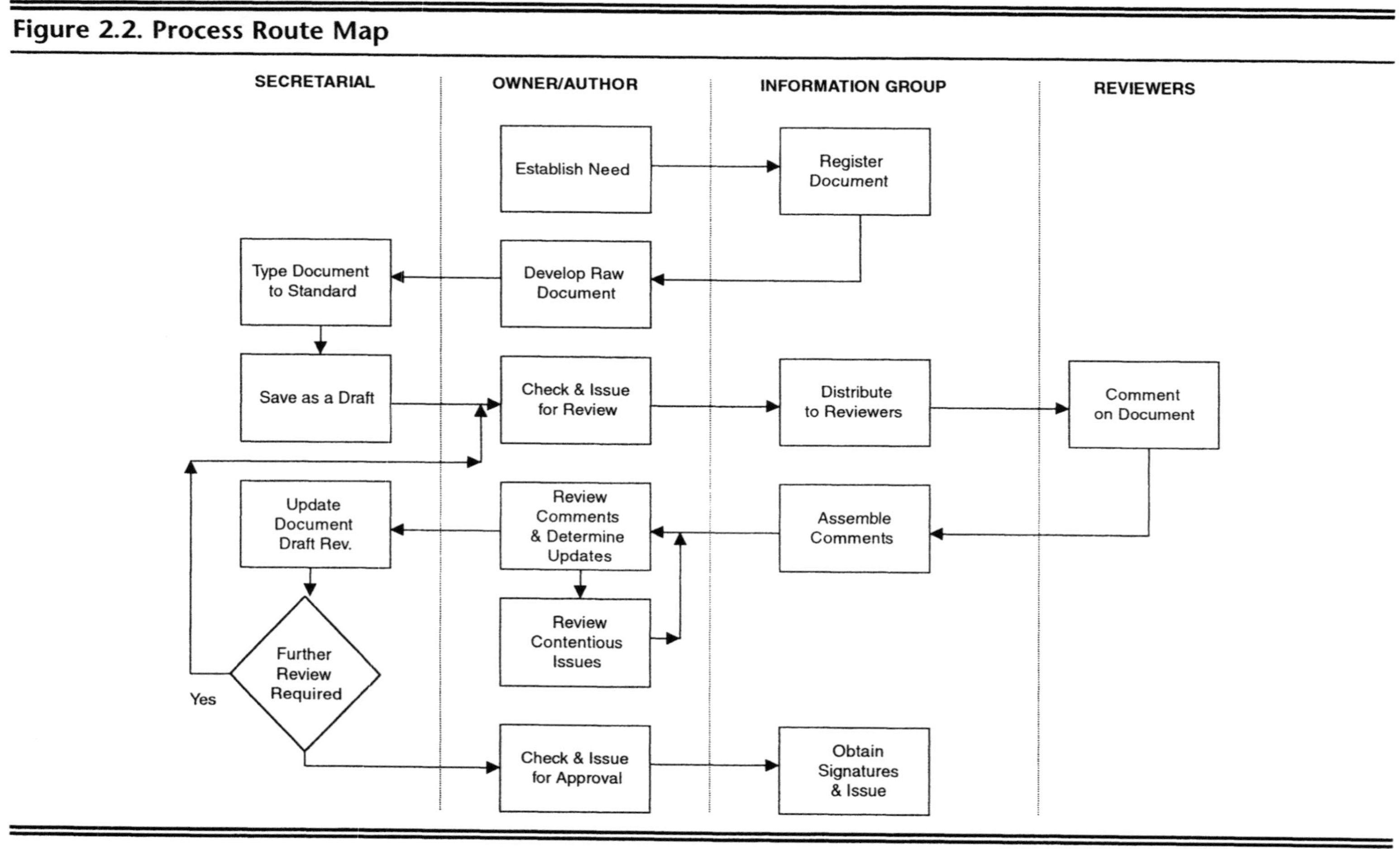

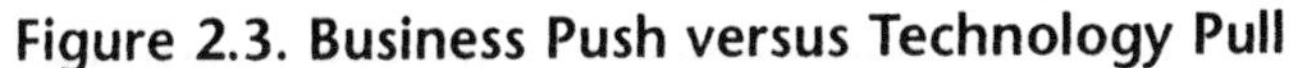

Figure 2.3. Business Push versus Technology Pull

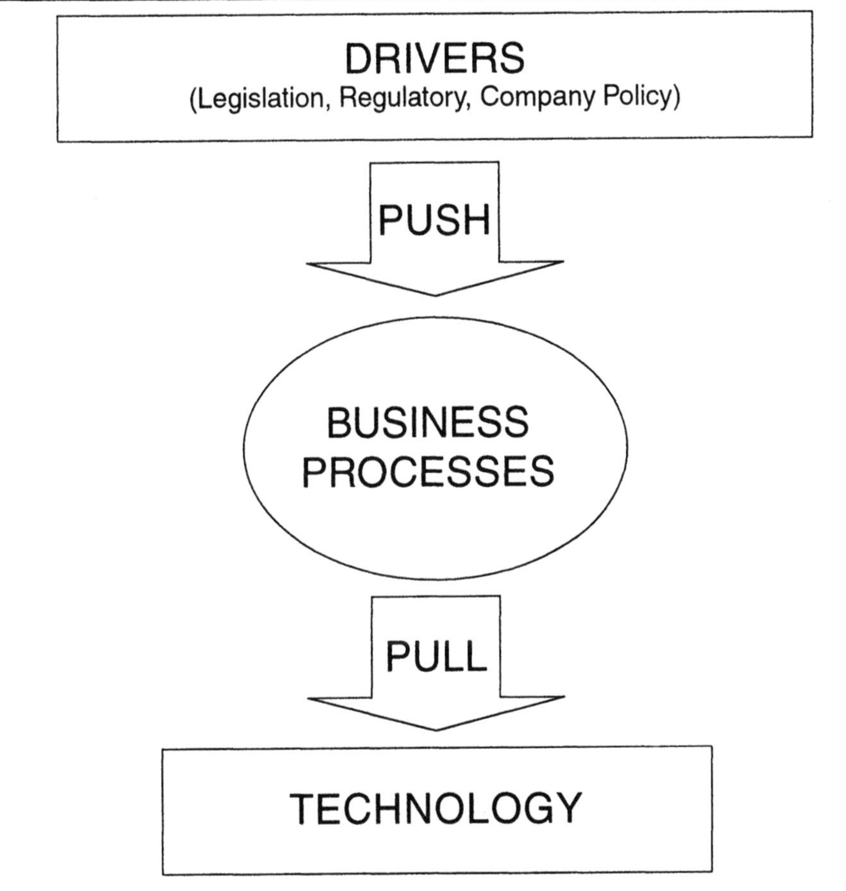

standardized is architectures and methods presents a business risk in its own right. For example, totally changing the way a business operates in order to avoid deviation from system architectures may present greater issues relating to cultural change, implementation of inefficient processes and training. On the other hand, resisting a technological solution that totally meets the objectives of the business process, simply because the process steps are implemented in a different order or adopt a slightly different terminology, will increase the risk of project failure. The balance, therefore, needs to favor the business process whilst recognizing the benefits of the standardized technological approach.

Once the process definition is advanced, functional user requirements will begin to evolve. It is important at this stage to understand the different levels of enterprise management [2]. As we move from the direct control systems to business management systems (Figure 2.4, the regulatory impact tends to be more indirect than direct. For example, a failure of the Distributed Control System (DCS) will immediately effect the quality of the drug product. However, a failure of the EDMS may result in the failure to carry out a procedure correctly, which may in turn lead to a failure to calibrate an instrument correctly, which may eventually result in product quality issues.

Once processes and functions are defined, the appropriate system can be mapped to the functionality. Figure 2.5 [2] provides an overlay of systems onto the four layers of enterprise management (sometimes referred to as the Computer Integrated Manufacturing hierarchy). The choice of system, however, may not always be problem free, particularly when operating within a highly regulated industry. Various products will provide overlapping functionality, as illustrated in Figure 2.6 [2]. The choice of system must therefore be based on the best approach to risk management. Three factors influence risk:

Figure 2.4. Enterprise Management

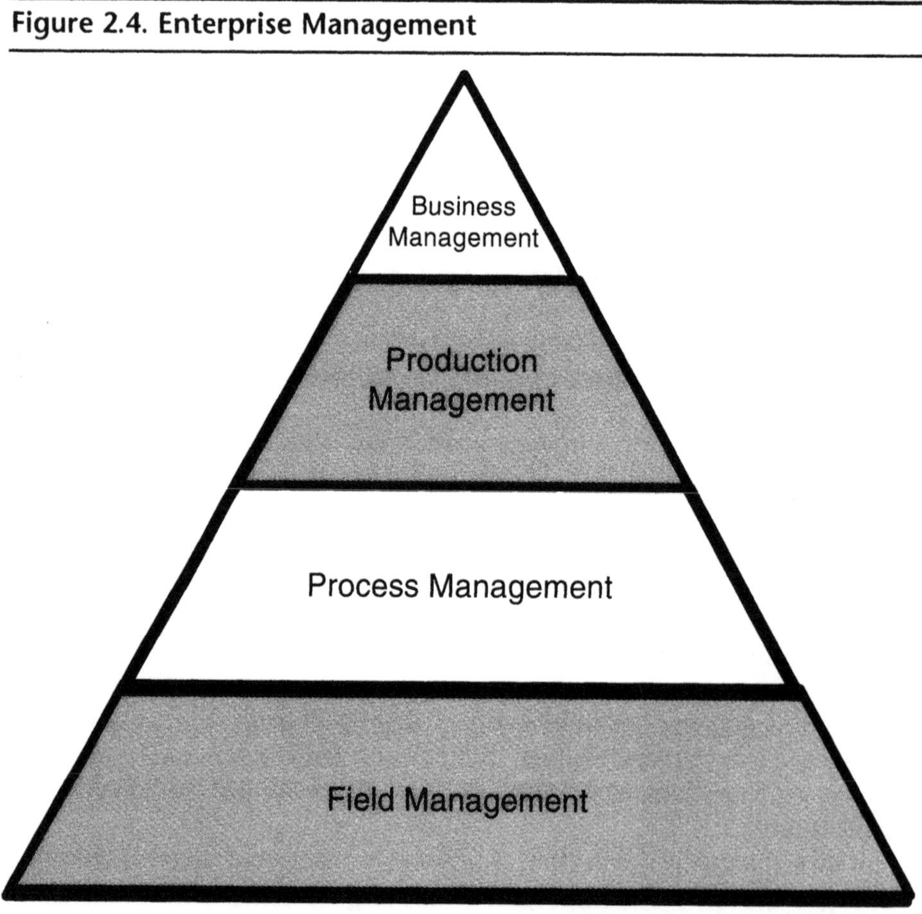

Figure 2.5. Business System Hierarchy [2]

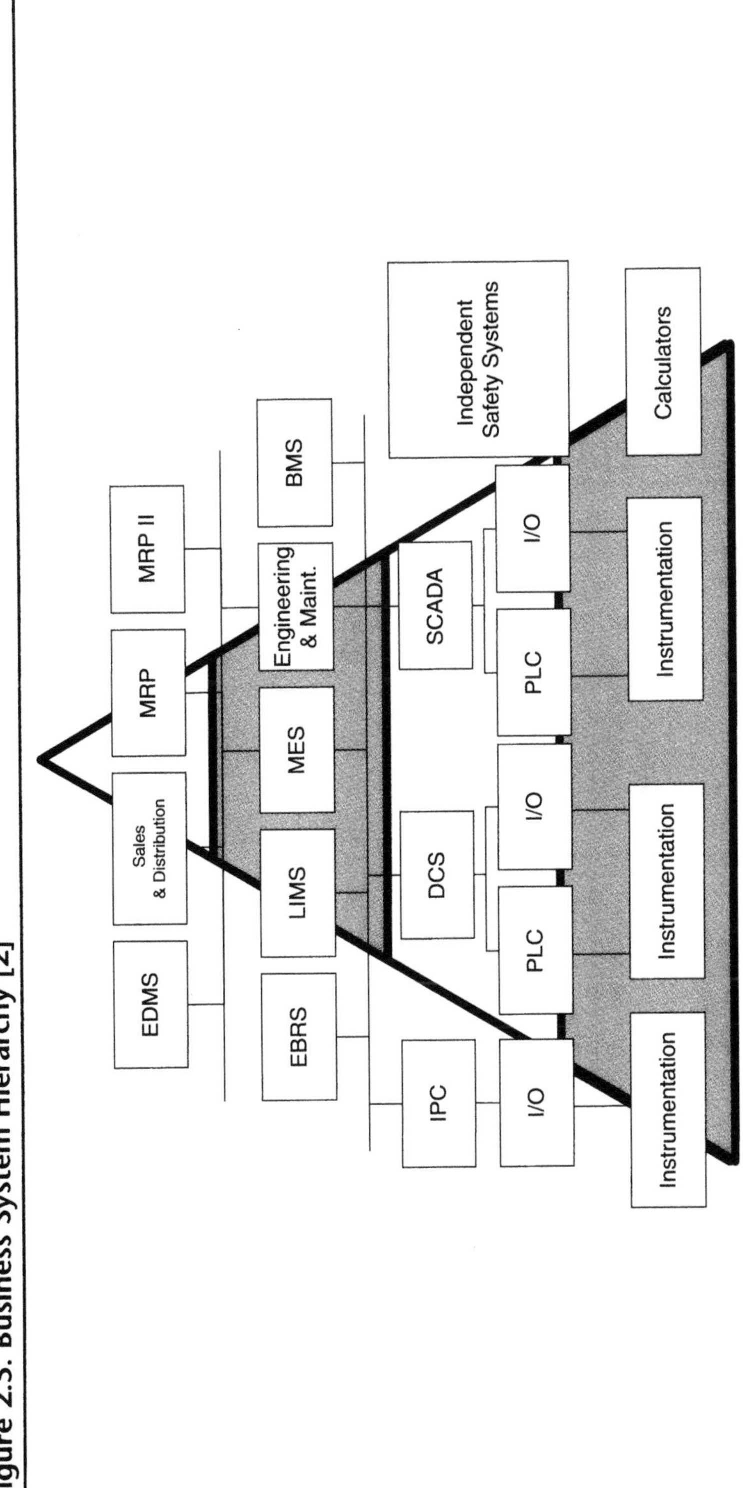

MRP = Materials Management System; EBRS = Electronic Batch Recording System; LIMS = Laboratory Information Management Systems; MES = Manufacturing Execution System; BMS = Building Management System; IPC = in-process control; DCS = Distributed Control System; SCADA = Supervisory Control and Data Acquisition; PLC = Programmable Logic Controller

Figure 2.6. Functional Overlap [2]

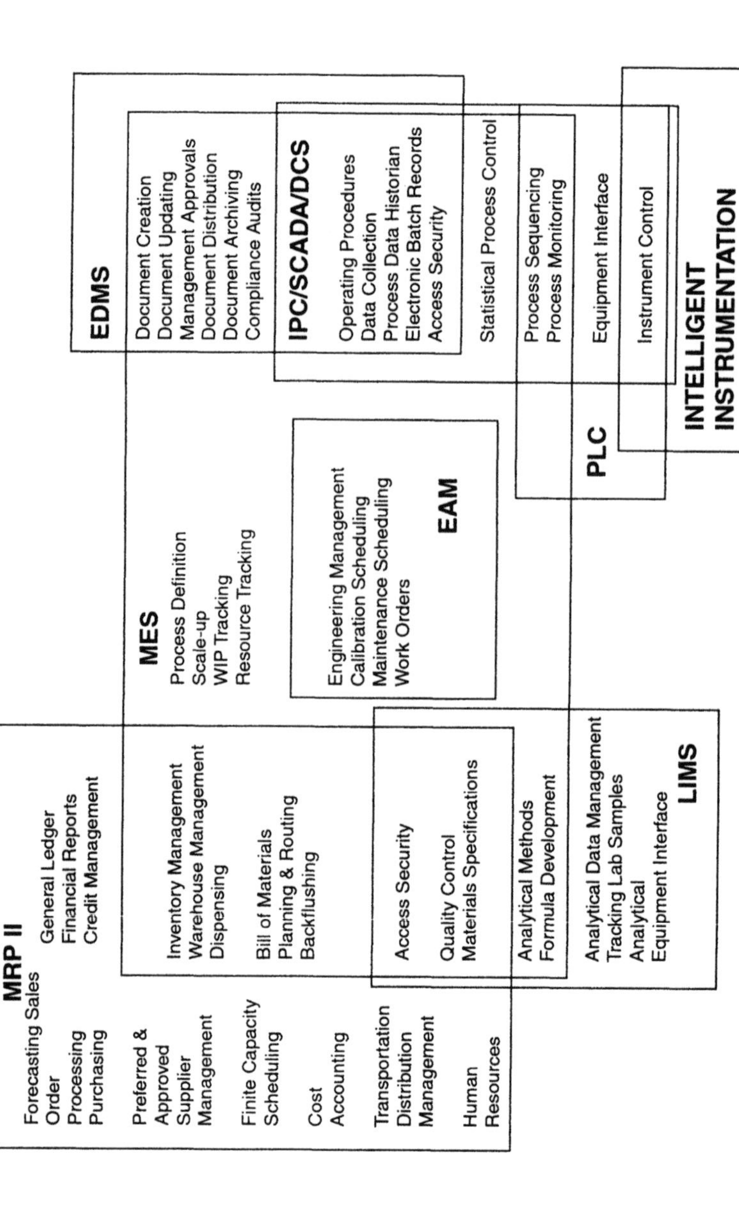

EAM = Enterprise Asset Management; WIP = Work in Progress

1. Compartmentalization of current GxP–related functionality into a minimal number of systems (limit validation to minimum number of systems)
2. Quality approach of suppliers
3. Best technical fit to defined business processes

The information strategy will normally be implemented over a number of years. If the culture is oriented toward continuous improvement, the strategy will continue to evolve.

As previously indicated, IS strategies will involve the implementation and integration of many computer systems. A map of the potential systems aids explanation and provides the analyst with a basic plan of what applications are in place, which need changing and which will have to be acquired. It also makes visible the most likely original sources of information circulating in the current system's portfolio.

Data are the building block of useful information and are the most valuable asset in the entire development of an IS suite. Data should be captured in as accurate a form as possible and securely stored. The key target of an IS strategy, once the sources of data have been identified and safeguarded, must be to ensure that progressively the same correct information is used every time that information is required. This is often described as "store once, use many". An associated benefit is that the master data source is clearly identified, which avoids issues with maintaining copies, an issue with particular relevance to the pharmaceutical industry. In terms of most physical asset data, this leads to a systems map in which systems are shown clustered around one or several linked databases accessing and, in the case of advanced design, being populated by centrally stored, approved and time/date-stamped information. This obviously constitutes a major step forward in the management of compliance-based information, as it ensures consistency of information and enables effective control.

The result is that whereas a truly accurate drawing is still a difficult aim in conventional Computer-Aided Design (CAD) applications, the drawing becomes one possible window on the data in such advanced systems. This clearly is a complex, expensive process and has only succeeded where an industry and its suppliers have agreed to cooperate in the development of the definitions. This transferability of information is essential to competitive survival in the car, oil and aircraft industries. Pharmaceutical companies are now beginning to follow suit.

KNOWLEDGE–BASED SYSTEMS

Considerable use of knowledge techniques is now occurring in the computer-based training area. Whilst there are undoubtedly technological and skill developments behind this use, they are based on the rigorous and structured capture of work experience in the use of recorded business processes and by database and multimedia techniques recording, retrieving and disseminating this information to the target personnel of the organization.

Knowledge is fundamental to all of the strategy that surrounds information, e.g., knowledge of the processes that create and demand data and the limitations of these data. This includes knowledge of the market in which you operate and the information resources that are needed to prosper. The drug submissions to regulatory authorities and the bringing of products to market more quickly are areas in which hard-won knowledge in laboratories, plants and engineering maintenance can repay the unavoidable investment costs many times over.

RELATIONSHIP BETWEEN STRATEGY AND IMPLEMENTATION

The implementation of the information strategy will usually be subdivided into a number of projects, each subject to normal, annual budgetary approvals. A project team will be established for each system or group of related systems. The project team should provide an appropriate blend of operational, technical, regulatory and other skills to meet the specific project needs.

The subdivision of the IS strategy into a number of manageable projects presents a number of issues, including the following:

- Consistent implementation of the strategy
- Integration of various projects
- Recognition of interdepartmental and section needs
- Integrated and consistent approach to validation

As such, a senior member of the organization must have responsibility for the evolution and implementation of the strategy. A typical relationship between strategy and project organization is presented in Figure 2.7.

Each project team will generally comprise the following individuals or groups (see Figure 2.8):

- Project manager
- Users
- Quality assurance
- Strategy manager
- System integrator
- Validation team
- Business process teams
- Internal technical IS representatives

Figure 2.7. IS Strategy Organization

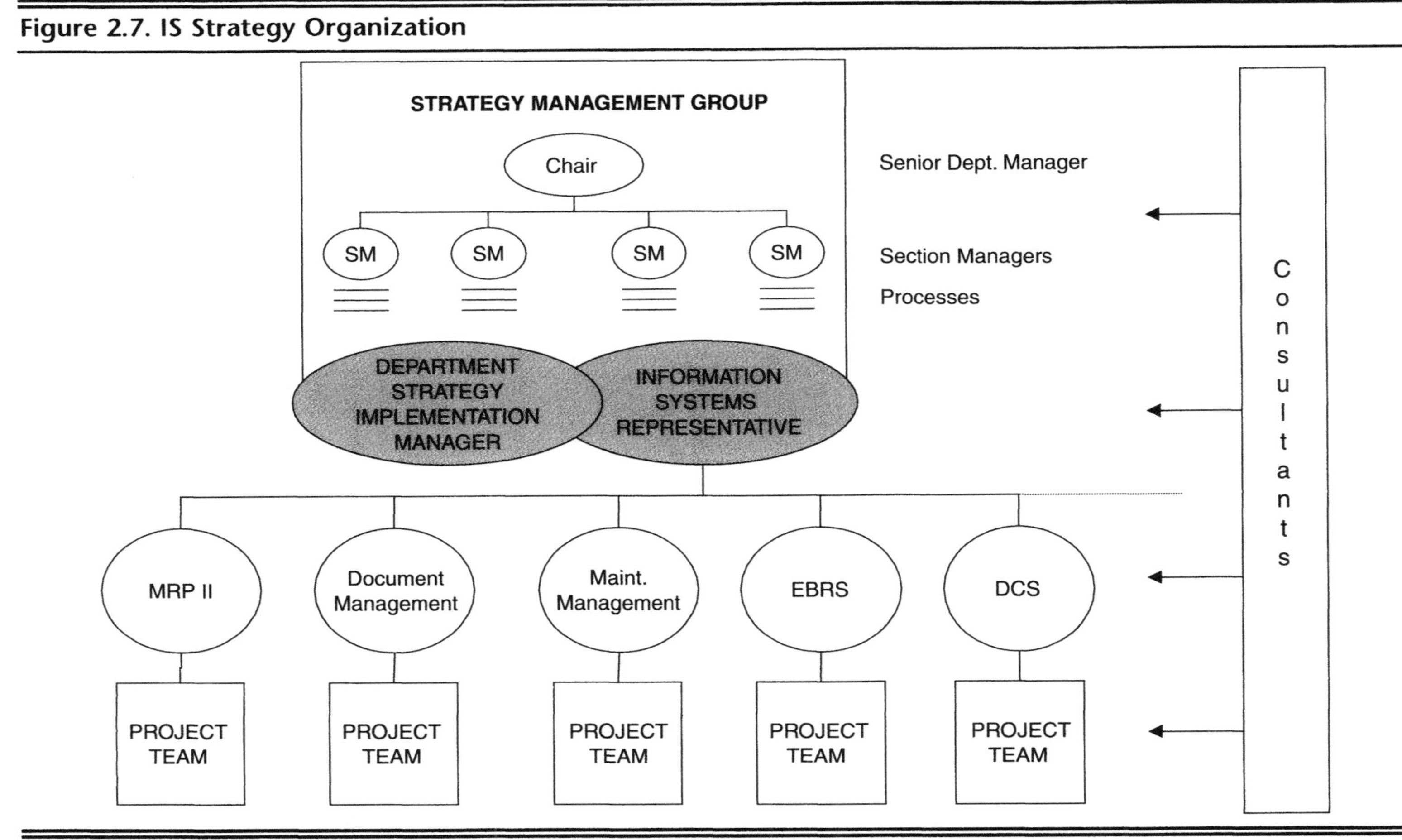

Figure 2.8. Project Organization

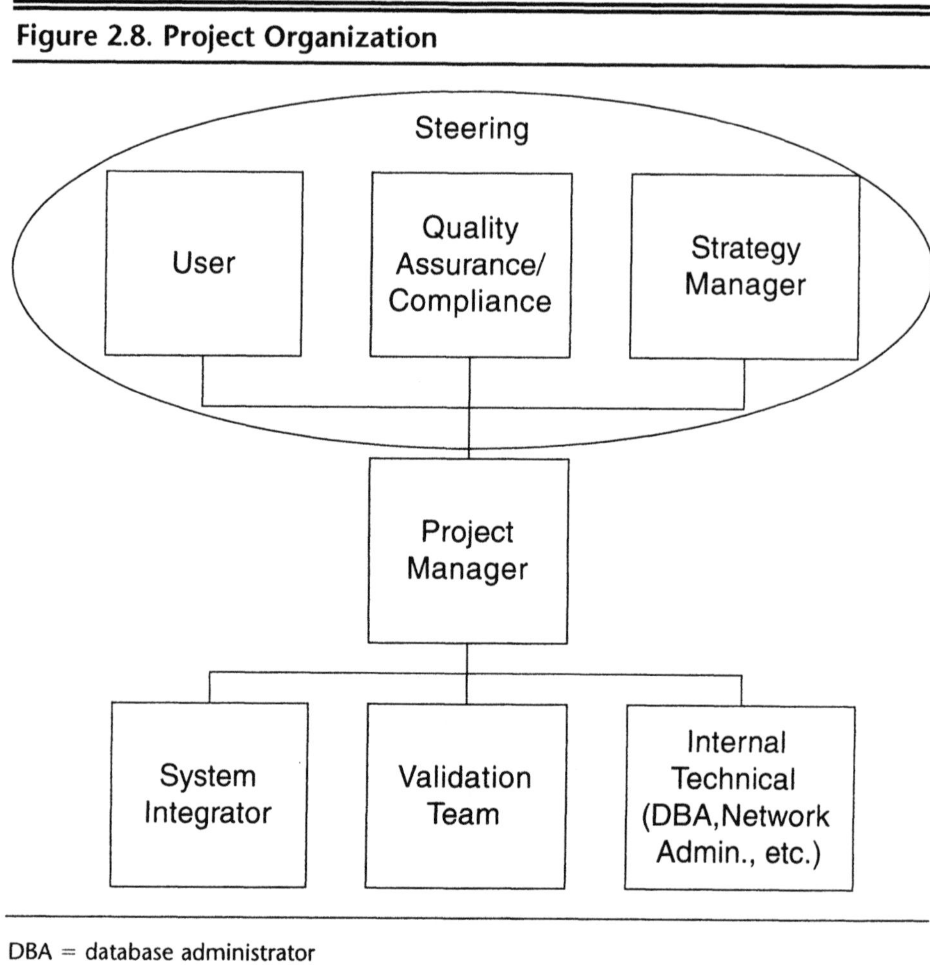

DBA = database administrator

The validation team will report to the project manager, who will ensure that validation tasks are conducted in accordance with the project program and budget. The validation team is normally independent of the system integrator in order to provide increased confidence in the outcome. Central validation steering groups are responsible for setting consistent validation policy and procedures that will provide a consistent and controlled approach to the validation.

SELECTING PROJECT PARTNERS

It is likely that the pharmaceutical company will have a continuing relationship for at least five years with the selected partner; after five years, the business processes may have been so drastically reengineered that the selection of new technology and partner becomes a viable option. No matter how well the system has been documented by the first supplier, their premature replacement will be burdened with the costs of bringing the new supplier up to speed and the extra management time needed to make the change.

In the selection of suppliers, a number of factors have to be considered. Weighting of the factors involved, by a Kepnor Tragor or structured process (see Table 2.1), is usually an essential tool used in the final selection of a supplier. The various attributes of the supplier are categorized into those which are essential, very desirable or merely nice, with appropriate weighting factors for each. The scores of the selection team are then applied to produce the resulting "best" selection. This is rarely the final answer.

A variety of internal and external organizations will support IS implementation, including but not limited to

- the hardware supplier,
- the core application provider,
- the systems integrator,
- consultants and
- internal IS functions (e.g., network and database administrators).

Before utilizing such organizations, the pharmaceutical company must be satisfied that the preferred organizations will meet the technical, regulatory, budgetary and program constraints defined in the project terms of reference. The vehicle most commonly adopted to assess the capability of an organization is the Supplier Audit.

Table 2.1. Supplier Selection Criteria

Selection Criteria	Score (1–10)	Weighting	Adjusted Score
Quality of technical proposal	8	90%	7.2
Locality of organization	5	70%	3.5
Structure of organization	7	80%	5.6
Pharmaceutical experience	8	95%	7.6
Availability of quality systems	9	100%	9
Application of quality systems	9	100%	9
Support infrastructure	6	85%	5.1
Cost	4	75%	3
Terms and conditions	6	60%	3.6
Total	62		53.6

The objective of the Supplier Audit is to determine

- technical competency,
- the experience profile against project requirements (technical, pharmaceutical, regulatory),
- quality systems effectiveness,
- organizational structure,
- support capability and
- financial stability.

There are potentially a number of outcomes from the audit:

- Organization suitable (usually with minor observations that are easily remedied)
- Organization suitable following satisfactory resolution of corrective actions
- Organization unsuitable

The audit must be structured in order to meet the specific requirements of the project and/or the generic requirements of the corporate pharmaceutical organization.

An audit team must be formed to ensure that all of the objectives of the audit are met. The lead auditor should be competent in technical issues and therefore can fulfill the technical role. The typical responsibilities of the audit team are shown in Table 2.2.

The audit scope must enable the establishment of a ". . . high degree of assurance . . ." in the supplier organization. A typical scope is presented in Table 2.3. Supplier audits are discussed in more detail in Chapters 14 and 15.

Demonstrable evidence in support of responses to audit questions should be sought whenever possible. Supplier responses should be challenged by asking how, where, when, what, why and who.

The audit observations and recommended corrective actions must be documented in an audit report. Observations should be categorized as *major* or *minor.*

- *Major:* A significant noncompliance which must be corrected and approved by the audit team before the services affected by the nonconformance can be provided.
- *Minor:* The nonconformance cannot immediately impact the process, product or safety. The corrective action is to be reviewed at the next scheduled follow-up audit.

Typical inclusions and exclusions from the audit report are given in Table 2.4.

Where nonconformances are identified, a date for a follow-up audit should be set. The timing of the follow-up audit is determined by the risk

Table 2.2. Audit Team Responsibilities

Role	Responsibilities
Lead auditor	• Open and steer the audit. • Ensure that the audit objectives and scope are met. • Ensure that questions are fully understood by those being audited, rephrasing questions when appropriate. • Ensure that questions are answered in full. • Ensure that questions are supported by documentary evidence where appropriate. • Summarize and close out. • Prepare report.
Technical representative	• Ask technical questions regarding the hardware, software, integration issues and so on and challenge responses.
User and/or project representative	• Relate discussions to the user environment and project constraints.

presented by the nonconformances. The objective of the follow-up audit will be to review progress against agreed corrective action plans and to ensure continued conformance to original audit findings.

VALIDATION STRATEGY

GxP Assessment

There are two questions most commonly faced by validation organizations: "What do we validate?" and "How does validation differ from normal project activities such as Factory Acceptance Testing and Site Acceptance Testing?" The answers are that we validate GxP critical aspects of the system, i.e., those aspects that can have a direct or indirect impact on product quality, safety and efficacy. *Validation* is essentially a term used to denote the additional rigor applied to the management and documentation of the each phase of the development and operational life cycle of the system.

Table 2.3. The Scope of a Supplier Audit

Subject	Typical Line of Questioning
Quality Management System	• Quality Management System accreditation (ISO 9001, TickIT) • Life-cycle approach • Established and approved procedures, standards and guidelines • Quality management organization (quality manager with executive responsibilities) • Internal audits planned and conducted • Methodologies and tools utilized and controlled • Change control applied • QMS continuous improvement process • Control of third-party product • In/out controls
Application of Quality	• Design specifications (URS, FDS, module design, hardware design) • Design review records • Test specifications (module, integration, upgrade) • Test records • Source Code Review/inspection records • Requirements traceability • Change control records • Release control records • Training records
Configuration Management Systems	• Software and configuration version control • Control of multiuser access and modifications • Security (access and modifications) • Control of build tools (compilers, linkers and assemblers)

Table 2.3 continued on the next page

Table 2.3 continued from the previous page

Subject	Typical Line of Questioning
	• Backup, archive, disaster recovery • Release control • Control of development bureau operating systems and hardware • Control of customer specific configuration/bespoke software • Command files • Control of packaged solutions, e.g., DocSolutions®
Competency	• Staff development plans • Experience profiles (resumes) • Training records
Support environment	• Operation and maintenance documentation • User training • Fault reporting • Upgrade management (number of versions supported) • Support facilities (e.g., help desk, etc.)
Organization	• Structure of Organization – Quality management (including reporting routes) – Technical development – Operational support
Experience	• Pharmaceuticals • Regulatory requirements (European and U.S.) • Validation
Commercial robustness	• Credit/debt ratios • Turnover • Gross/pre-tax profits • Three-year profile

Table 2.4. Audit Report Inclusions and Exclusions

Inclusions	Exclusions
• Name of organization audited • Audit date • Audit location • Standard against which audit was conducted (ISO 9000, corporate or department questionnaire, etc.) • Personnel involved • Observations made • Recommendations for selection, rejection, corrective action and follow-up	• Deficiencies discovered and corrected during the audit • Confidential information without express permission of organization • Subjective opinion • Ambiguous statements • Antagonistic words or phrases

The GxP Assessment process identifies

- each function of the system,
- failure modes of each function (the way each function can fail) and the
- consequence of failure to the business.

If the consequence is an impact on product quality, safety or efficacy or violates GxP regulations, then the function is deemed to be GxP critical and must be validated. A less rigorous approach to review, testing and documentation can be applied to non-GxP–critical functions.

Level of Validation

The validation strategy must take into consideration the structure of a typical information system (see Figure 2.9). The UK GAMP Forum [1] has defined five fundamental categories of systems and, consequently, the level of validation required. These are reviewed below.

Operating System

The operating system (e.g., VMS [Virtual Memory System], Windows NT, etc.) provides the platform for the product. Where the operating system utilized is an "established, commercially available operating system . . .", the functionality of the operating system is considered validated as part of the application software.

Figure 2.9. System Components

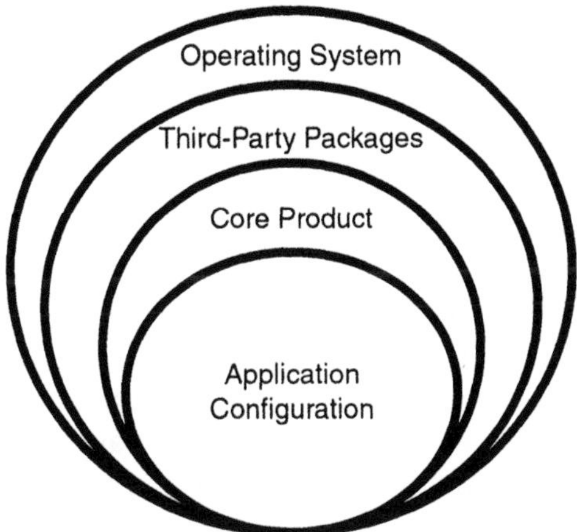

Essentially, version control is applied and upgrades controlled in order to minimize potential GxP impact. The impact of operating system upgrades on third-party software, application software and application configuration must be reviewed and issues addressed following all upgrades.

Firmware

Firmware is electronically programmable read-only memory (EPROM) acting as system software. Version control must be applied.

Standard Third-Party Packages

Standard third-party software such as Oracle®, standard communication interfaces and so on are often utilized by application software. This software provides the environment for the application software and as such will be implicitly validated when validating the application software. As with operating systems, rigorous upgrade management is essential.

Configurable Application Software

Configurable application software usually includes custom configurable packages such as Manufacturing Resource Planning (MRP II), Documentum®, Maximo® and so on. Validation of such systems requires an audit of the supplier and validation of any bespoke application software.

Bespoke Application Software

Bespoke application software is specifically coded to meet the needs of the pharmaceutical organization. The application configuration must be fully validated. In addition, we must consider functional criticality of the core product being configured, i.e., the risk to the business posed by any single or group of functions. Figure 2.10 shows that as functional criticality increases, so does the degree of rigor applied to proving the system. The term *validation* as used within the pharmaceutical industry is therefore indicative of a highly rigorous approach.

There is great debate regarding the inclusion of operational safety within the scope of validation. In principle, there can be no question that safety of personnel is as critical if not more critical than product safety. With the advent of IEC 61508, there is no doubt that industry regulators, e.g., the U.S. Food and Drug Administration (FDA) and UK Medicines Control Agency (MCA), are interested in demonstrable evidence of system quality and performance.

The GxP Assessment process described later in this chapter is a common vehicle for establishing functional risk.

Figure 2.10. Risk and Rigor

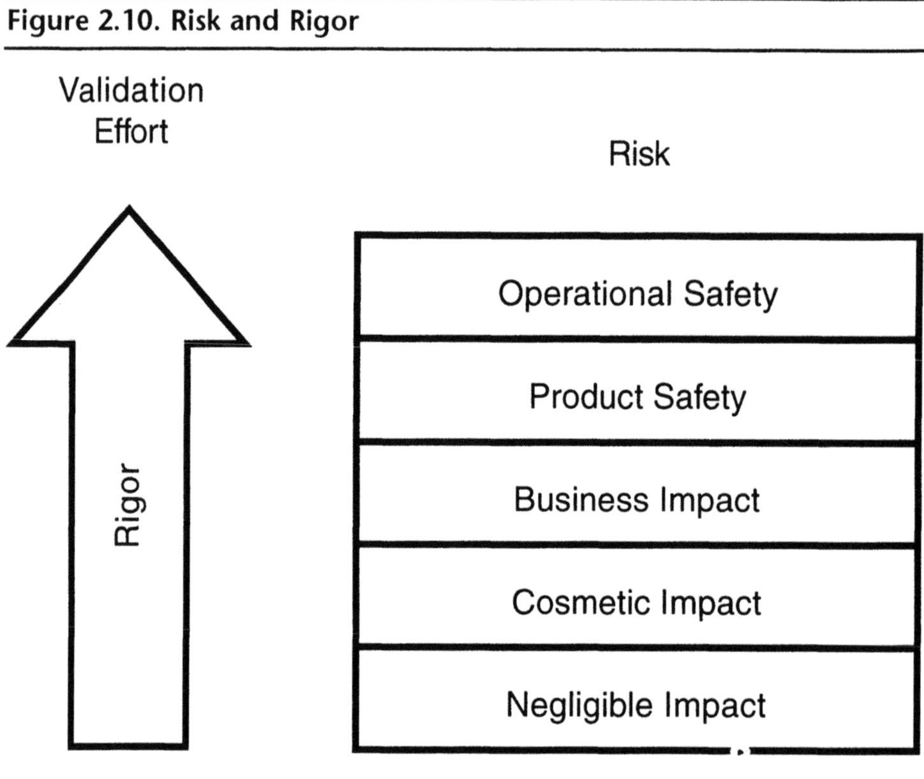

Life-Cycle Methodology

Validation life cycles (Figure 2.11) provide a methodical approach to ensuring that quality is "built in" to the software and hardware development from the User Requirements Specification (URS) through to final testing and ongoing operation and maintenance. A Validation Master Plan (VMP) must be developed to define the validation strategy, organization and roles and responsibilities for project participants.

User Requirements

The user requirements phase allows the pharmaceutical company to define its business needs. This is often communicated by pictorially defining the business processes, as illustrated in Figure 2.12; each business process is accompanied by a descriptive narrative describing the functional requirements of each process step.

Figure 2.11. Life-Cycle Methodology

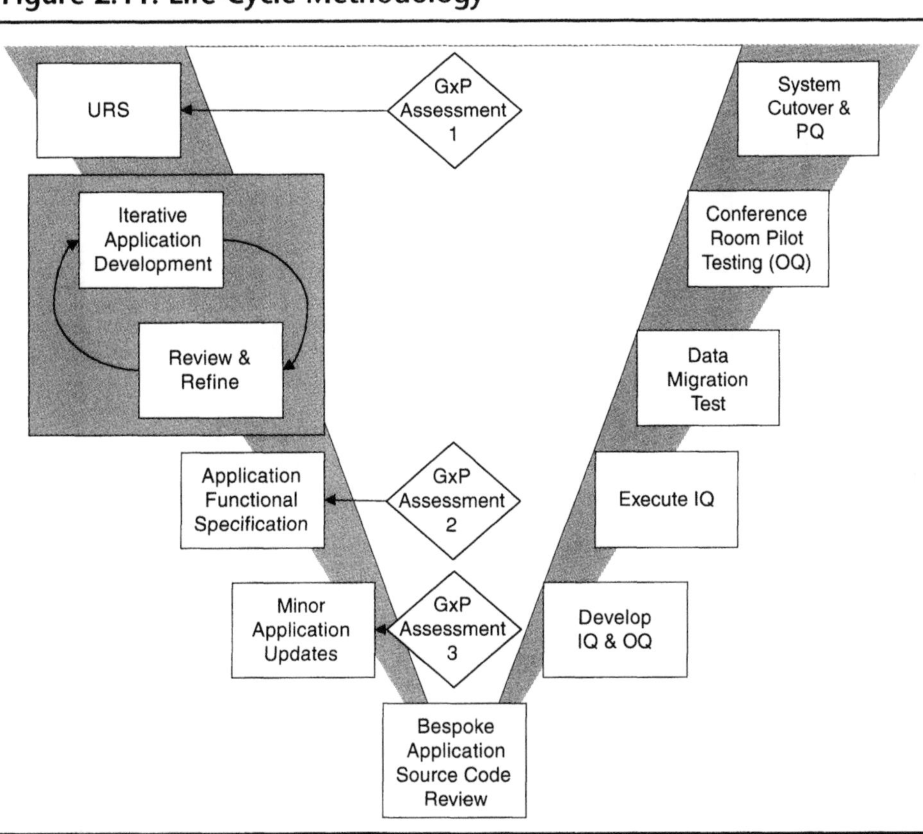

Figure 2.12. Process Requirements

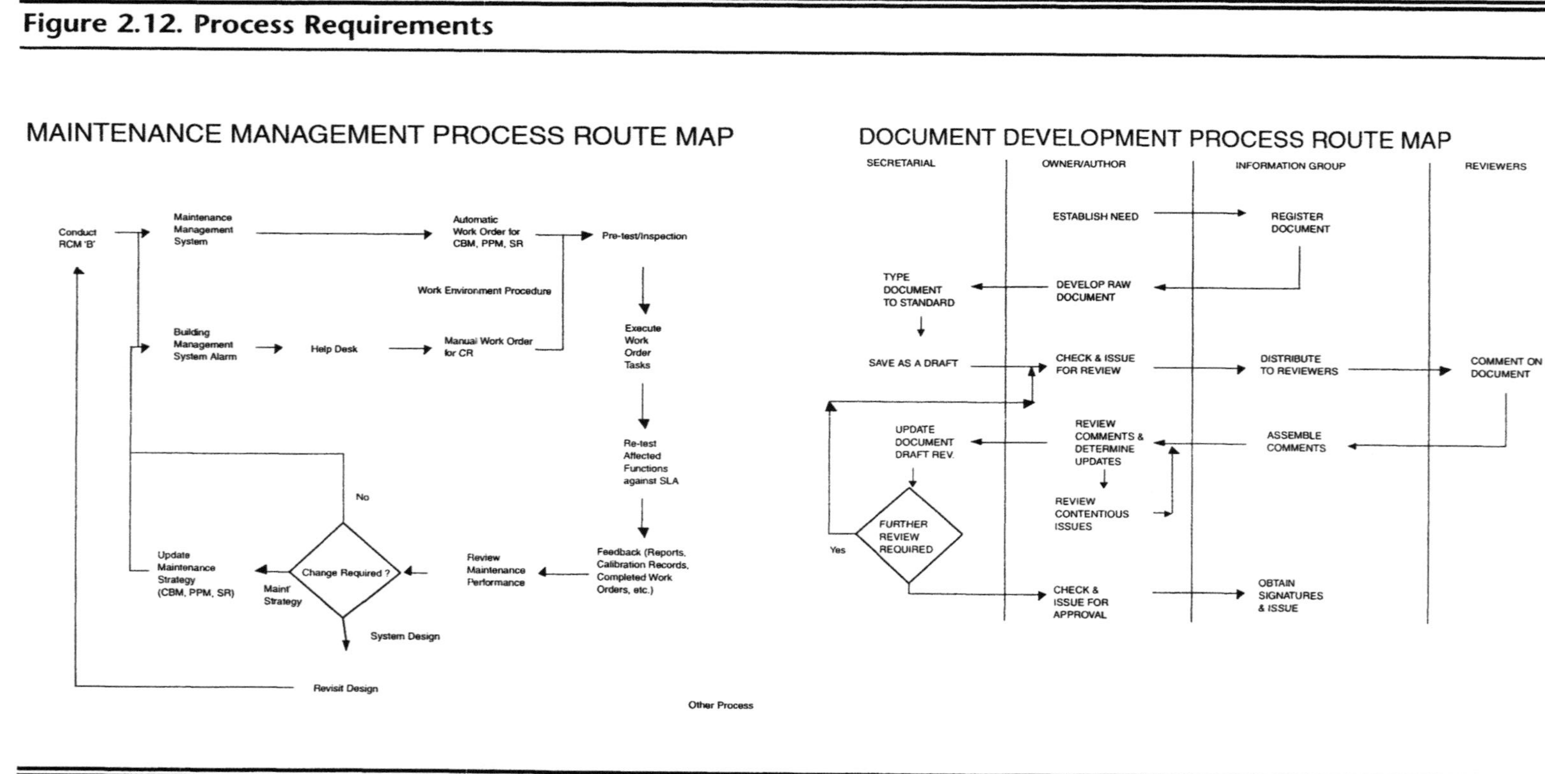

The user requirements must define the following elements:

- Business processes
- Interfaces between business processes (internal and external)
- Critical business information
- Consequences of business process failure
- Implementation constraints (IT infrastructures, details of legacy systems)
- Specific performance criteria (including tolerances)
- Data migration from legacy or manual systems

Technical requirements or constraints such as corporate software and hardware requirements, network configurations, and so on must also be stated by the IS department.

The URS should be written in a clear and unambiguous manner that enables full "requirements traceability" through each phase of the project life cycle.

Level 1 GxP Assessment

The level 1 GxP Assessment is conducted in order to determine the impact of each stated user requirement on GxP regulations. This assessment is conducted by a small team comprising representatives from validation/regulatory compliance/quality, users and technical and operational groups. The objectives of this assessment are to ensure that

- relevant GxPs have been implicitly or explicitly stated in the URS,
- GxPs have not been contravened by statements within the URS,
- design considerations (backup systems, design constraints, etc.) have been stated,
- the validation scope can be determined and
- the resource requirements are defined.

The impact of a stated requirement on GxP is determined by studying the consequence of the function failing, e.g., What happens if we corrupt a batch record? Where the consequence cannot be tolerated, the business risk is high and the rigor applied to the validation of that function increases. The process often adopted to determine failure consequences is Failure Modes and Effects Analysis (FMEA).

The findings and recommendations of the GxP Assessment must be clearly documented for reference throughout the project.

Rapid Application Development

Iterative or Rapid Application Development (RAD) is frequently employed in the development of configurable information systems. Where this approach is used, a team is normally formed comprising system integrators and key users who collectively develop the preferred solution. The benefit of such an approach is that users are able to quickly visualize the system and therefore confirm its requirements at an early stage. Further, the system integrator is able to demonstrate the key features of the system and propose implementation options that meet both business and technical needs. The pitfalls of such an approach can be many if this phase of the project is not correctly managed. Typical pitfalls include the following:

- Requirements are invented on the fly and not documented.
- Appropriate controls, e.g., reviews, are not applied to the design.
- Implementation is ad hoc and does not conform to configuration and coding standards.
- Regulatory issues are not considered.
- The end solution is difficult to maintain.

In order to overcome such issues, it is essential to control this phase of the project in order that

- deviations, modifications and extensions to the original URS are captured;
- configuration is formally documented and reviewed;
- bespoke software is reviewed against coding standards; and
- the level 1 GxP Assessment can be refined.

A formal process of recording configurations and changes during the RAD phase is essential in order to record

- configurations;
- deviations, modifications and extensions to the URS;
- issues requiring further consideration; and
- rationales, assessments and assumptions.

Functional Specification

The Functional Specification has two primary objectives:

1. Retrospectively document configuration that can be immediately finalized.
2. Prospectively document the design basis for more complex functionality.

The primary inputs (Figure 2.13) to the Functional Specification are the records established in the RAD phase, structural design for bespoke software development and standard system documentation. The Functional Specification should clearly demonstrate that all requirements stated in the URS have been accommodated. Where there is a nonconformance with the URS, it should be documented and formally approved by the project sponsor.

The Functional Specification forms the basis for subsequent Operational Qualification (OQ). The Functional Specification should be sufficiently detailed to enable test objectives and acceptance criteria to be set. Detailed design may need to be progressed before detailed test methodologies can be developed.

Level 2 GxP Assessment

The level 2 GxP Assessment is conducted once the Functional Specification has been developed and reviewed. The assessment compares the Functional Specification with the original URS in order to establish the differences and

Figure 2.13. Creating the Functional Specification

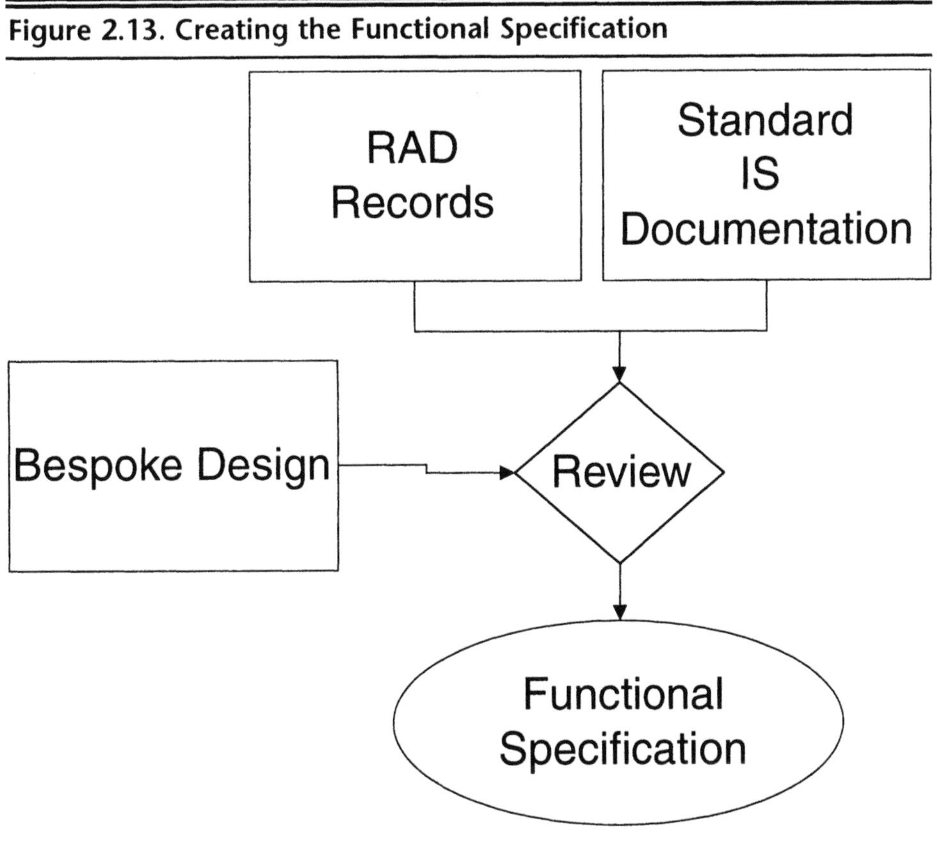

determine the impact on the original GxP Assessment. Following the review, the level 1 GxP Assessment is revised and the scope and approach to the validation amended accordingly.

Detailed Software Design

Where bespoke software needs to be developed to meet the URS, a detailed design is required. Often, however, bespoke software is developed within the RAD phase in the absence of software coding standards and configuration controls. This does not overcome the need for a fully documented design.

A Software Design Specification (Figure 2.14) must be developed to document the requirements of the bespoke software application. Typically, the Software Design Specification defines the software architecture in modular terms.

For each software module, the specification defines module architecture, descriptions, interfaces and relationships (events, timers, handshaking). Each module can be further decomposed into

- the module input parameter definition (integer, real, character)
- global and local data definitions,
- the parameter passing mechanism (pass by value, pass by reference),
- the detailed functional definition and
- function returns.

Detailed Hardware Design

Enterprise-wide information systems, by their definition, must be networked. The physical hardware and configuration supporting the applications must be specified, which is a combined effort between the hardware supplier and the internal IS function.

The system supplier will define the hardware requirements required to support the application within the framework of standards set by the internal IS group. The Hardware Design Specification will define the following parameters:

- Performance (CPU [central processing unit], bus, cache, clock, etc.)
- Capacities (RAM [random access memory], hard disk, floppy disk, DAT [digital audio tape], CD-ROM [compact disk, read-only memory], etc.)
- Peripherals (MMI [man machine interface], printers, keyboards, mouse, bar codes, etc.)
- Interfaces (I/O [input/output] cards, dedicated/network, cabling, speed)

Figure 2.14. Software Design

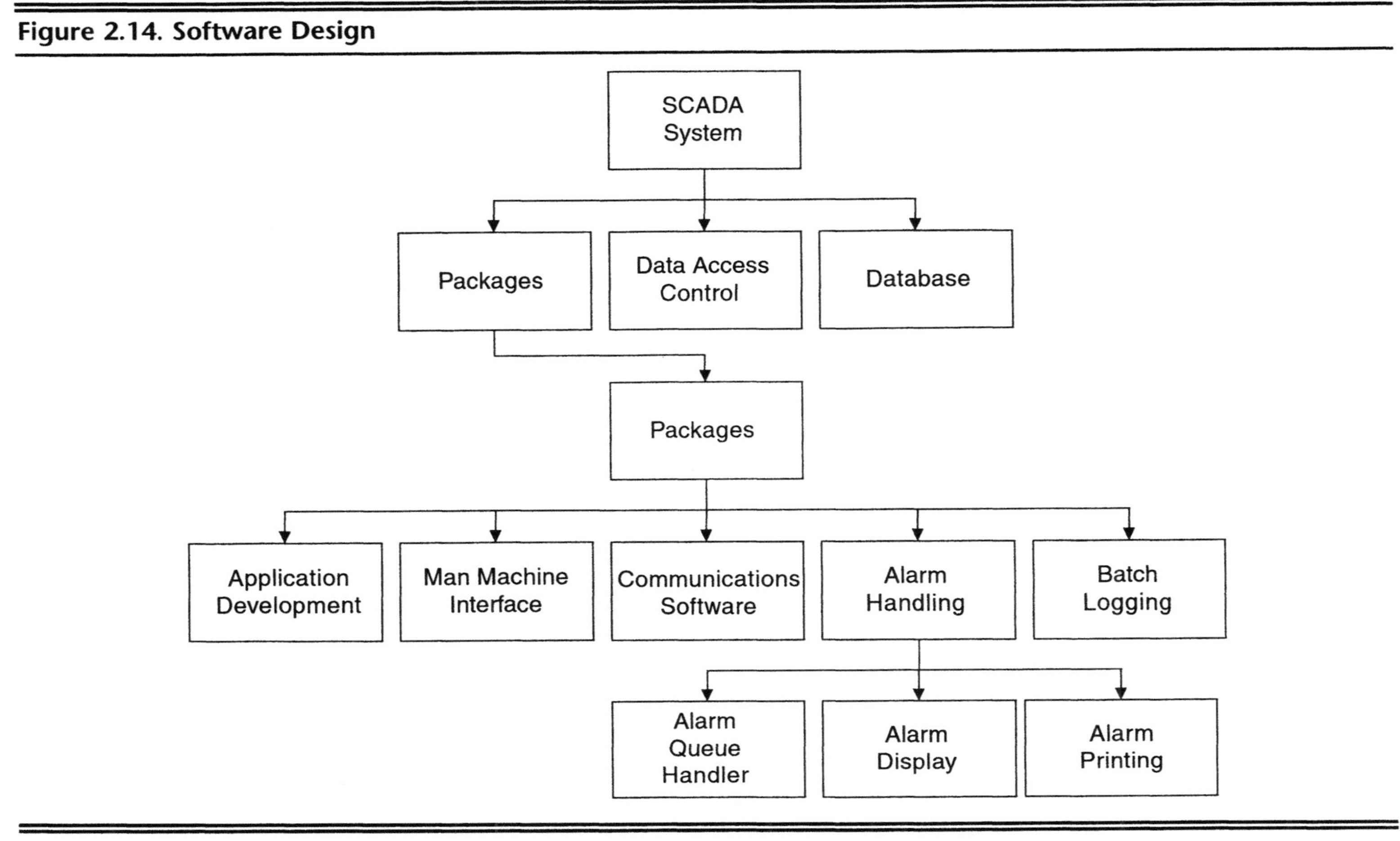

- Settings (switch settings, firmware configuration)
- Environment (temperature, humidity, RFI [radio-frequency interference], UV [ultraviolet], electromagnetic)
- Electrical supplies (UPS [uninterruptible power supply], earthing, filters, etc.)
- Define relevant standards (safety, electrical, etc.)

In addition, a Configuration Specification should be developed to document network addressing and routing, switch settings, communications set-up and so on.

Implementation

When only minimal bespoke software development is required, the implementation phase is integrated into the RAD phase. This means that the final solution (excluding minor modifications) will be the output of the RAD phase.

Where there is a greater need for bespoke software development, the implementation (coding) must be controlled by coding standards. A typical coding standard will address the following:

- Data scoping (global, local, parameter passing)
- Module size
- Module layout
- Module cohesion and coupling
- Naming conventions (modules, procedures, data, etc.)
- Use of control blocks (if-then-else, while-do, repeat-until)
- Module structure (indentation, control block nesting, etc.)
- Commenting (annotation)

As bespoke software presents the greatest risk of system failure, an independent Source Code Review must be conducted on the critical aspects of the development. Critical aspects will generally cover

- calculations,
- status messages,
- synchronization of system task,
- interrelationships between business processes,
- start-up and shutdown,
- data storage and
- critical functions determined by GxP Assessments.

Corrective actions arising from the software code review may lead to remedial software development or, more likely, additional testing.

Level 3 GxP Assessment

The level 3 GxP Assessment is the final assessment intended to capture any modifications applied during the implementation phase. Such modifications may arise from tidying up activities following the RAD phase or as a result of technical issues that arise during the implementation phase. The scope of the qualification phase must be reviewed and refined in accordance with any issues raised.

Qualification Protocols

Installation Qualification (IQ) and OQ protocols will be developed to verify the installation and functional operation of the system. IQ verifies the software and hardware installation against the Software and Hardware Design Specifications, ensuring that the correct software and hardware components have been installed, identified, connected, configured and documented. OQ is functional testing of the application software and configuration focusing on GxP critical functionality.

Inspections and functional tests defined within the qualification protocols must meet pharmaceutical industry standards. These standards dictate that each test protocol must include the items listed in Table 2.5.

Table 2.5. Qualification Protocol Requirements and Objectives

Requirement	Objective
Objective	A clear and unambiguous statement of what the test aims to demonstrate.
Traceability	The test is traceable to Functional, Hardware or Software Design Specifications in order that the scope of testing can be verified.
Prerequisites	Test sequencing and test environment setup requirements are clearly defined.
Equipment	Equipment required to conduct the test such as measurement devices and simulators are available and calibrated.
Methodology	Stepwise approach to conducting the test in order to meet the test objective.
Acceptance criteria	Clear statement of success criteria for test.
Results/observations	Provision of forms/space for recording of test outcomes.
Pass/failure	Status of inspection or test, i.e., Do results satisfy acceptance criteria set forward?
Approval	Approval of testers, witnesses and/or reviewers.

Data Migration

In addition to the installation of new systems, the IS strategy will also involve the replacement of legacy systems—systems that are no longer supported or do not meet the current business needs. When replacing a system, data migration becomes a critical issue. Indeed, even when the legacy system is a manual, paper-based system, there are still data migration issues to be considered. Table 2.6 defines outline methods for addressing such issues.

Table 2.6. Data Migration Issues

Migration Category	Outline Method for Addressing Issues
Obsolete data	• Identify obsolete data. • Archive data for regulatory inspection, product review and so on. • Delete obsolete data from legacy system before transfer.
Inaccurate data	• Review data to determine inaccuracy. • Prioritize risk of inaccurate data in accordance with criticality. • Correct inaccuracies in accordance priority. Note: Low priority (i.e., low criticality) data may be migrated prior to rectification; however, a plan must be in place to address the issue.
Overlapped data	• Review data to establish overlaps. • Ensure consistency between overlaps. • Remove overlaps where there is a high business risk. • Develop plan to improve information structures.
Omissions	• Omissions will be addressed on a business and system needs basis. The business need relates the criticality of the information in the operation of the business. System needs are where new features provided by the system require the population of certain data that do not currently exist. • Determine omissions. • Prioritize on business and system basis. • Populate new system following legacy data transfer.
Transfer integrity	• Develop and validate automatic data migration routines. • Validate data migration routines. • Conduct migration. • Verify new database against old.

Conference Room Pilot Testing/Operational Qualification

The OQ protocol is executed once the system has been loaded with the operational data, accepting that a final data migration may be required to capture updates occurring during the test period. This process may often be referred to as "conference testing" or the "conference room pilot".

This testing further presents an ideal opportunity for training. Personnel who will be involved in the operation of the system after cutover often form part of the test team. Working through detailed test methods and contributing to the analysis of test failures is a valuable learning vehicle.

Cutover Testing/Performance Qualification

Risk Management Plans should be established prior to cutover in order to define the fallback measures to be taken in the event that the system fails to perform within the operational environment. Risk management strategies may include parallel running of manual or legacy systems, interim manual systems or fallback to the previous versions. Whenever possible, a phased cutover from manual or legacy systems is recommended.

Following successful cutover, the Performance Qualification (PQ) is conducted. PQ monitors the ongoing performance of the system within the operational environment. Changes and faults are monitored in order to establish a measure of the success of the system design and implementation.

Creation of an Operational Environment

The greatest threat to the validated status of an information system is the failure to establish a controlled environment for system management, operation, modification and failure (Figure 2.15). An overview of the operational environment is given below; more detail can be found in Chapter 13.

Change Control. Change control systems must be established to ensure that critical changes are appropriately defined, planned, authorized, implemented, tested and approved. In essence, the regulatory and business impact of changes must be validated.

Service Level Agreements. Service Level Agreements must be established between the user and supplier to ensure that the appropriate expertise and facilities are available to assess and implement changes, either as a result of defects or a change in business need.

Operational Procedures. SOPs must be developed to cover system operation. Such procedures may be implemented in electronic form, providing they are suitably controlled.

Figure 2.15. Operational Environment

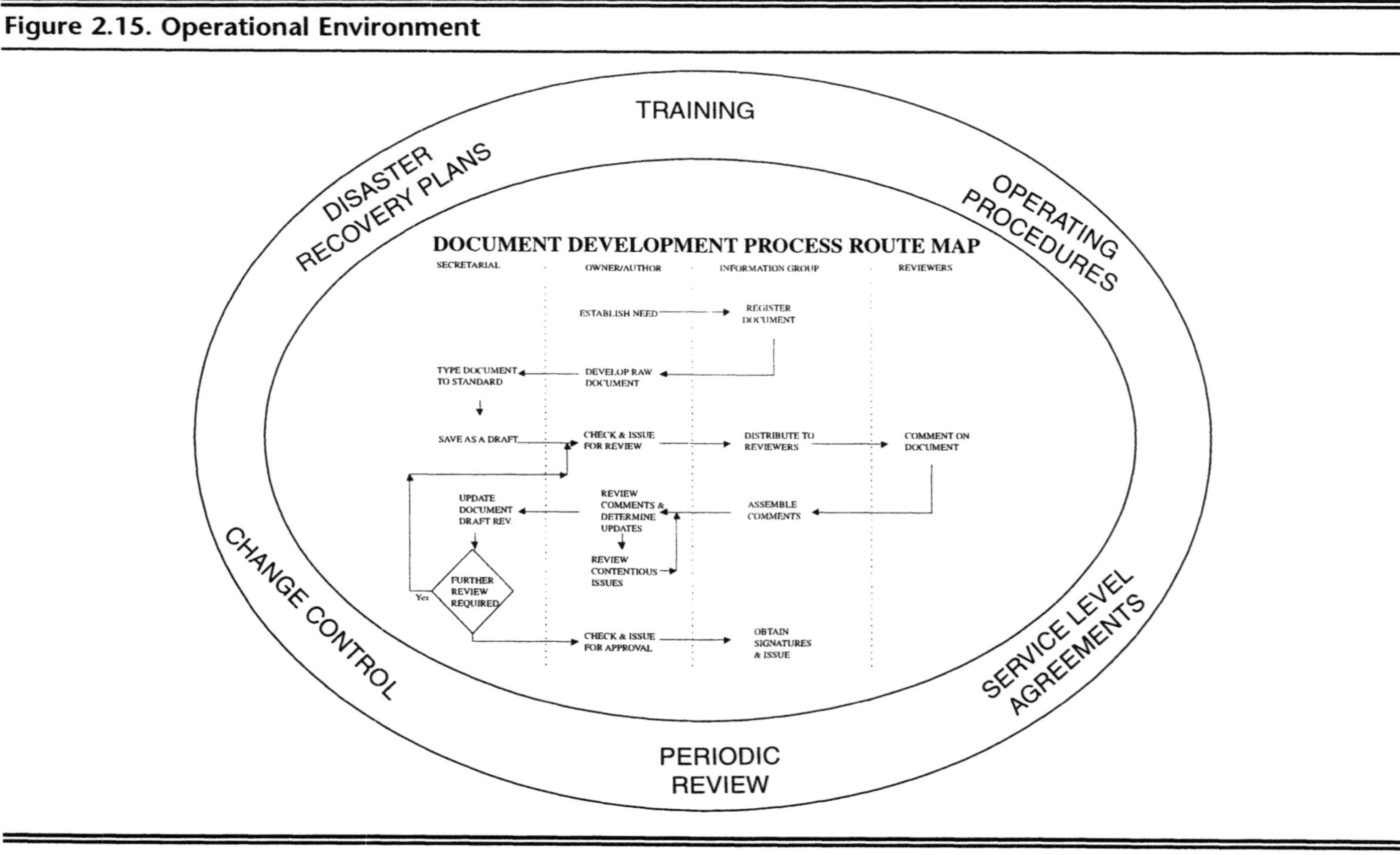

Disaster Recovery. SOPs must be developed to ensure that it is possible to recover from system failures in a controlled manner to a known state. Such procedures will cover backup, archive and restoration, check-pointing and so on. In addition to being able to recover to a known state, it is also necessary to consider backup processes to ensure GMP compliance during the period of system outage and also to assess and address potential noncompliance arising from the failure.

Training. All personnel involved in the operation or maintenance of the system must be appropriately trained or experienced to conduct their duties as stated in GMP regulations. The project must determine the level of training required in order to establish the appropriate awareness, understanding and/or expertise required to exercise these tasks.

Periodic Review. Periodic validation reviews should be conducted at approximately 12-month intervals. The periodic review has a number of objectives:

- Review change controls to ensure that appropriate testing has been conducted to maintain validated status. Where change controls are focused in a common area of functionality, the original validation should be reviewed for its adequacy.
- Determine the impact of any changes to regulatory requirements on the validation package.
- Review fault reports to establish the adequacy of the original validation.

Validation Report

The Validation Report must respond to the VMP, clearly summarizing the approach taken against the original intent. All documentation produced should be listed, including those generated in support of the operational environment.

Any deviations from the approach proposed by the VMP and any failures encountered during testing must be documented in the report, including a rationale of the impact of the deviation.

The report should conclude with a Corrective Action Plan and a statement of suitability of the system to move into operational life. Any restrictions on use of the system must be clearly stated.

SUMMARY

There is an explosion in the reliance on information systems to support the management of pharmaceutical research, development, manufacture and distribution. The benefits of exploiting the latest technology include improved definition and control of business processes, improved efficiency and greater regulatory compliance. As the level of dependence on automated systems increases, risk strategies must shift from people-based processes to automated processes. Understanding such risk is an essential ingredient to establishing a rational validation strategy. The use of quality and robust organizations in conjunction with adopting industry-standard system development life cycles is the means by which the risks can be demonstrably managed.

REFERENCES

1. UK GAMP Forum. 1998. *Suppliers Guide for Validation of Automated Systems in Pharmaceuticals Manufacture,* Version 3. Available from the International Society for Pharmaceutical Engineering.

2. Wingate, G. A. S. 1997. *Validating Automated Manufacturing and Laboratory Applications—Putting Principles into Practice.* Buffalo Grove, Ill., USA: Interpharm Press, Inc.

BIBLIOGRAPHY

European Union Guide to Directive 91/356/EEC. 1991. *Directive Laying Down the Principles of Good Manufacturing Practice for Medicinal Products for Human Use, European Commission.*

FDA. 1998. U.S. Code of Federal Regulations Title 21, Part 210: Current Good Manufacturing Practice in Manufacturing, Processing, Packaging, or Holding of Drugs (General); Part 211: *Current Good Manufacturing Practice for Finished Pharmaceuticals.* Washington, D.C.: National Archives and Records Administration.

Rules and Guidance for Pharmaceuticals Manufacturers. 1997. London: Her Majesty's Standards Office.

3

Regulatory Expectations

Heinrich Hambloch
GITP
Kredfeld, Germany

SURVEY OF REGULATIONS AND GUIDELINES

Since 1982, a number of guidelines applicable to information technology (IT) and network applications have appeared for the validation of computer systems. Between 1995 and 1998, five comprehensive regulations and guidelines were published. It is therefore timely to compare these guidelines in regard to their scope and contents.

This chapter reviews the regulations and guidelines that

- were published after 1990,
- were published before 1990 and represent a milestone in the validation of computer systems
- apply to the development and manufacture of conventional drugs (no medical devices, blood products etc.).

Table 3.1 lists the regulations and guidelines discussed in this chapter, and Table 3.2 presents an overview of the number, origin and scope of the guidelines. Many of these documents have been only recently published. These regulations and guidelines differ in their respective scope and fall into three types:

Table 3.1. Regulations and Guidelines for the Validation of Computer Systems (1982–1998)

Title of Regulation/Guideline	Country/ Field	Authors/Source	Regulatory Authority (RA)/ Pharmaceutical Industry (PI)/ Binding for the Industry (Y/N)	Year of Publication
Compliance Policy Guides on Computerized Drug Processing: • Input-Output Checking • Identification of Persons—Batch Control Records • cGMP Applicability to Hardware and Software • Vendor Responsibility • Source Code for Process Control Application Programs	USA GMP	FDA	RA (Y)	1982–1987
Guide to Inspection of Computerized Systems in Drug Processing ("Blue Book"): Reference Materials and Training Aids for Investigators	USA GMP	FDA	RA (N)	1983
Validation Concepts for Computer Systems Used in the USA GMP Manufacture of Drug Products	USA GMP	PMA's Computer System Validation Committee (*Pharm. Technol.* May 1986, pp. 24–34)	PI (N)	1986
Guide to the Inspection of Software Development Activities: Reference Materials and Training Aids for Investigators	USA GMP	FDA	RA (N)	1987

Table 3.1 continued on the next page

Table 3.1 continued from the previous page

Title of Regulation/Guideline	Country/ Field	Authors/Source	Regulatory Authority (RA)/ Pharmaceutical Industry (PI)/ Binding for the Industry (Y/N)	Year of Publication
Computerized Data Systems for Nonclinical Safety Assessment—Current Concepts and Quality Assurance	USA GLP	Drug Information Association, (Maple Glen, Penn.)	RA + PI (N)	1988
EC Guide to Good Manufacturing Practice for Medicinal Products (The Rules Governing Medicinal Products in the European Community, Volume IV)—Annex 11, Computerized Systems (ISBN 92–826-3180-X)	EC GMP	European Community	RA (N)	1992
Use of Computers—Australian Code of GMP for Therapeutic Goods—Medicinal Products—Part 1, Section 9	Australia	GMP Therapeutic Goods Administration	RA (Y)	1993
Guideline on Control of Computerized Systems in Drug Manufacturing	Japan GMP	Matsuda, T. (*J. Pharm. Sci. & Technol.* 48:11)	RA (Y)	1994
The Application of the Principles of GLP to Computerized Systems	OECD Countries GLP	OECD Environment Directorate (Paris)	RA + PI (Y)	1995
Technical Report No. 18—Validation of Computer-Related Systems	USA GMP	PDA	PI (N)	1995
The APV Guideline "Computerized Systems", based on Annex 11 to the EU GMP Guide	Germany GMP	APV	PI (N)	1996

Table 3.1 continued on the next page

Table 3.1 continued from previous page.

Title of Regulation/Guideline	Country/ Field	Authors/Source	Regulatory Authority (RA)/ Pharmaceutical Industry (PI)/ Binding for the Industry (Y/N)	Year of Publication
The GMA/NAMUR Guides for the Validation of Control Systems: NE58, NE68, NE71, NE72	Germany GMP	GMA/NAMUR	PI (N)	1997
21 CFR Part 11—Electronic Records; Electronic Signatures	USA GCP GLP GMP	FDA	RA (Y)	1997
GAMP Guide for Validation of Automated Systems in Pharmaceutical Manufacture, Version 3.0	UK GMP	The GAMP Forum	PI (N)	1998
Guidance for Industry: Computerized Systems Used in Clinical Trials (Draft)	USA GCP	FDA	RA (N)	1998

*The shaded guidelines are part of the "Body of Knowledge".

Table 3.2. Number of Validation Regulations and Guidelines for Computer Systems Published Between 1982 and 1998 (Excludes 21 CFR Part 11)

	Authorship			
	Regulatory Authority	Pharmaceutical Industry	Both	Total
GCP	1*	0	0	1*
GLP	0	0	2	2
GMP	6	5	0	11
Total	7	5	2	14

*Draft

1. *Industry guidelines:* the compliance is voluntary in nature.
2. *Guidelines for inspectors:* not binding for the industry, however, they do present regulatory expectations and, as such, pharmaceutical manufacturers are wise to ensure that their systems and procedures are equivalent to these guidelines.
3. *Official regulations:* binding for inspectors and industry.

BODY OF KNOWLEDGE

Some of the organizations responsible for the industry guidelines are preparing a collective review of their guides into what is likely to be known as the "Body of Knowledge". Abstracts of these guidelines are presented below.

FDA Software Development Activities

The U.S. Food and Drug Administration (FDA) published a technical report on software development activities in 1987. While under revision, its content is still very relevant and covers basic expectations for software development practices. A simple life cycle is described: the requirements phase, the design phase, the implementation phase, the test phase, the installation and checkout phase and the operation and maintenance phase. The technical report also has some appendices covering firmware and software test methods.

FDA Electronic Records and Electronic Signatures Regulation

Electronic Records/Electronic Signatures, 21 CFR 11, was a long time coming, but it became fully effective on 20 August 1997. It has a potential huge impact, if rigorously enforced, mandating the validation of just about every computer system used in research and development (R&D), manufacturing and distribution sites with the exception of some financial systems. Chapter 16 discusses the practical implications of 21 CFR 11.

GAMP Guide

The GAMP Guide (GAMP 3) was produced by the Good Automated Manufacturing Practice (GAMP) Forum and presents the user and the supplier with some examples of best practice. It describes an overview of computer validation and provides a framework of procedures to support validation projects. The GAMP Guide currently includes copies of the APV (Arbeitsgemeinschaft fur Pharmazeutische Verfahrenstechnik, an international association for pharmaceutical technology) interpretation of Annex 11 of the European Union Good Manufacturing Practice (EU GMP) regulation and the GMA/NAMUR (Gesellschaft Meβ-und Automatisierungstechnik/Normenarbeitsgemeinschaft fü Meβ-und Regelungstechnik) validation documents.

PDA Technical Report 18

Technical Report No. 18 of the Parenteral Drug Association (PDA) presents the computer system validation life cycle and includes consideration of cycle refinement during the development phase. Some specific checklists are given for system requirements, vendor evaluations and specifications. The document is currently being considered for revision.

GMA/NAMUR Control System Validation Guides

The Control System Validation Guides of GMA/NAMUR were produced by user and supplier committees representing 11 international pharmaceutical companies and 11 international control system suppliers and were published between 1995 and 1997. The topics covered include prospective validation (NE58), retrospective validation (NE68), operating and maintaining (NE71) and validation support (NE72). The work aims to give a more prescriptive view of the practical requirements for validating process control systems. The various guides, which have been accepted as German standards, are largely checklist driven and cover development and operational compliance.

APV Guide to Annex 11—Computerized Systems

The APV Guide is an interpretation of Annex 11 that specifically concerns the requirements of computerized systems within the EU GMPs. This work was conducted by the IT specialist group within the APV. The interpretation aims to break down the top-level requirements laid out in Annex 11 into practical guidance that can be easily understood and applied at a hands-on level.

How the Body of Knowledge Fits Together

This chapter focuses on the regulations and guidelines making up the so-called Body of Knowledge. These regulations and guidelines were highlighted as shaded rows in Table 3.1. A detailed examination looking at the life-cycle activities addressed by these regulations and guidelines is presented in Table 3.3.

The comparison shows a high degree of agreement in terms of software engineering. All regulations and guidelines demand a software life cycle such as the waterfall or the V model (a natural exception being 21 CFR 11 which covers the use of electronic signatures and electronic records). The essential difference between these regulations and guidelines lies in the degree of the detail. The following categories were formed in order to classify the degree of the detail:

- Level 0: Guideline does not mention the topic.
- Level 1: Guideline indicates *what* is required.
- Level 2: High-level general guidance about how to approach compliance.
- Level 3: Guidance which supplies practical advice on *how* to comply.
- Level 4: Guidance to allow a framework for an SOP (Standard Operating Procedure) to be drawn up.

One interesting observation from this analysis is that the operation of computer systems besides change control is covered in only a few guidelines. Table 3.4 summarizes the results of Table 3.3.

EXAMPLES OF NONCOMPLIANCE REPORTS

What happens if a regulatory inspector discovers a violation against the regulations and guidelines during an inspection? If the non-compliance is significant, then the regulatory authority will send an inspection report to the pharmaceutical manufacturer, detailing the findings and requesting or

Table 3.3. "Body of Knowledge" Comparison

Guideline	Version	Applies to	Software Development													Operation of the Computer System							
			Project Outline	Supplier Audits	Quality Planning	Requirements Specification	Design Specification	Risk Assessment	System Construction (HW and SW)	All Phases of Testing	User Documentation	Configuration Management and Control	Installation	System Acceptance Testing	Release to Production	Monitoring and Maintenance	Change Control	Error Handling	Data Storage, Archiving and Retrieval	System Failure and Disaster Recovery	Access Control	Data Input Control	Periodic Reviews and Revalidation
FDA Software Development Activities	1987	All computerized systems																					
EU Guide Annex 11	1992	All computerized systems																					
PDA Report 18	1996	All computerized systems																					
APV Guide to Annex 11	1996	All computerized systems																					
GMA / NAMUR Guide	1997	PLCs and DCSs																					
21 CFR Part 11	1997	All computerized systems																					
GAMP 3	1998	All computerized systems																					

Level 0—Does not mention the topic.
Level 1—Official regulations indicating **what** is required.
Level 2—High-level general guidance about how to approach compliance.
Level 3—Guidance which supplies practical advice on **how** to comply.
Level 4—Guidance to allow a framework for an SOP to be drawn up.

PLC = Programmable Logic Controller; DCS = Distributed Control System; HW = hardware; SW = software

Table 3.4. High Level Summary of the "Body of Knowledge"

Guideline (Short Title)	Level of Detail	
	Software Development Level	Operation
FDA Software Development Activities	4	0
EU Guide Annex 11	1	1
PDA Report 18	2	1
APV Guide to Annex 11	2	3
GMA/NAMUR Guide	4	4
21 CFR Part 11	0	4
GAMP 3	4	0

demanding a resolution. The confidentiality of this information, however, is handled differently in the United States and the European Union (EU):

- In the European Union, this information is absolutely confidential and is not published.
- In the United States, there is the so-called "Freedom of Information Act", which states that everyone with a justifiable interest in the information has a right to receive it. In addition, there are publication firms, for example, GMP Trends Inc., that regularly publish extracts from the latest GxP inspection findings released into the public domain. The publications include extracts covering computer systems validation.

U.S. FDA Form 483 and Warning Letters

The U.S. FDA first describes its findings on a form, the so-called "Form 483". This form is completed immediately after the inspection and is sent to the company so that it can initiate the necessary corrective actions. Some corrective actions may have been taken in response to FDA observations during the audit, and the inspector will normally take account of this proactive response. A quick response to the formal issue of a Form 483 is required by the pharmaceutical manufacturer to demonstrate management commitment to the resolution and close out issues identified by the FDA.

Between 1984 and 1995, 163 Form 483s regarding the validation of computer systems were submitted to pharmaceutical companies. These

Table 3.5. Classification of the Form 483s Related to the Validation of Computer Systems (1984–1995)

Class	Number of Findings
Testing and/or Qualification	132
Development Methodology	101
Validation Methodology and Planning	95
Change Control/Management	91
Quality Assurance and Auditing	89
Operating Procedures	51
Device Manufacturing	19
Security	25
Hardware, Equipment Records and Maintenance	23
Training, Education and Experience	11

companies were drug, medical device and blood product manufactures. These forms contain 627 individual findings, which are grouped into classes in Table 3.5 [1].

Table 3.6 presents some examples of Form 483s. They have been published in the "GMP Trends" in the years 1996 to 1998 and involve findings from both manufacturing facilities and analytical control laboratories.

The next level of alert after the Form 483 is the Warning Letter. The FDA defines a Warning Letter as follows:

> *A written communication from FDA notifying an individual or firm that the agency considers one or more products, practices, processes, or other activities to be in violation of the Federal FD&C Act [Food, Drug and Cosmetic Act], or other acts, and that failure of the responsible party to take appropriate and prompt action to correct and prevent any future repeat of the violation may result in administrative and/or regulatory enforcement action without further notice.*

More background information about the nature of Warning Letters can be found at http:/www.fda.gov.

In the field of computer systems validation, by far the largest proportion of Warning Letters is issued to the medical device industry. For conventional pharmaceutical drugs, Warning Letters are the absolute exception. Table 3.7 shows a typical Warning Letter from 1998 on computer systems validation for an IT system used by a pharmaceutical manufacturer.

Table 3.6. Examples of Form 483s from 1996 to 1998

There are no detailed specification documents for any of the computerized process control systems that contain sufficient details on how these systems/software were represented and developed. The only specification documents made available and referred to as the design document were the 'System Specifications" However, these documents only provide a high level explanation of what the systems do. They lack sufficient detailed description of specific and complete data structure, data control flow, design bases, procedural design, development standards and so on to serve as the model for writing code and to support future changes to the code.

Failure to comply with computer and network system backup procedures in that not all required backup processes were documented as scheduled, showing lack of documented evidence that tape replacements were done. The current backup logbook has entries only for backup and shows several weekly backups using the "manual contingency tape backup procedures" (for reverification purposes), yet there is no documented evidence this is done.

The following items were observed in relation to the validation of Quality Tracking System (QTS) software: Software validation packages lack detailed specifications to test against. Without detailed specifications to test against, thorough testing cannot be and is not performed. Without thorough testing, proper validation cannot be accomplished.

QTS software validation specifications lack details such as the hardware platforms the software should run on (model number, CPU [central processing unit], RAM [random access memory], hard drive size, free hard drive space, etc.), the software operating system, software database specifications (such as number of records accommodated, number of fields, whether those fields are character or numeric or date or alphabetical fields), the number of concurrent users allowed, the development platform used to write the application, network information, report contents and other information needed to perform testing properly.

No written specifications for the computer system used to generate batch records and print/reconcile labels, including descriptions of such items as hardware and software to be used, system functions (all activities, operations and processes to be performed), all error and alarm points, measures to verify accuracy of the system during use, measures to protect the system from accidental or intentional abuse and testing to be conducted to validate the operation after installation.

No validation of the computer system used to generate batch records and print/reconcile labels to include such items as source code inspection, testing before installation, use and testing after installation involving nominal (normal range) input values, boundary input values at and outside of normal input limits, invalid or unexpected input values and data volume and environmental extremes.

Complete data derived from all tests are not maintained by the Quality Control Laboratory. The computer automated laboratory system, which calculates the percent purity and potency of products released for distribution and tested for stability by HPLC (high performance liquid chromatography), is programmed to overwrite chromatographic data. The firm does not electronically store all integration parameters and chromatograms, nor can the firm determine if reintegration of a sample occurred once or several times. Each reintegrated chromatogram replaces the previous chromatogram.

Table 3.6 continued on the next page

Table 3.6 continuedfrom the previous page

There are no SOPs for determining the degree of testing necessary to assure the proper function of the system following any hardware or software modifications. There are no SOPs in place to periodically revalidate and challenge the software program to assure data acquired on the system are accurate and reliable for determining the purity and potency of products.

The software error checking system is not periodically challenged and evaluated. Software validation does not include the challenge of critical decision paths and error routines within the program to determine the effect on program execution. Testing has not been conducted simulating worst-case conditions.

The right to publish these examples is acknowledged to GMP Trends Inc.

Table 3.7. Warning Letter from the Validation of Computer Systems for the Manufacture of Conventional Drugs*

Warning Letter 03/98 (Extract)

1. Inadequate retrospective validation of the . . . network computer system used for overall quality assurance/quality control operations and for the control and release of raw materials, in-process materials and finished products. For example,
   a. There were no original planning or systems design documents included in the validation materials for the programs.
   b. There were no structural and functional designs included in the validation materials for the programs.
   c. Only a small fraction of each program's code underwent detailed review.
   d. Sections of code lacked annotations (e.g., the meaning of variables) and contained "dead" or unused code.
   e. There were inconsistencies in the review of the validation documentation.
   f. There was inadequate software version control.

Your response indicates that your firm is looking torward replacing the system rather validating the existing system. The new system is expected to be in place and validated prior to . . . The first phase of installation and qualification of the new system, affecting production and warehousing, is to be completed in . . . The new system should not be used until it is satisfactorily validated. The current system is unacceptable for current use because of the deficiencies listed above. During the telephone conference on March 18, 1998, you agreed to submit a plan to bring the system into compliance with validation requirements. We agreed to review this corrective action plan.

*published on http://www.fda.gov

EU Notifications of Noncompliance

As already pointed out above, the results of inspections in the EU are confidential and, therefore, only few published findings are available. Table 3.8 presents some findings that an UK Medicines Control Agency (MCA) inspector has published in abstract form [2].

Consequences of Noncompliance

The severity of a noncompliance citation depends on the

- nature and extent of the noncompliance(s),
- impact on product quality and data integrity associated with drug and
- compliance history of the pharmaceutical manufacturer with the regulatory authority.

The pharmaceutical manufacturer should quickly establish a response to the regulator's findings, which is commensurate with the severity of the findings. Adequate and timely corrective action plans put into place by the pharmaceutical manufacturer in response to regulatory inspection findings are key to reestablishing a level of confidence between the pharmaceutical regulator and the pharmaceutical manufacturer. If the regulatory authority is not satisfied with the response, it is likely to escalate its action, possibly resulting in the revocation of the license to market a drug within the regulator's home market. The consequence of lost sales revenue from a market can be devastating to a pharmaceutical manufacturer. There is a clear financial incentive for compliance.

Table 3.8. Inspection Findings Published by an MCA Inspector

- Suppliers have not been audited . . .
- Responsibilities for the system . . . were not defined.
- Detailed written descriptions of the computer systems were not kept up to date.
- The disaster/breakdown arrangement did not work.
- No routine system audits were performed by Quality Assurance.
- . . . the database recorded theoretical rather than actual weighings.

INFRASTRUCTURE FOR THE VALIDATION OF COMPUTER SYSTEMS

How can a pharmaceutical company prevent a complaint from the regulatory authorities? The answer is to establish a quality system infrastructure to govern the acquisition and development of software and the acquisition and operation of hardware. Such a system should be implemented for both regulatory compliance and cost savings:

- Equivalent work is performed in the same way everywhere.
- The work is done correctly and completely; nothing will be omitted.
- New employees are trained faster.

Chapter 4 contains an example validation policy, which describes a quality management system for the hardware and software components making up an IT system and its support environment. Chapter 2 also discusses IT strategies with due consideration to validation compliance.

The term *validation* is often misunderstood by IT professionals not familiar with pharmaceutical industry terminology. Validation has often erroneously been seen as limited to a testing activity at the end of a software/hardware project. Validation in the pharmaceutical industry is closely aligned to the principle of quality assurance, which is a holistic approach that aims to build quality attributes into a system rather than verifying quality through testing. A system that is built according to

- well-structured requirement and design specifications prepared according to well-developed checklists,
- well-written state-of-the-art development guidelines for programming/parameterization,
- strict configuration management using modern tools and
- thorough review techniques applied to all documents and source code/parameterization

will have the necessary quality assured, and testing just confirms that. Testing does not add any quality to a system.

CONCLUSION

This chapter has reviewed the pharmaceutical industry's guidelines and regulations governing computer systems validation. Special emphasis has been placed on regulatory expectations and inspection practice. The aim has not been to present a detailed examination of inspection practice but to highlight key issues affecting the validation of IT and network applications.

REFERENCES

1. Wyrick, M. 1998. State of the Industry—Current Events. Presentation made at the ISPE Seminar: Computer Systems for the New Millennium, in Orlando, Fla. on 18 November 1998.

2. Trill, A. 1994. Computer Validation Projects, Problems, and Solutions: An Introduction. In *Good Computer Validation Practices,* edited by T. Stokes, R. C. Branning, K. G. Chapman, H. Hambloch, and A. J. Trill. Buffalo Grove, Ill., USA: Interpharm Press, Inc.

4

IT Validation Policy

Heinrich Hambloch
GITP
Kredfeld, Germany

In this chapter, an example company policy for quality assurance of IT (information technology) systems is presented. It provides an overview of Good IT Practices that would normally be expected from internal IT project and support groups within a pharmaceutical manufacturing company. These IT practices are equally applicable to supplier organizations providing project and support services.

The terminology used this chapter is not prescriptive—different organizations will naturally use different terminology. This is not an issue in its own right as long as the terms used are defined, and the basic concepts and principles outlined in this chapter are covered.

The chapter appendices are example procedures for risk assessment and performance monitoring of IT systems. These procedures would not normally be included as appendices to an actual validation policy but are included here as useful reference material.

GOOD IT PRACTICES

This chapter describes an IT Quality Assurance System (ITQA System) for software and its environment (server, networks, personal computers [PCs]), giving the basic procedures and general responsibilities involved. The detail measures of the ITQA System would be described in associated Standard Operating Procedures (SOPs).

ITQA is not to be regarded as a separate activity; it is an integrated part of each IT project. Therefore, the term *validation* is generally avoided in this chapter because it was and unfortunately is often seen as an activity performed at the end of a project rather than as an integrated activity which goes along with the project.

Good IT Practices are applicable for QA activities of all IT systems which acquire and process data from processes subject to GxP (x = C, D, L, M; hence GCP [Good Clinical Practice], GDP [Good Distribution Practice], GLP [Good Laboratory Practice], GMP [Good Manufacturing Practice]). IT systems in this field are subject to QA if they

- may have an impact on the pharmaceutical-technological quality of a product;
- may affect the safety of a patient; and
- acquire, process or store information, which may become part of approval documents submitted to a regulatory authority or may be subject to inspection by that regulator.

The guidance presented in this chapter only applies to the IT system and its interfaces to other pharmaceutical systems (Figure 4.1). The whole computerized system can only regarded as being "validated" if all the parts of the IT system and its interfaces to other pharmaceutical systems have undergone quality assurance.

QA APPROACH

All of the activities described in this chapter must be documented, whether or not this is explicitly stated.

IT Objects/Activities

All objects and activities that influence the quality status of an IT system, should be put under QA (see Figure 4.2), including

- hardware;
- operating system and tools (system software);
- application software kernels;
- configured application software, new bespoke developments; and
- the system environment and operating procedures.

Figure 4.1. Parts of a Computerized System

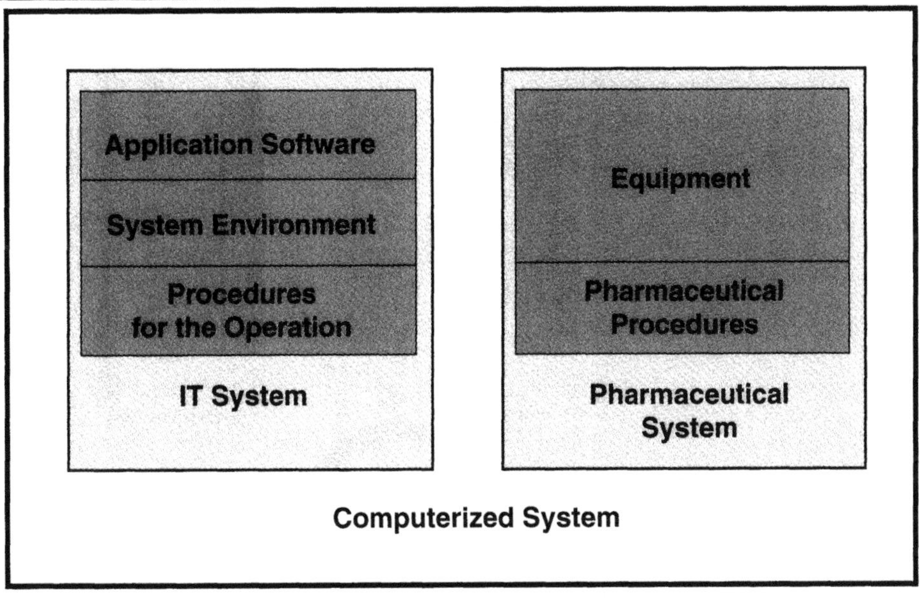

Quality Criteria for IT Objects/Activities

The following criteria concerning quality should be assessed for hardware and software suppliers:

- The manufacturer has been established on the market for a long time.
- There are a high number of installations and users.
- The manufacturer has an ISO 9000 certificate or a corresponding Quality Assurance System/Quality Management System (QMS).
- The manufacturer offers services (e.g., helpdesk).
- The vendor is authorized by the manufacturer.
- There is long-term guaranteed performance.
- The product line is mature.
- There are good references within the pharmaceutical sector.
- Use of the product was not objected to during inspection by regulatory authorities.
- Tools for installation, simulation, configuration, monitoring, documentation and introduction are supplied with the product.

Figure 4.2. Objects and Factors Affecting the Quality of IT Systems

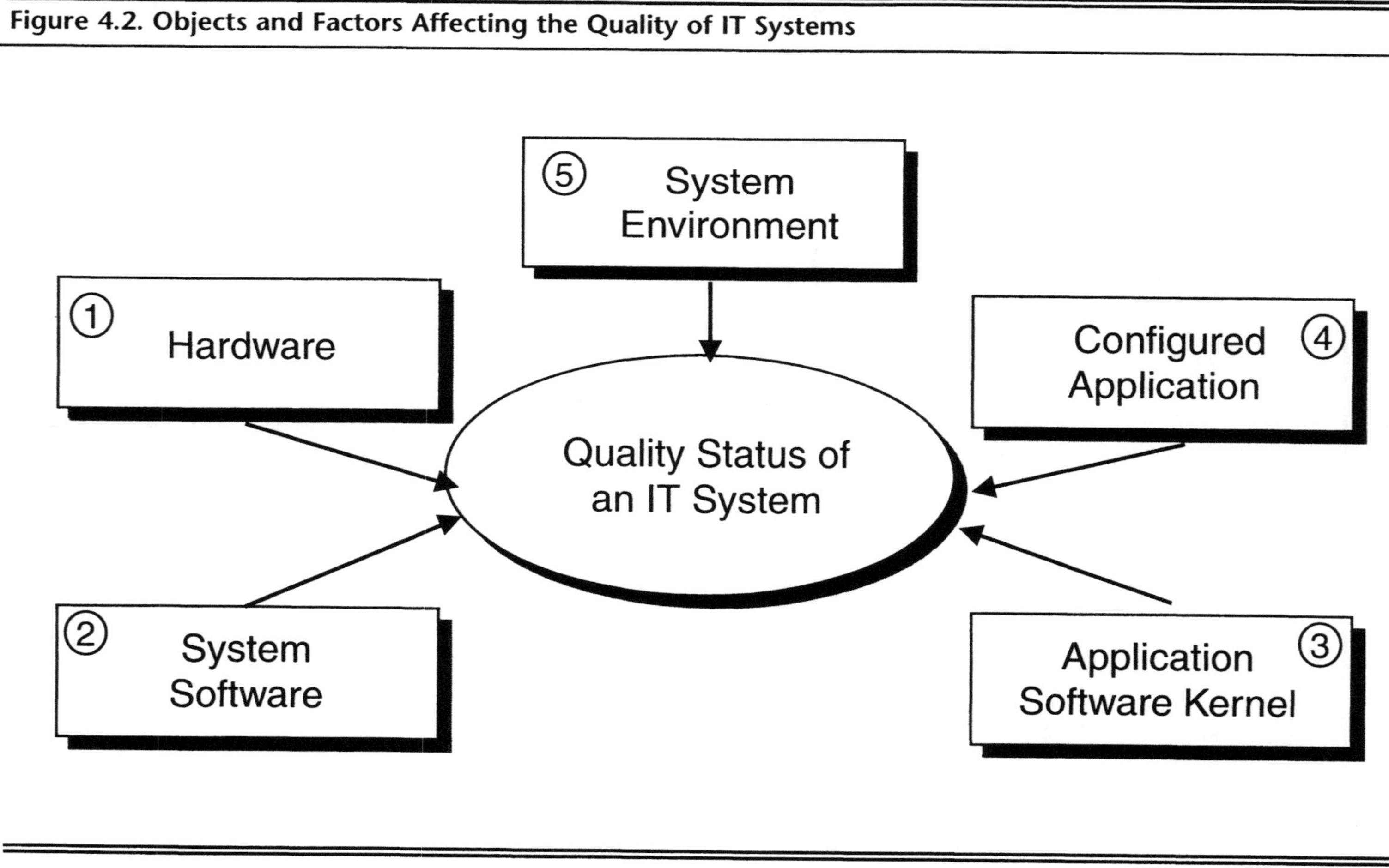

If the quality of hardware and software objects is not adequately covered through this evaluation, it should be checked before procurement via a questionnaire that contains questions regarding the Quality Assurance System of the supplier. If the questions are not answered satisfactorily and cannot be demonstrated with examples, an audit should be conducted.

An audit is a detailed check of the Quality Assurance System by IT specialists. An audit is carried out on the basis of a SOP containing a checklist with the respective questions. Audits conducted by other companies may be referenced. (Supplier audits are discussed in more detail within Chapters 14 and 15.)

Hardware

When planning an IT system, only reliable hardware should be considered for implementation.

Operating System and Tools

Operational software and tools are defined as

- operating systems (e.g., UNIX®, VMS® [Virtual Memory System], VM/VSE®, DOS®, MS Windows®, Windows NT®);
- network software (e.g., Novell®, Pathworks®);
- system management and diagnosis tools (e.g., Norton-Commander®, Novell-NMS®);
- programming languages and appropriate development tools;
- database software (e.g., Oracle®, query languages);
- statistics software (e.g., SAS®, Statgraphics®);
- office communication (e.g., All-in-One®, Teamlinks®, Lotus Notes®); and
- end user tools (e.g., MS Excel®, MS Word®, ABC Flowcharter®, Harvard Graphics®, Project Manager®).

Only proven operational software and tools should be considered.

Application Software Kernels

Application software kernels are predefined software packages which are adapted to the requirements of the company by configuration/parameterization (e.g., LIMS [Laboratory Information Management Systems], MRP [Materials Management System], PLC Programmable Logic Controller]). Only reliable application software kernels should be selected.

Configured Application Software and New Bespoke Developments

Configuration/parameterization and new bespoke development of application software should be carried out in accordance with a software Quality Assurance System (see Figure 4.3). The diagram shows the following relationships:

- The left side of the diagram consists of the documents in the System Development Life Cycle (SDLC).
- The right side of the diagram shows generally applicable SOPs of the QA System, describing the form and the content of SDLC documents, mainly in form of checklists. The quality of these SOPs is directly related to the quality of the software.
- The central part consists of project-specific instructions based on the SOPs of the right part, which result in the documents represented on the left.

The individual phases and the phase-independent activities of the SDLC are explained below:

- Phase-related activities/documents:
  - Create initial concept.
  - Create a project plan.
  - Create User Requirements Specification (URS)/functional description.
  - Perform risk assessment.
  - Create System Design Specifications.
  - Develop source code/configuration/parameterization.
  - Plan and execute module/integration/system test.
  - Develop the user and system manuals.
  - Installation.
  - Plan and execute acceptance test.
  - Release.
  - Decommission.
- Phase-independent activities/documents:
  - Problem management.
  - Change management.
  - Configuration control.
  - Training.
  - Project audits.
  - Document control.

Figure 4.3. QA System for Application Software

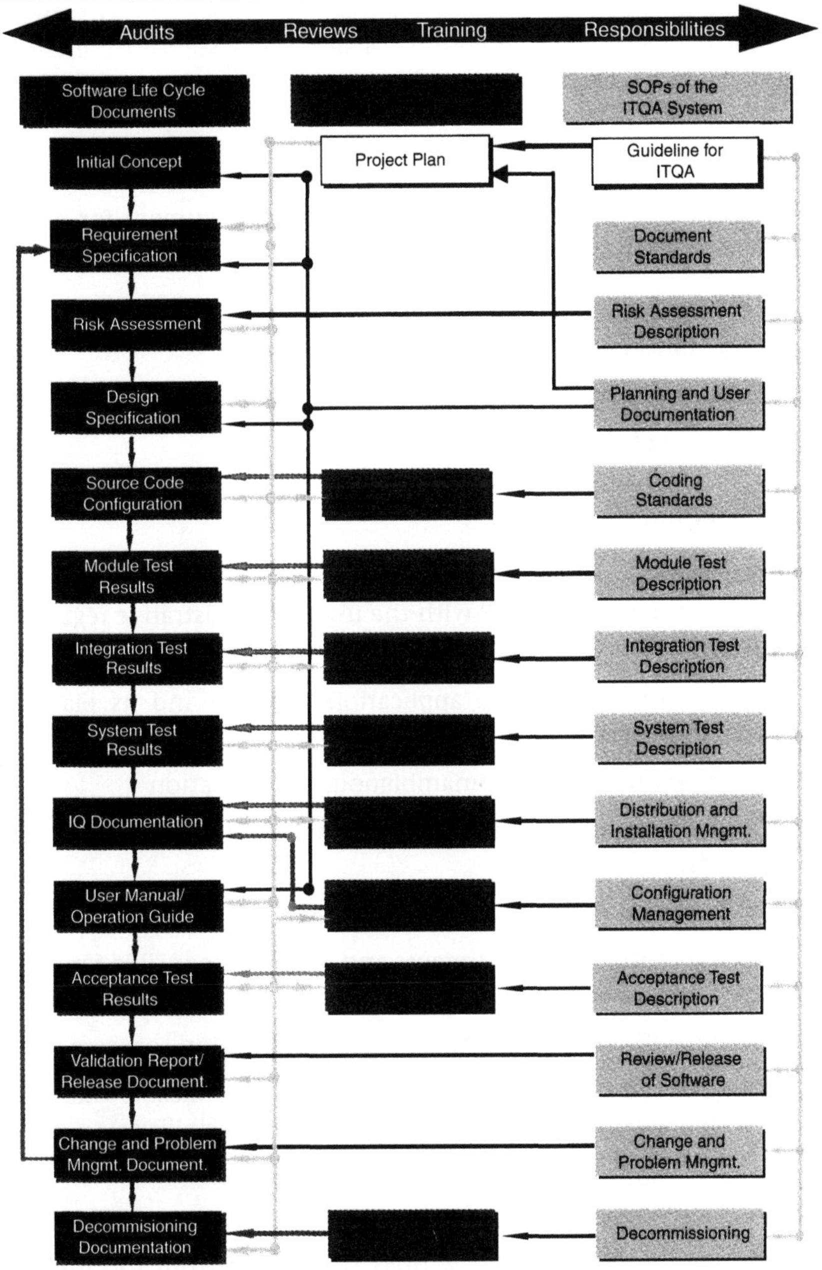

Audits Reviews Training Responsibilities

Software Life Cycle Documents

SOPs of the ITQA System

Initial Concept

Project Plan

Guideline for ITQA

Requirement Specification

Document Standards

Risk Assessment

Risk Assessment Description

Design Specification

Planning and User Documentation

Source Code Configuration

Coding Standards

Module Test Results

Module Test Description

Integration Test Results

Integration Test Description

System Test Results

System Test Description

IQ Documentation

Distribution and Installation Mngmt.

User Manual/ Operation Guide

Configuration Management

Acceptance Test Results

Acceptance Test Description

Validation Report/ Release Document.

Review/Release of Software

Change and Problem Mngmt. Document.

Change and Problem Mngmt.

Decommisioning Documentation

Decommissioning

These activities and their documentation represent a principle only and may be either combined (e.g., for a small spreadsheet macro application) or further subdivided according to the complexity of a specific IT system (e.g., for an MRP II [Manufacturing Resource Planning] system). Different parties (internal or external) may carry out these activities.

Initial Concept

The initial concept document serves as a decision document for project approval and should describe the following:

- Current practice
- Description of the benefits of a new system
- Outline of possible solutions
- Cost/benefit evaluation

Project Validation Plan

A project Validation Plan should be established for each IT application. The plan has the character of an SOP with the usual administrative regulation for these documents. The plan comprises the following elements:

- Characterization of the application software and its place in the overall process
- Manufacturer, version, unambiguous identification
- Short description of the application software, application scope and interfaces
- Project team and liaison responsibilities
- Listing of the SDLC documents and who is responsible for preparing them and adhering to work schedules
- List of SOPs used for development of application software
- Filing of the documents (according to the applicable GxP rules for documentation, type of storage, place, responsibilities)

Requirement Specifications

The URS/functional description describes the system from the user's point of view. It should be established by the users in cooperation with IT specialists and has to be detailed and structured in a way that it can be used as a basis for

- design specification,
- risk assessment,

- the establishment of acceptance test plans and
- the user manual.

Risk Assessment

The risk assessment examines the GxP impact of all the functions described in the specification. Its output will be the System Design Specification as well as the scope and depth of testing. An example SOP is presented in Appendix A of this chapter.

System Design Specification

The System Design Specification describes the application software from the application developer's point of view. It should be possible to set up the system exclusively using this specification. The System Design Specification is the basis for the module, and integration test plans should be developed with this aim in mind. The requirements defined in the functional description must be fully implemented unless otherwise agreed.

Source Code/Configuration

To ensure uniformity and safe maintenance of the source code, there should be guidelines for its development/parameterization/configuration. These should, for instance, comprise the following elements:

- Program header
- Declaration of variables
- Commenting
- Form and contents of the instructive part
- Format, such as indents of logic structures and use of capital or small letters
- Permitted and forbidden types of data and commands

Module/Integration/System Test Planning and Execution

Module/integration/system tests verify that the software works correctly and as defined in the URS/functional description and System Design Specification.

Modules, including interfaces, should generally be tested against the System Design Specification. Testing may include black box tests, white box tests and a Source Code Review.

The overall system should generally be tested against the URS/functional description. Black box tests are most appropriate here.

A test plan should be established for all of the methods used. After the tests have been carried out, the documented test results are compared to the expected results or the acceptance criteria. Errors should be resolved accordingly.

Installation

An Installation Plan should be developed for each application. The plan must address the following items:

- Description of all preparatory work
- Execution of installation instructions
- Description of all subsequent work
- Description of how to reestablish the initial state in case the installation goes wrong

Raw data must be collected as evidence of a successful installation. An Installation Qualification (IQ) protocol and an IQ Report summarizing the test results should determine whether the system has been successfully installed and is ready for use.

User and System Manuals

The user manual describes all the functions of the system in a way the user can understand. It should enable the user to operate the system correctly and exclusively using this documentation. Users should establish the user manual in cooperation with IT specialists.

The system manual describes all the work necessary for the operation of the application. IT specialists (system administrators) generally carry out the work described in the system manual. The system manual may make extensive use of references to vendor-supplied documentation.

User Acceptance Test Planning and Execution

User acceptance testing (UAT) demonstrates that the application software meets the expectations of the user, based on the URS/functional description. It checks if the application software works in practice and is demonstrated by test cases/scenarios.

UAT is sometimes referred to as consisting of Operational Qualification (OQ) and Performance Qualification (PQ). Raw data must be collected as evidence of a successful operation. Protocols and reports summarizing the test results and concluding whether the system is fit for purpose should be prepared.

Release

Application software can be released for use if the project Validation Plan has been satisfactorily completed. This needs to be documented in a Validation Report.

Application software can be released if it contains errors, as long as these correspond to critical functions determined in the risk assessment. If this situation exists, then it needs to be documented in the Validation Report, together with appropriate and justified workarounds.

Decommissioning

Decommissioning ends the life cycle of application software. On decommissioning, the following points should be taken into account:

- Evaluation of side effects on other systems
- Data migration/archiving
- Storage of system documentation
- Cancellation of servicing contracts/licenses
- Physical disposal

Problem Management

The following points should be documented during problem management:

- Evaluate the GxP risk.
- Conduct an error analysis.
- Implement additional organizational measures (workarounds).
- Commission and carry out the repairs.
- Test the repaired objects.
- Check stored data.
- Reconstruct data if necessary.
- System release for reuse.
- Documentation instructions.
- Counter-measures to prevent the problem in the future.

Change Control (Management)

Changes in application software must be managed and documented. All documents which are affected by the change should be defined. The documentation of the change then comprises the

- affected documents,
- SOPs used for working on these documents and
- persons who applied the SOPs.

The decision on the revalidation of the overall system should be examined in each individual case, depending on the complexity of the application software and the risks involved with the change.

Software Configuration Control

A Configuration Plan should be established for

- software objects of the application software and their versions,
- tools and procedures which allow the application software to work in the desired versions in a specific environment (e.g., compilers, linker scripts) and
- the environment for which the application software has been configured and in which it can run.

Training

A Training Plan should be set up for the IT professionals involved in an IT project. Training records must be maintained for contractors and permanent staff.

Project Audits

The quality status of a project should be monitored by regular audits using a prepared checklist. Corrective actions resulting from the project audit should be included in a report so that progress of their resolution can be tracked and managed.

Document Control

All documents of the life cycle (including software programs) should be reviewed before their issue. The checking criteria are to be established in the SOPs describing document content. A competent person must check the document structure and content. The Source Code Review of software programs should be carried out only for those parts of the source code that were categorized as particularly critical in the risk assessment.

System Environment and Operation

Apart from the documentation of the life cycle for software, a QA System should be available for the IT system environment and operation. The IT-System environment comprises of the following:

- Closed data centers and distributed servers
- Clients/workstations (e.g., PCs)
- Networks
- The IT part of a computerized system (e.g., the PCs of an HPLC [high performance liquid chromatography] system)

The software necessary for operation should also be included. In this respect, the non-IT part of a computerized system (e.g., scales, HLPC instrument, sterilizer) does not form a part of the IT system environment. The description and the operation of these parts need to be treated separately in a different quality system.

The QA System for the IT system environment and operation is shown in Figure 4.4. It is designed in the same way as the ITQA System for software (Figure 4.3). The diagram shows the following relationships:

- The left side of the diagram consists of the documents of the Environment/Operation Life Cycle (EOLC).
- The right side of the diagram shows generally applicable SOPs of the QA System, describing the form and the content of EOLC documents, mainly in form of checklists. The quality of these SOPs is directly related to the quality of the environment and the operation.
- The central part consists of object-specific instructions based on the SOPs of the right part, which result in the documents represented on the left.

The individual phases of the EOLC are explained below.

- Phase I (purchase/installation)
  - Create initial concept.
  - Create a project plan.
  - Create requirement specifications.
  - Perform risk assessment.
  - Create design specifications.
  - Delivery control.
  - Install and test the environment objects.
  - Assign responsibilities for the operation.
- Phase II (operation)
  - Problem management.
  - Change control.
  - Maintenance.
  - Backup/restore.

Figure 4.4. QA System for IT System Environment and Operation

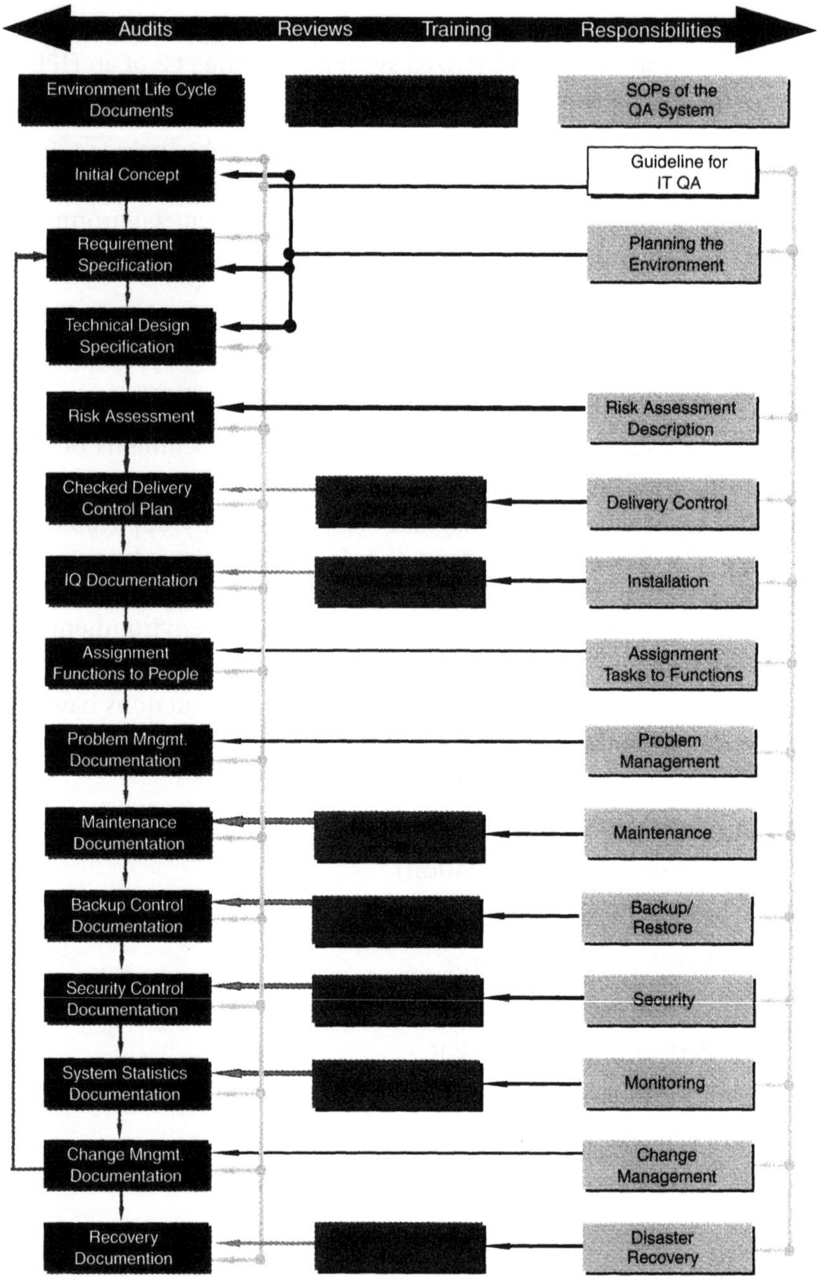

Audits Reviews Training Responsibilities

Environment Life Cycle Documents

SOPs of the QA System

Initial Concept

Guideline for IT QA

Requirement Specification

Planning the Environment

Technical Design Specification

Risk Assessment

Risk Assessment Description

Checked Delivery Control Plan

Delivery Control

IQ Documentation

Installation

Assignment Functions to People

Assignment Tasks to Functions

Problem Mngmt. Documentation

Problem Management

Maintenance Documentation

Maintenance

Backup Control Documentation

Backup/ Restore

Security Control Documentation

Security

System Statistics Documentation

Monitoring

Change Mngmt. Documentation

Change Management

Recovery Documention

Disaster Recovery

  - Security.
  - System performance monitoring.
  - Disaster recovery.
- Phase-independent activities
  - Training.
  - Periodic audits.
  - Document control.

The individual phases of the life cycle for the IT system environment and operation are described in more detail below.

Initial Concept

The initial concept document should describe the following:

- A link from the application software to enhance/change the environment
- The reason why the new application software cannot run in the current environment

Project Validation Plan

The project Validation Plan should take into account the following elements:

- Project team and liaison responsibilities
- Listing of the EOLC documents and who is responsible for establishing them and adhering to schedules
- List of SOPs used for installation and operation of the environment objects
- Filing of the documents (according to the applicable GxP rules for documentation, type of storage, place, responsibilities)

Requirement Specifications

The requirement specifications for environment objects describe the demands from the user's point of view and are usually derived from a requirement specification/functional description for application software. These specifications should address the following points:

- Number of concurrent users
- Maximum system response time
- Added/generated data volume within capacity of system
- Number of pages printed within a time interval

- Minimum uptimes
- Interfaces
- Backup/archiving cycles

Design Specifications

The design specifications for environment objects describe in detail the architecture as well as all components. They are directly derived from the requirement specifications for environment objects. The design specifications should be written at a level of detail that allows the objects to be ordered at a vendor. The following items should be addressed:

- Manner of combined action of the components
- Integration into the existing infrastructure
- Scalability
- Location
- Exact technical specification with type of equipment and manufacturer

Risk Assessment

The risk assessment for the environment examines the critical effects which may emerge from the objects listed in the design specifications and their combined action. New technologies are assessed here. In addition, the risks determined in the assessment of the application software are evaluated again with the view on their possible impact on the environment.

Delivery Control

The delivery is to be checked for completeness and correctness against the design specification for environment objects according to a predefined plan.

Installing and Testing the Environment Objects

The following points should be considered when environment objects are installed and tested:

- Description of all preparative tasks prior to installation
- Description of the detailed steps for executing the installation
- Checking for correct installation according to predefined plans
- Checking the functions of the individual components according to predefined test plans

- Description of all subsequent work after the installation
- Description of re-establishing the initial state in case the installation goes wrong

An IQ protocol and report should be prepared as described for application software. Evidence of testing needs to be retained. Configuration documentation will require updating.

Problem Management

The following elements should be documented during problem management:

- Determine the GxP risk.
- Conduct an error analysis.
- Implement additional organizational measures (workarounds).
- Commission and carry out the repairs.
- Test the repaired objects.
- Check stored data.
- Reconstruct data if necessary.
- System release for reuse.
- Documentation instructions.
- Counter-measures to prevent the problem in the future.

Change Control

The following changes should be approved and documented by the person responsible for the environment object:

- Hardware including peripheral devices
- System software
- System parameters (e.g., configuration files)
- Object location
- Network connection

Maintenance

Internal and external maintenance measures should be described in a Maintenance Plan, which should make use of one or more of the following maintenance models:

- Object is still under warranty and is replaced by the vendor.
- Maintenance is guaranteed in a maintenance contract.
- A spare part is in stock for immediate exchange.

In addition, all periodic maintenance activities, such as cleaning, should be described in the plan.

Backup/Restore

The following points should be considered in a backup/restore plan:

- Description of the backup intervals
- Backup control
- Storage/location/duration
- Restore method
- Handling of media

Security

The security of physical and logical access has to be described. This includes

- administration of user accounts,
- password management,
- periodic security checks (viruses, intrusions etc.) and
- access barriers for rooms/cabinets with environment objects.

Environment Performance Monitoring

A Monitoring Plan should be established for each environment object which describes the measures for monitoring. The following items should be considered when preparing a Monitoring Plan:

- Monitored item
- Warning limit
- Frequency of observation
- Monitoring tool
- Alarm system
- Documentation method
- Storage period of the results

An example SOP is presented in Appendix B of this chapter.

Disaster Recovery/Business Continuity Planning

For each system, contingency measures which are to be taken in the case of system failure for an extended period of time should be developed and regularly tested. The circumstances of the use of these measures should be defined with respect to disaster recovery deadlines. Management of a shorter system or environment outage should also be defined in SOPs that are regularly tested.

Training

A Training Plan should be set up for the IT professionals involved in the planning and operation of environmental objects. Training records must be maintained for contractors and permanent staff.

Periodic Review

The quality status of the operation of environmental objects should be monitored by regular audits using a prepared checklist. Corrective actions resulting from the project audit should be included in a report so that progress of their resolution can be tracked and managed.

Document Control

All documents of the EOLC should be reviewed, during which their form and content are checked by a competent person. The checking criteria are to be established in writing in the SOPs describing document content.

ROLES AND RESPONSIBILITIES FOR QUALITY ASSURANCE

The following roles should be defined for the implementation and operation of each IT system and its environment:

- ITQA committee
- Owner of the IT system
- Project manager
- Project team
- System/network/PC operator
- Application manager

Other roles may be required depending on the organization of the IT department.

ITQA committee

Chairman:	Nominated by the members, responsible for one year
Members:	Research, manufacturing, QA, logistics, sales, IT, engineering
Nominated by:	Company management
Meetings:	At least every three months
Reports to:	Company management
Required profile:	• Overviews the use of IT systems within the company • knows the relevant IT regulatory and industry "validation guidelines" • knows the field of IT and pharmaceutical quality assurance
Tasks:	• Keeps the company ITQA guideline up to date • Coordinates and controls the ITQA System • Arranges the designation of the owners of the IT systems in agreement with the respective departments • Possesses all the relevant regulatory and industry guidelines on the use of IT • Maintains contact with authorities and other companies

System Owner

Nominated by:	ITQA committee
Reports to:	ITQA committee (technical)
Required profile:	Is responsible for the results generated with the IT system and knows the GxP requirements for his or her area
Tasks:	• Nominates the head of the project team and the team members in agreement with the ITQA committee or takes on the function of project leader himself • Triggers the initial QA of the IT system • Provides the personnel and financial resources • Approves the IT system for routine use • Has the right of veto concerning the functioning and quality of the IT system for GxP–relevant applications

Project Leader

Nominated by:	System owner in agreement with the ITQA committee
Reports to:	System owner (technical)
Required profile:	• knows the IT system with regard to the application and the IT sides • Knows the relevant subordinate SOPs in detail
Tasks:	• Establishes the project plan • Estimates resources • Carries out the project professionally according to the project plan • Manages the project documentation

Project Team

Chairman:	Project leader
Members:	People working on the user side, IT specialists (internal and/or external)
Nominated by:	System owner
Reports to:	Project leader
Required profile:	• Knows the IT system in detail from the IT side as well as from the user's point of view • Knows the relevant IT guidelines and the subordinate SOPs in detail
Tasks:	• Develops the IT system according to the ITQA guideline and the subordinate SOPs • Performs all defined QA tasks

System/Network/PC operator

Nominated by:	IT department or owner of the IT system
Reports to:	IT department or owner of the IT system
Required profile:	Knows how to operate the basic software and hardware for environment objects
Tasks:	Carries out all the work necessary to maintain the operation of the environment objects according to the respective SOPs

Application Manager

Nominated by:	IT department or owner of the IT system
Reports to:	IT department or owner of the IT system
Required profile:	• Knows how to develop and maintain application software • Masters the SOPs for the software life cycle
Tasks:	• Develops, changes and operates application software • Supports users in using the application software • Conducts training for the application software

Training

The personnel involved in the development/operation of an IT system must prove that they have knowledge in their fields and keep up to date by means of further training. Table 4.1 shows the fields of training that are to be taken into account for the individual groups of people.

Table 4.1. Training Requirements

	ITQA Committee	System Owner	Project Leader	Project Team	System/Network/ PC Operator	Application Manager	User
Guidelines by Authority & Industry	X						
Quality Assurance (GxP)	X	X					
Quality Assurance (IT)	X						
SOPs for ITQA	X		X	X	X	X	
IT Audit Methods						X	
Software Application Development Maintenance			X	X		X	
Operation of the Environment					X		
Use of the IT System		X	X	X	X	X	X

APPENDIX A: EXAMPLE SOP FOR RISK ASSESSMENT OF IT SYSTEMS

1. Purpose

This SOP describes how a risk assessment for an IT systems is conducted. A risk assessment is applied to discover the GxP–relevant functions of an IT system in order to

- build in plausibility checks and other safety measures into the System Design Specification,
- determine an adequate test width and test thoroughness and
- decide about retesting ("revalidation") of the application after changes.

The risk assessment is carried out on the basis of the URS/functional description.

2. Responsibilities

The owner of the IT system is responsible for the investigation and evaluation of risks that emerge from the GxP operational process. Application managers/system operators are responsible for the investigation and evaluation of the risks that emerge from the IT technical realization.

3. Procedure

3.1. Overview of a Risk Assessment

The following representation gives an overview of the risk assessment. The parts belonging to the risk assessment are in **bold.**

URS/functional description:

- Functions
- Sub-functions
- Further sub-functions if necessary
  - **Investigation of GxP relevance**
  - **Investigation of possible impacts**
  - **Recognition of impact in the further process**
  - **Investigation of the level of the risk**
  - **Establishment of measures for risk minimization**
    - Building in plausibility checks and other safety measures into the system design specification
    - Determining an adequate test width and test thoroughness
    - Applying organizational measures

3.2. Risk Assessment Steps

3.2.1. Collection of the Functions

The functions and sub-functions and if necessary further sub-functions are derived from the URS/functional description and are recorded in the first columns of Table A1 for the risk assessment. Additionally, a reference is made to the corresponding position in the URS/functional description. If no GxP relevance is determined for the investigated functional level, then affiliated sub-functions are not examined further.

3.2.2. Investigation of GxP Relevance

For each risk assessment, GxP–relevant criteria that can be influenced by the IT system are to be determined.

- The pharmaceutical quality of a product
  - Composition
  - Ingredient mix-ups
  - Packaging mix-ups
  - Batch status
  - Storage conditions
  - Batch recalls
- The security of patients and consumers
  - Adverse reactions
  - Mix-ups of samples used in clinical trials
- Correctness of the information that becomes part of registration records
  - Statistical evaluations
  - Calculations
  - Composition of dossier

These criteria have to be determined for each IT system individually. For each function and subfunction, the influence of these criteria must be determined. This is documented in the risk assessment table (Table A1). If no influence is determined, then this has to be justified.

3.2.3. Investigation of Possible Impacts

For the further investigation of a risk determined above, all relevant impacts are evaluated. According to the cause of the risk, they are divided into impacts emerging from

- the operational process and the pharmaceutical GxP procedure and
- IT technical realization.

Impacts are documented in the table for the risk assessment (Table A1).

3.2.4. Recognition of the Impact in the Further Process

Whether an impact will be discovered at its source or as a knock-on effect will be determined from the

- URS/functional description and/or
- description of the tandem-arranged organizational surroundings.

This is documented in the risk assessment table (Table A1).

3.2.5. Investigation of the Level of the Risk

The length of time it takes to recognize an impact together with the significance of the impact will affect the level of risk posed. Risks can be classified as

- high (e.g., impairment of persons, damage to equipment, batch recalls, loss of company image, high costs) or
- low (e.g., failure/breakdowns without damage, low costs, short production standstill without negative quality effects).

3.2.6. Establishing Measures for Risk Minimization

If a high risk was determined for a function, one or several of the following measures are be taken:

- Additional risk-decreasing measures, for example, plausibility checks built into the System Design Specification
- Adequately high test width and test thoroughness
- Organizational measures outside IT (e.g., double checks)

The persons responsible for introducing these measures are to be documented.

3.3. Review of the Risk Assessment

The following items are checked during review of the risk assessment:

- The structure of the risk assessment is in accordance with this SOP.
- Correct judgment of GxP relevance.
- The list of observable impacts is as complete as possible.
- Correct judgment of the risk.
- Appropriateness of the countermeasures introduced into the design.
- Thoroughness of testing to identify if potential impacts exist and how they are managed by design enhancements or operational SOPs.

Table A1. Example of a Risk Assessment Table

Function with Cross-Reference to Specification Document	Subfunction	GxP Relevant (Y/N); Short Justification	Potential Impact Cause: OP (Operation) IT (Design)		Impact Recognition (S—Source) (K—Knock-on Effect)	Level of Risk Without Risk Management	Risk Management • Design Enhancement • Test Thoroughness • Operational Measures (e.g., SOPs)
Preparation for weighing (Section 5.4 in the functional specification)	Identification of an ingredient by barcode	Y; Mix-up of ingredients, wrong potency in finished product	OP IT	1. Barcode read incorrectly.	S (Checksum test will fail.)	low	None beside the usual careful treatment of this type of equipment according to SOP.
			OP	2. Wrong ingredient has been delivered into the weighing cabin.	K	high	Application software compares the substance ID with the recipe. Source code review of that software part. Responsibility: programming and testing group.
			OP	3. Right ingredient, but not yet released.	K	high	Application software compares the delivered lot with the pool of released batches. Source code review of that software part. Responsibility: programming and testing group.
			OP	4. Right label, wrong ingredient.	K	high	No IT measure possible; double check at labeling instead. Responsibility: labeling group.

APPENDIX B: EXAMPLE SOP FOR PERFORMANCE MONITORING OF IT SYSTEMS

1. Purpose

This SOP describes the performance measures for monitoring an IT system.

2. Responsibilities

The system operators are responsible for planning and executing the monitoring measures.

3. Procedure

3.1. Parameters to Be Monitored

Depending on the GxP risks of the applications running in the environment, the following system conditions should be continuously checked with suitable tools:

- Servers/workstations/PCs
  - CPU (central processing unit) utilization
  - Interactive response time
  - Number of transactions/time unit
  - Average job waiting time
  - Disk filling grade
  - System error messages
  - Existence of critical batch jobs
  - Existence of critical processes
- Network
  - collisions
  - Network load
  - Broadcasts
- Applications
  - application dependent

3.2. Notification Mechanisms

Depending on the risk of the monitored parameter, one or more of the following notification mechanisms should be chosen:

- Message on the system console
- Mail to system operator
- Mail to external services
- Pager message to system operators
- Printed lists

3.3. Structure of a Monitoring Plan

The Monitoring Plan should be in the form of a table with the following columns:

- The monitored parameter
- Warning limit
- Frequency of observation
- Monitoring tool
- Notification mechanism
- Documentation of the monitoring results
- Duration of storage of the results

An example Monitoring Plan is shown in Table B1.

3.4. Review of the Monitoring Plan

The following items are checked during review of the Monitoring Plan:

- The structure of the Monitoring Plan is in accordance with this SOP.
- The risks determined in the risk assessment have been appropriately addressed.
- The time intervals and the warning limits for the observed parameters are adequate.
- The method of notification allows timely alert.
- The storage of the monitoring results is adequate.

Table B1. Example Monitoring Plan for VMS Server Running a Laboratory Information Management System

Monitored Parameter	Warning Limit	Frequency of Observation	Monitoring Tool	Notification Mechanism	Documentation of the Monitoring Results	Storage Duration
CPU Utilization	> 25% at the average over 24 hours	Every 10 minutes	System procedure	System console	File with 24 hour CPU statistic	6 month
Disk Filling Grade	> 90%	Hourly	System procedure	Mail to system operator	File with mail	30 days
System Error Message	Error count increased by severe system error (defined in the tool)	Every second	"CheckSys" Tool	Message to pager with error number	According to SOP "Problem Management"	According to appropriate GxP regulations
Critical Batch Jobs: • **all Monitor Jobs** • **FULLBACKUP.COM** • **DIRCHECK.COM** • **CHECK_PRINTQUEUES.COM** • **STOP_ORACLE.COM** • **LIMS**	If batch job is lost	Every 10 minutes	System procedure	Mail to system operator Automatic restart of batch jobs by system procedure	File with mail	30 days
Critical Processes: • **LIMS** • **PATHWORKS** • **ORACLE** • **PERFECT DISK** • **UCX** • **DECnet** • **Security Audit**	If process is not running	Every minute	"CheckSys" Tool	Mail to system operator	File with mail	30 days

5

Demonstrating GxP Compliance

Guy Wingate
Glaxo Wellcome
Barnard Castle, United Kingdom

Pharmaceutical manufacturing sites work under a raft of what has become known in the industry as Good Practice regulations. Examples of these regulations include Good Clinical Practice (GCP), associated with clinical trials of medicinal products; Good Distribution Practice (GDP), associated with the distribution of medicinal products after manufacture; Good Manufacturing Practice (GMP), associated with the manufacture of licensed medicinal products; and Good Laboratory Practice (GLP), associated with laboratory operations. Other Good Practice regulations include those for medical devices. The philosophy behind the various Good Practices is broadly the same: to ensure that medicinal products are consistently produced and controlled to the quality, safety and efficacy standards appropriate to their use. Collectively, they are known as the GxPs (because there are so many variants of the three-letter acronyms).

GxP VALIDATION OF COMPUTERIZED SYSTEMS

Advances in computer technology have made available innovative systems to improve business operations. Twenty years ago, most computer systems found in a pharmaceutical business involved stand-alone financial systems and limited process control. Since then, the use of computer systems has blossomed through process and production management systems to enterprise systems. Pharmaceutical manufacturers may consider embarking on a number of enterprise projects, including the Enterprise Asset Management (EAM) System, Laboratory Information Management Systems, (LIMS), and Manufacturing Resource Planning (MRP II).

The broad nature of enterprise systems and their interconnectivity means that the isolated application of specific Good Practice regulations is not always possible. For instance, the MRP II system, whilst supporting the manufacture of medicinal products (hence GMP is appropriate), also supports LIMS (hence potential impact of GLP) and distribution (hence impact of GDP). If clinical trial material is handled, then GCP regulations will also apply. Validating these systems can be extremely complex if the aim is to apply particular Good Practice regulations solely in their stated domain of application. The boundaries between the application of these regulations are fuzzy at the best of times. An alternative approach that many pharmaceutical manufacturers are now adopting is "GxP validation". GxP validation denotes a single level of validation complying with all the Good Practice regulations by achieving the most stringent compliance criteria from each regulation.

REGULATORY EXPECTATIONS

It is important to bear in mind that the regulatory expectations for computer validation are embedded in numerous regulatory documents issued across the Good Practice domains. For instance, the most recent issue of U.S. Food and Drug Administration (FDA) guidance on software validation came through the medical devices arena but describes the FDA's general expectation for computer systems validation to satisfy the Good Practice regulations (see Figure 5.1). GxP validation is therefore applicable across a range of industry sectors, including research and development, biotechnology, bulk pharmaceutical chemicals (BPCs), cosmetic products, medical devices, diagnostic systems, and finished pharmaceuticals for both human and veterinary use. Indeed, the UK Medicines Control Agency (MCA) uses the term *GxP* in official guidance. The various supply chain operations of a pharmaceutical business encompass these Good Practice regulations, and it is not unreasonable to expect inspection questions framed using the term *GxP*.

Figure 5.1. Regulatory Timetable

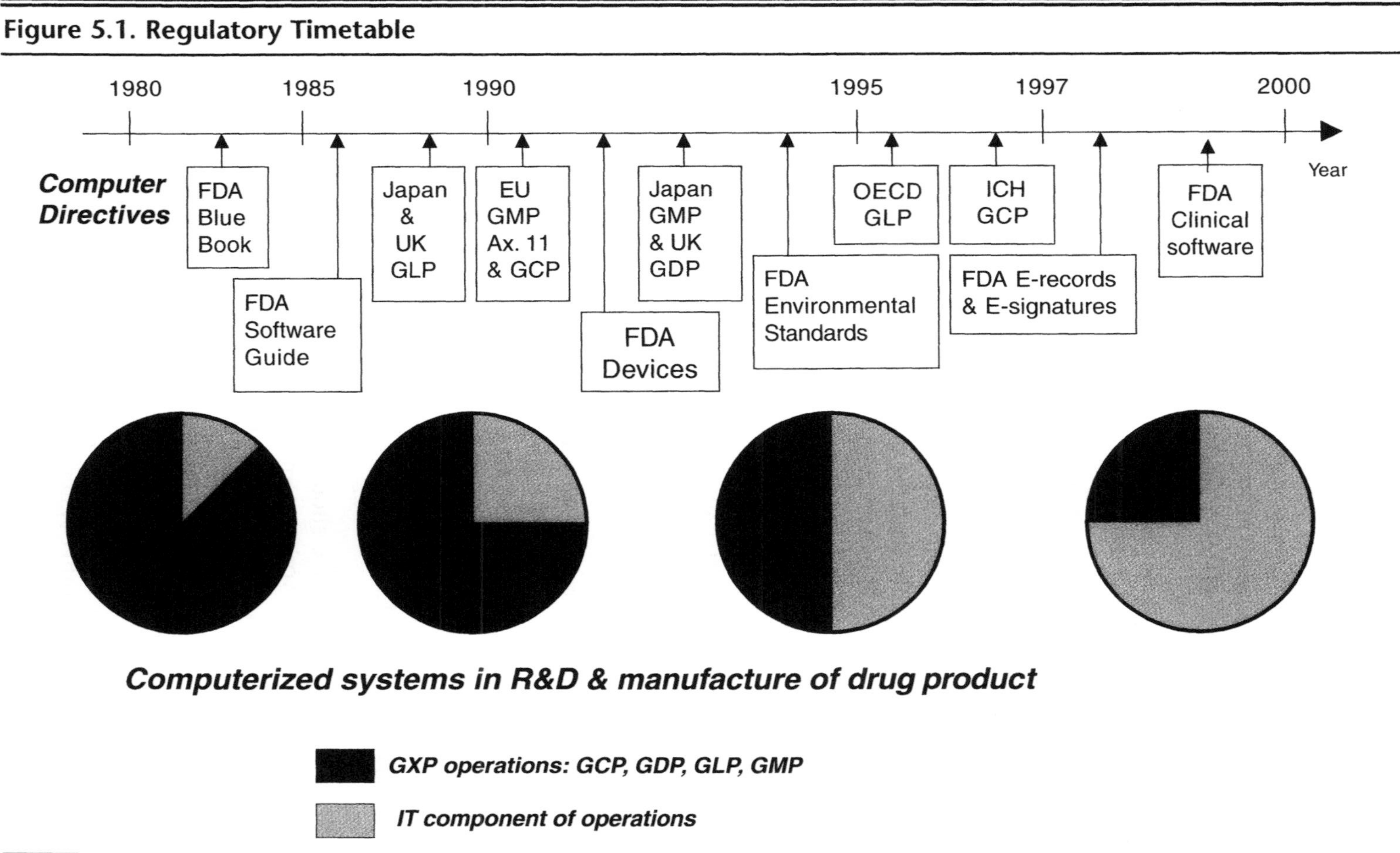

BENEFITS OF THE GxP APPROACH

GxP validation should deliver tangible benefits.

- GxP validation is more manageable in large organizations because it brings consistency and standardization. The overall cost of GxP validation should be lower than the cost of implementing different levels of validation to different systems, although it is acknowledged that there may be limited instances where costs may appear higher under GxP validation.
- Suggested compromises to the level of validation from projects and support functions will be more transparent. Noncompliant practices should be reduced.
- Less time will be spent defining the boundaries and defending different levels of validation to regulators. Those residual noncompliances that slip through the net should be easily discovered through an internal audit before a regulator discovers them.
- Validation skills are more transferable between different computer systems—a key issue where specialist computer validation resources are rare.
- A standard approach also allows the impact of new and developing regulations and computer technology on the standard GxP validation approach to be more easily assessed and necessary corrective actions taken in a consistent and timely manner.

Pharmaceutical manufacturers are generally moving toward using the term *GxP* validation when describing the validation requirements for new corporate computer projects and initiatives.

SYSTEMS PERSPECTIVE

A broad systems perspective should be taken when validating information technology (IT) and network systems. The GxP regulations are aimed at the ways of working in an organization, and this spirit should be applied to the use of IT and network applications. A system is composed of not only computer hardware and software components but also associated desktop infrastructure. Hence, servers, clients and networks are included during validation. Other system aspects, which should be considered during validation, are procedures and training for users of the system, system documentation, and any interdependencies with other systems sharing data. Validation must include due consideration of data management and integrity that can be complex in large distributed systems. In addition, the general operating environment will need to validate and consider security measures, virus controls, and the physical environment (humidity, power stability, electromagnetic interference, radio-frequency interference, dust, and hygiene).

COMPUTER VALIDATION FOR GxP

A validation life cycle should be adopted to fulfill the quality assurance (QA) expectations of the GxP regulations. Chapters 2, 4 and 13 outline a life cycle framework suitable for pharmaceutical manufacturers. Life cycles can vary and still fully support GxP. A typical life cycle might include the following:

- Validation Plan
- Supplier audit
- Specification
- Design
- Hazard Analysis
- Source code review
- Data load
- Installation Qualification (IQ)
- Operational Qualification (OQ)
- Performance Qualification (PQ)
- Validation Report
- Handover

Projects implementing corporate computer systems might expect the following four activities to soak up most effort, in descending order: testing, data load, configuration, and specification. It could be argued that documentation should appear in this league table, and indeed it would if treated as an activity in its own right instead of an underpinning "given" in support of pharmaceutical projects.

Table 5.1 reviews the primary U.S. and European Union (EU) GMP regulations with reference to that part of the validation life cycle where particular regulatory clauses can be addressed. The table is not definitive and is given as an illustrative example only. Like many other GCP, GDP and GLP regulations, there are few direct references within the GMP regulations to computer systems validation. For instance, 21 CFR 211 [1] has only one clause directly applicable to computer systems.

Clause 211.68: Automatic, Mechanical, and Electronic Equipment

(a) Automatic, mechanical, or electronic equipment or other types of equipment, including computers, or related systems that will perform a

Table 5.1. GMP Compliance for Computer Systems

Topic of Interest	Key Issues	U.S. GMP (21 CFR 211)	EU GMP (91/356/EEC)	Validation Project Activity
Organization and Personnel				
Project and validation team	Organizational chart, job descriptions, training records	211.25, 211.28	2.1, 2.2, 2.9, A11.1	Validation Plan
Maintenance and operational compliance team	Organizational chart, job descriptions, training records	211.25, 211.28	2.1, 2.2, 2.9, A11.1	Validation Plan, Validation Report, handover
Quality Management System	Validation procedure	NSR	5.21, 5.24, A11.2	Policies, procedures, working instructions
Contract agreement	Equipment, support services and contractors/consultants	NSR	7.13, A11.18	Validation Plan and IQ
Contractors and consultants	Name, address, qualifications, service used	211.34	7.6, A11.1	Validation Plan, training records
Audits	Self-inspections and audits of suppliers/vendors	NSR	7.13	Supplier audit, SOPs, Handover
Building and Facilities				
Computer room and locations for field/distributed equipment including instrumentation	Design for ease of installation and operation	211.42, 211.63	A11.4	Specification, design, DQ
Computer room and locations for field/distributed equipment including instrumentation	Operating environment monitoring and records	211.44, 211.46, 211.58, 211.194(d)	3.41, 6.5	Specification, IQ, PQ

Table 5.1 continued on the next page

Table 5.1 continued from the previous page

Topic of Interest	Key Issues	U.S. GMP (21 CFR 211)	EU GMP (91/356/EEC)	Validation Project Activity
Equipment				
Equipment identification	Unique identification (make, model, serial number/software version) of software, firmware and hardware components	211.105	A11.4	IQ
Equipment design	Design of computer room, system and field/distributed equipment including instrumentation, hazard analysis, Source Code Review	211.42, 211.63	A11.4	Specification, design, Hazard and Operability Study, SCR
Data entry	Challenge invalid operator input	NSR	A11.6	Data load, OQ
Double-checking	System verification of critical user input	NSR	A11.9	Specification, design, data load, OQ
Calculations	Including conversion factors, yield mixes(?), units of measure, expiry dates, shelf life, etc	211.68(b) 211.137, 211.194(a)(5)	6.16–17	SCR, OQ
User identification	Record activities accessed by user(s)	21 CFR 11	A11.10	Specification, design, OQ
Security access	Challenge passwords	21 CFR 11	A11.8	OQ

Table 5.1 continued on the next page

Table 5.1 continued from the previous page

Topic of Interest	Key Issues	U.S. GMP (21 CFR 211)	EU GMP (91/356/EEC)	Validation Project Activity
Security practice	Security facilities	21 CFR 11	A11.4	Specification, design, SOPs, OQ
Exception reports	Management notification of exceptions	211.180(f)	5.39, 5.56	Specification, design, OQ, PQ
Printers	Test clarity of print for production and laboratory records	211.68(a), 211.186, 211.194	A11.12, 5.21, 6.7, 6.8, 6.9, 6.10	IQ, PQ
Equipment location	Physical location of equipment should be accessible	211.58	4.28–29, 6.7	Specification, design, IQ
Environment check	Monitoring operating environment of computer room and field equipment (e.g., power fluctuations, RFI, ESD, EMI)	NSR	3.41, 6.5	Specification, IQ, PQ
Calibration schedule	Instrumentation, analytical equipment and computer clocks	211.68(a) 211.67(b)	3.41, 4.28, 6.7	SOPs, IQ
Testing/qualification	Test records	211.68	A11.7	SOPs, IQ, OQ, PQ
Operating instructions	User manuals and procedures	211.100	4.27	SOPs, OQ
Maintenance and self-inspection	Maintenance and inspection schedules for computer room and field equipment	211.67(c), 211.68, 211.180–182	3.35–36, 3.43, 4.26, 5.19(e), 6.7	SOPs, handover

Table 5.1 continued on the next page

Table 5.1 continued from the previous page

Topic of Interest	Key Issues	U.S. GMP (21 CFR 211)	EU GMP (91/356/EEC)	Validation Project Activity
Backup and storage	Procedures and storage locations	211.68(b)	A11.13–14	SOPs, OQ, handover
Periodic review	Performance monitoring and analysis	211.192	9.1–3, A11.17	SOPs, handover
Revalidation	Periodic reappraisal of validated status and any required revalidation	NSR	5.24	SOPs, handover
Control of Materials				
Receipt of materials	System logging of materials receipts	211.80–82	5.2–3, 5.5	Specification, OQ
Inventory records	For each material; reconciliation	211.184(c)	5.56	Specification, OQ
Inventory monitoring	For each material; reconciliation	211.184(c)	5.56	Specification, PQ, handover
Material sampling	Notification of QC worksheets	211.84(c)	4.22, 6.11	Specification, OQ
Sample size	Statement of measure used	211.194(a)(3)	6.11	Specification, OQ
Retesting materials	System notification and tracking	211.87	5.64	Specification, OQ
Material allocations	Records of materials assigned to batches	211.860	5.32–33	Specification, OQ

Table 5.1 continued on the next page

Table 5.1 continued from the previous page

Topic of Interest	Key Issues	U.S. GMP (21 CFR 211)	EU GMP (91/356/EEC)	Validation Project Activity
Production and Process Controls				
Material	Lot disposition (distinctive status)	211.80(d)	5.31	Specification, design, OQ, PQ
Material specifications	References for materials specifications	211.84(d)(2)	4.10–12, 5.26	Data load
Batch records	Integrity of references to master records	211.188(a), (b)	4.17–18, 5.2	Specification, design, OQ, PQ
Batch closure	Identify person closing records	211.84, 211.184	A11.19	Specification, design, OQ, PQ
Packaging and Labelling Controls				
Labeling specification	Reference, identification, quantity, supplier, lot number, date (including format)	211.184(a)	4.19–21, 5.2, 5.13, 5.29	Specification, design, OQ, PQ
Labeling procedure	User procedure	211.112(a), 211.120	5.2, 5.34, 5.44, 5.47, 6.11	SOPs, OQ, PQ, handover
Label issuing	Initiation and control	211.125(f)	5.41	Specification, design, SOPs, OQ, handover
Label examination	For conformity with specifications and correctness	211.130(c), 211.184(d)	4.18, 5.61	SOPs, OQ, PQ, handover
Disposal of labels	Record of destruction	NSR	5.43	SOPs, PQ, handover

Table 5.1 continued on the next page

Table 5.1 continued on the next page

Topic of Interest	Key Issues	U.S. GMP (21 CFR 211)	EU GMP (91/356/EEC)	Validation Project Activity
Holding and Distribution				
Warehouse procedures	Use of system for tracking	211.150	1.3, 5.2	SOPs, OQ, handover
Exception reports	Perpetual inventory	211.180(f)	5.39, 5.56	Specification, design, PQ, handover
Laboratory Controls				
Sampling	Sampling calculation	211.160–167	4.22	Design, SCR, PQ
Records and Reports				
Change control	Control of all changes to system	211.68(b)	A11.11	SOPs, PQ, handover
Review of records	At least annual review	211.80(e)	NSR	SOPs, handover
Deviations/justifications	Records of deviations from production procedures	211.100(b)	4.17(i), 5.15, 5.39	SOPs, PQ, handover
Deviations/justifications	Records of deviations from laboratory procedures	211.160(a)	1.4, 6.2	SOPs, PQ, handover
Deviations/justifications	Time limits on production	211.111	4.15, 4.17(i)	Specification, design, OQ, PQ

Table 5.1 continued on the next page

Table 5.1 continued from the previous page

Topic of Interest	Key Issues	U.S. GMP (21 CFR 211)	EU GMP (91/356/EEC)	Validation Project Activity
Returned and Salvaged Drug Products				
Returned product records	Records of returned or re-called materials	211.204	5.65, 8.9	Specification, OQ
Recall records	Use of computer system to facilitate a recall	211.204–208	8.1–15	Specification, design, Hazard Analysis, SOPs, OQ, handover
System recovery	Computer and records	211.68(b)	5.63, A11.15	Specification, Business Continuity Plans, OQ, handover
Recovered materials	Records of salvaged batches or batch reprocessing	211.21, 211.115(a)	5.62	Specification, OQ

Notes:

(1) NSR means "not specifically referenced" in GMP regulation.

(2) SOPs (Standard Operating Procedures) usually required for all activities by default. One SOP may cover more than one activity.

(3) 21 CFR 11 refers to the Title 21, *Code of Federal Regulation* on Electronic Signatures and Electronic Records, which is discussed in more detail in Chapter 16.

(4) Business Continuity Plans cover both Contingency Plans and Disaster Recovery Plans.

(5) Handover signifies only that this topic should be included for operational compliance; its absence does not indicate that an activity is not required for operational compliance.

(6) SCR = Source Code Review; DQ = Design Qualification; RFI = radio-frequency interference; ESD = electrostatic discharge; EMI = electromagnetic interference; QC = quality control.

function satisfactorily, may be used in the manufacture, processing, packing, and holding of a drug product. If such equipment is so used, it shall be routinely calibrated, inspected, or checked according to a written program designed to assure proper performance. Written records of those calibration checks and inspections shall be maintained.

(b) Appropriate controls shall be exercised over computer or related systems to assure that changes in master production and control records or other records are instituted only by authorized personnel. Input to and output from the computer or related system of formulas or other records or data shall be checked for accuracy. The degree and frequency of input/output verification shall be based on the complexity and reliability of the computer or related system. A backup file of data entered into the computer or related system shall be maintained except where certain data, such as calculations performed in connection with laboratory analysis, are eliminated by computerization or other automated processes. In such instances a written record of the program shall be maintained along with appropriate validation data. Hard copy or alternative systems, such as duplicates, tapes, or microfilm, designed to assure that backup data are exact and complete and that it is secure from alteration, inadvertent erasures, or loss shall be maintained.

Meanwhile, European Directive 91/356/EEC[2] on GMP includes an Annex that specifically addresses the validation of computerized systems. Table 5.2 lists the aspects of computer validation covered.

Table 5.2. Validation of Computerized Systems

Aspect of Computer Validation	Clauses in Annex 11, 91/356/EEC
Personnel	1
Development life cycle	2, 5
Change control	2, 11
Specification and design	2, 3, 4, 6
Acceptance testing and handover	7
Access authorization	8, 10
Data input	6, 8, 9, 10, 19
Data storage and backup	12, 13, 14
Error handling and system failure	15, 16, 17
Third-party services	5, 18

Computer validation requirements are not limited to the development and operation of a system but also include how it is used within a GMP environment. Table 5.1 includes other clauses in 21 CFR 211 and 91/356/EEC that are applicable to wider interpretation and affect the use of computer systems.

Not all computer applications will be regulated by 21 CFR 211 and EEC/356/91. Table 5.1 can be further extended to include other GxP regulations as applicable to specific projects. Some sample regulatory requirements that may also need to be addressed from the Japanese Ministry of Health and Welfare (MHW) and the Australian Therapeutic Goods Administration (TGA) are given below.

- Australian Code of GMP for Therapeutic Goods—Medicinal Products—Part 1, Use of Computers

A change to an existing computer system should be made in accordance with a defined change control procedure which should document the details of each change made, its purpose and its date of effect and should provide for a check to confirm that the change has been applied correctly.

Persons with appropriate expertise should be responsible for the design, introduction, and regular review of a computer system.

- Japanese Guideline on Computer Systems in Drug Manufacturing—Scope of Application

This guideline shall apply to the drug manufacturing plants governed by Drug GMP and other regulations where any of the following systems are introduced. However, the system with the limited function and are operated simply by inputting the parameters to the programme fixed by the supplier of the hardware (including every device and equipment controlled by a computer) as general functions of the devices shall be exempted from this guideline.

(1) *The system for manufacturing process control and management including that for recording these data.*

(2) *The system for production control such as storage and inventory of starting materials and final products including intermediates.*

(3) *The system for documentation of manufacturing directions, testing protocols or records.*

(4) *The system for quality control and recording these data.*

- Japanese Guideline on Computer Systems in Drug Manufacturing—System Designing

The manufacturer shall document a system development manual which shall normally include:

> *(A) Procedures of system development,*
>
> *(B) Documents to be prepared during each step of the system development and the methods of maintaining these documents,*
>
> *(C) Procedures for evaluation and approval of each step of system development,*
>
> *(D) Procedures for amending or repeal of the system development manual.*

- Japanese Guideline on Computer Systems in Drug Manufacturing—Operation Control 4.1.3 (A). *The manufacturer shall prepare a system alteration standard code which includes, in principle, the following:*
  - *(a) Procedures for alteration application and approval of alteration,*
  - *(b) Examination after system alteration,*
  - *(c) Amendment of operation procedures,*
  - *(d) Notification of the amended parts to the persons concerned,*
  - *(e) Title or name of the manager and person/s in charge. (B) The provisions [. . . during . . .]"development of the system" shall generally apply to the system alteration when associated with programme change after its introduction.*

- Japanese Guideline on Computer Systems in Drug Manufacturing—Accidents

The manager shall ensure that the person/s in charge take immediate actions when accidents occur in the system in accordance with the provisions of the operation procedures and let him/her conduct a cause finding study as well as take necessary actions for the prevention of these accidents.

- Japanese Guideline on Computer Systems in Drug Manufacturing—Self-Inspection

The manufacturer shall ensure that the operation of the system is consonant with this guideline through periodical inspection by its manager.

- 21 CFR 58.61—Equipment Design

Equipment used in the generation, measurement, or assessment of data . . . shall be of appropriate design and adequate capacity to function according to the protocol and shall be suitably located for operation, inspection, cleaning, and maintenance.

- 21 CFR 58.63—Maintenance and Calibration of Equipment
    - *(a)* *Equipment used for the generation, measurement, or assessment of data shall be adequately tested, calibrated, and/or standardized.*
    - *(b)* *The written standard operating procedures required under § 58.81(b)(11) shall set forth in sufficient detail the methods, materials, and schedules to be used in the routine inspection, cleaning, maintenance, testing, calibration, and/or standardization of equipment, and shall specify, when appropriate, remedial action to be taken in the event of failure or malfunction of equipment. . . .*
    - *(c)* *Written records shall be maintained of all inspection . . . operations. . . .*

- 21 CFR 58.130—Conduct of a Nonclinical Laboratory Study

In automated data collection systems, the individual responsible for direct data input shall be identified at the time of data input. Any change in automated data entries shall be made so as not to obscure the original entry, shall indicate the reason for the change, shall be dated, and the responsible individual shall be identified.

- Japanese Guideline on Computer Systems in Drug Manufacturing—System Designing 3.1.1 (A)

The manufacturer shall designate a manager and person/s in charge of each step of the system development ranging from "system engineering" to "installation and operation test". The same person may also assume the responsibility for several steps of the system development. The manager shall specify individual duties of person/s in charge.

- Japanese Guideline on Computer Systems in Drug Manufacturing—Operation Control 3.1.1 (A)

The manufacturer shall designate a manager and person/s in charge of each step of the system development ranging from "operation of hardware" to "self-inspection". The same person may also assume the responsibility for several steps of the system development. The manager shall specify individual duties of person/s in charge.

- Japanese Guideline on Computer Systems in Drug Manufacturing—Others

All or part of the activities described in "development of the system" can be entrusted to other firms on condition that the manufacturer concerned shall obtain and maintain the relevant documents of the activities from the contracting firm or otherwise ensure that these documents are properly retained by the contracting firm. However, the contracted activities which fall within "system performance test" or "installation and operation test" shall be directly supervised, evaluated, and accordingly approved by the responsible person designated by the manufacturer concerned.

ELECTRONIC RECORDS AND ELECTRONIC SIGNATURES

Pharmaceutical manufacturers are discovering that compliance with 21 CFR 11 can be very difficult. Table 5.3 outlines the topics covered by the regulation. Most supplier systems (hardware and software) do not fully support the use of electronic records and signatures as specified in the regulation. Typical problems include

Table 5.3. Electronic Records and Electronic Signatures

Aspects of Electronic Records and Electronic Signatures	Clauses in 21 CFR 11
Electronic records	2(a), 30(a, b)
Data integrity	10(c, e, f, g, h), 30
Access control	10(d), 300(b, c, d)
Physical security	10(b, c, g)
Training/operating procedures	10(i, j, k), 100(b), 300(e)
Electronic signatures	2(b), 50, 70, 100(a, b), 200(a), 300(a)
Biometrics	200(b)
Validation	10(a), 300(e)

- system synchronization,
- password aging,
- password size,
- log of attempted and successful security violations and
- full audit trails for electronic records.

Until suppliers embody 21 CFR 11 compliance within the products, demonstrating an application's compliance will almost inevitably involve the use of hybrid systems. The term *hybrid system* is used to describe the management procedures defining how to use a computer system in combination with the computer system's embedded functionality. Where hybrid systems are used, it is vital that actions managed by procedures are contemporaneous with their associated electronic record or signature.

It is interesting to note that the EU does not have an equivalent regulation to 21 CFR 11 but rather claims its basic requirements are already embodied in existing EU GxP directives and standards, such as British Standard (BS) 7799. Indeed, the MCA has stated at conferences that they do not hold to all the requirements prescribed in 21 CFR 11. Harmonization may alter this situation (see Chapter 17). A requirement for corporate computer systems used in the EU that is not specified in 21 CFR 11, however, is the need to define the role of the qualified person in terms of electronic approvals and authorizations. This is especially important for critical GxP actions like batch release, recall, and annual product reviews. Further information on the practical issues affecting 21 CFR 11 can be found in Chapter 16.

FAIR WARNING

The UK Good Automated Manufacturing Practice (GAMP) Forum reported a threefold increase in the number of regulatory inspections affecting computer systems validation during the first six months of 1999 compared to previous six month periods in recent years. Significant computer system inspections during this period have involved international pharmaceutical manufacturers in the United Kingdom, United States, Denmark and Finland by both the FDA and the MCA. These inspections have been conducted under the remit of GDP, GLP and GMP.

There would appear to be an appreciation amongst regulators that computer systems are extensively used throughout the drug manufacturing, packaging and distribution process. The computer systems covered by the inspections included MRP II, LIMS, spreadsheets, process control systems and network applications. Between these inspections, a large number of noncompliance observations were made and are summarized below. This summary is based on several hundred noncompliance observations made between the inspections.

It is interesting to note that these noncompliance observations were pretty consistent between inspections and that the noncompliance inspections were made by a number of inspectors. It appears that the regulators and their inspectors are building a common understanding of GxP requirements for computer systems validation. The standards being promoted for computer systems validation are not, as it might have been positioned in the past, being set by individual inspectors but by the respective regulatory authorities.

- *Systems inventory:* name, versions, platform, hardware identifier, operating system and version, versions of utility software, details of system developer (vendor, implementer)
- *SOPs and policies for computer validation:* exist and adopted
- *Computer system:* system overview, architecture design, organization charts, list of all programs, software flow diagram, software application versions, production data, system descriptions, list of all hardware, version of operating system and compilers, diagram showing interfaces, proof software listings were accurate, QA approval for use of specific version of system being used
- *System overviews:* simple/short, required for all systems and networks, architecture design required for larger systems, systems overviews also required for quality contract research organizations
- *Hardware:* maintenance logs, mislabelled computer cabinet, concerns over disconnected wires in computer cabinets, cabling schedules
- *Testing records:* no unit/module test records available, user test documents
- *Validation records:* version of software to be recorded on documents
- *Change control and configuration management:* concern about large numbers of changes, potential functional ripple effect of changes within software, lack of version identifiers on software, all software changes to be approved by QA including software product upgrades and patches (bug fix releases)
- *Preventative and emergency maintenance logs:* records required
- *Disaster recovery:* plans exist and have been tested
- *Periodic reviews:* are conducted
- *Organization charts:* should be archived
- *Training records:* exist and are up to date, include contractors
- *Security access:* no more than four super users on major IT systems, shared user identifiers not allowed, should not allow anybody to log on under another person's identifier, QA approval of lists of users given system access
- *Data center audits:* location of information in distributed databases understood, diagram of physical layout of data center, hardware

and network diagrams (including telecomms layout), SOPs, operating system versions, compilers, access privileges, layered products, applications running, maintenance logs, vendor maintenance contracts, internal audits

- *Archive records:* archivists must have deputies, log of archive requests required, control required between main and satellite archives, legacy data not in archives should be deposited there
- *Electronic records and electronic signatures:* backup of electronic records, plan to bring 21 CFR 11 compliance

An important feature of these inspections to note is that the inspectors often wanted to be taken on tours of areas and meet operational staff. These tours were of a spontaneous nature; the inspectors did not want them to be planned or managed by their hosts.

Individually, these observations may not be considered as major concerns. However, the collective nature of the observations together with the depth of the computer systems audit (in two instances 15 man-days of inspection effort were dedicated to computer systems) indicate what could well be described in the future as a pivotal increment in regulatory expectations. A similar situation exists with UK MCA inspections.

Where these observations did not directly result in an FDA 483 citation for noncompliance, they were positioned as fair warning of what the inspectors thought were reasonable expectations and that would be inspected as outright requirements in the near future.

INSPECTION READINESS

No matter how well a pharmaceutical manufacturer believes it conducts validation, it will count for nothing unless during an inspection the regulator understands what has been done and can easily find his or her way around supporting documentation. To this extent, a key feature in any validation exercise is inspection readiness.

Seven key elements for being inspection ready are listed below; others can be added as appropriate to show a pharmaceutical manufacturer wishes to manage regulatory inspections.

1. Inventory of systems
2. System/project overviews
3. Validation Plans/Reports and Reviews
4. Presentation slides
5. Internal position papers
6. Document map
7. Trained personnel

Inventory of Systems

An inventory of systems and knowledge of which ones are GMP critical must be maintained and available for inspections. A MCA preinspection checklist includes the inventory of systems as one of its opening topics. The availability of this information is a clear indicator of whether management is in control of its computer systems validation. The use of an inventory need not be limited to inspection readiness; it could also be used for determining Supplier Audits, periodic reviews and so on. Many pharmaceutical manufacturers use a spreadsheet or database to maintain these data. Where a site's inventory is managed between a number of such applications (perhaps one per laboratory, one for process control systems, one for IT systems), care must be taken that duplicate entries are avoided, and, equally, some systems are missed and not listed anywhere. It should be borne in mind that where spreadsheets and databases are used to manage an inventory, then these should be validated just like any other GxP computer application.

System/Project Overviews

Management overviews should be available for systems and projects to give a succinct summary of the scope of the system, essentially drawing boundaries and identifying functionality and use of the system/application concerned. Top-level functional diagrams and physical layout diagrams are highly recommended. It is also worthwhile to consider developing some system maps showing various links between systems and dealing with both manual and automatic interfaces. Care must be taken to keep system maps up to date as new systems are introduced, old systems are decommissioned and as the use and interfaces of some systems are modified to meet evolving user demands. Regulators are often interested in system interfaces, manual and electronic, and the validation status of connected systems. As a rule of thumb, all systems providing GxP information (data, records, documents, instructions, authorizations, or approvals) to a validated computer system should themselves be validated together with the interface.

Some regulators have requested guidance from pharmaceutical manufacturers on what is of particular relevance in terms of GxP functionality within their corporate computer systems. Such GxP Assessments often fit neatly in the system overview. The reason for this request by regulators is to help them concentrate on the key aspects of the system during an inspection without getting bogged down in aspects of the system which are not of a prime concern. It is easy for a regulator who is unfamiliar with a corporate computer system to get lost in its extensive and complex functionality (information overload). Needless to say, any GxP Assessment information presented to a regulator must be understood and carefully justified.

Validation Plans/Reports and Reviews

It is likely that during a GxP inspection a regulator will ask whether or not a particular system has been validated. This line of investigation may stop with a yes/no response from a pharmaceutical manufacturer. The line of investigation may, however, lead to a follow-up request to see the Validation Plan and Report for a system described as validated. Many of the computer systems used today have been in use for many years, and the regulator may also ask for any evidence of any Validation Reviews. These documents are, not too surprisingly, vital in demonstrating GxP compliance. It is not very clever to let a regulator discover a system in use with a Validation Plan but an incomplete or nonexistent Validation Report. Equally, if the system has been used for many years, it is more than reasonable to expect a recent Validation Review. Validation Plans, Reports, and Reviews should be checked to make sure they exist, are approved, and meet current regulatory expectations. In some instances, pharmaceutical manufacturers, when considering this point, may put in place a review program to check that the items discussed above are complete and in place.

Presentation Slides

It is useful to prepare a small presentation of each system that may be subject to an inspection, perhaps four or five slides and certainly less than a dozen. The presentation slides should not be too detailed but provide a broad picture describing a system/application and facilitate discussion. It is worthwhile letting the legal department look over the slides because there may be a danger of too high a level of information being interpreted as misleading if the detail of a system/application is examined. There is a careful balance to be struck between too much information and concise clarity. The slides should be in a suitable state to provide the inspector with a copy if requested. The slides are a GxP document and should be treated as such.

Internal Position Papers

Position papers, for internal consumption only, are very useful when preparing for and receiving an inspection. In practice, computer systems are not perfect, and projects implementing applications will typically raise many management issues—that's life in the real world! The validation of any system/application will present its own special problems and solutions. Rationales need to be prepared and documented to demonstrate how problems and solutions have been managed. In essence, the position papers should provide a brief to enable the system/application to be presented in a positive light. Knowing how to effectively position problems and solutions will dramatically enhance the overall perception of the standard of validation on a system/application. The aim must be not to mislead an inspector, rather to present validation issues in the vein of a glass half full rather than a glass half empty. If all reasonable endeavors have been taken by a pharmaceutical manufacturer to validate a system/application, then this should normally be sufficient to satisfy an inspector, remembering that reasonable endeavors might include replacement where an original system/application cannot be validated to meet current regulatory expectations.

Document Map

It is vital to be able to easily locate documentation. Validation documentation that exists but cannot be retrieved as required during an inspection is worthless—it might as well have not been prepared in the first place. To this end, an index to documentation should be produced. The index can take many forms: a document tree, a straight listing or a Requirements Traceability Matrix (RTM). An RTM is a very valuable tool during and after a validation project. By tracking requirements (more detail than just tracking documents) in a project through design to testing, a project can verify that its needs are incorporated into the system/application and that they have been fully tested. After validation, the RTM can, if designed correctly, provide a very effective means of demonstrating how a particular function of a system/application has been validated through design (including any Design Qualification [DQ]), programming (including Source Code Review), and testing (prequalification, IQ, OQ, PQ). Table 5.4 illustrates the basic concept of an RTM. Typically, multiple tables will be necessary to track (provide an audit trail) requirements as they progress into implementation and testing. Only in the simplest systems will a single table suffice the needs of an RTM.

Trained Personnel

Last but by no means least, the availability and use of trained personnel to "front" inspections is key. Those who make presentations to an inspector should be permanent employees, otherwise there may be an impression of dependence on quality from temporary staff whose loyalty and long-term commitment to a pharmaceutical manufacturer could be questioned. Presenters need to be knowledgeable about systems/projects they are asked to front. They need to understand the validation approach and appreciate why certain project and validation decisions were taken. The position papers, slide packs and Validation Plans/Reports/Reviews should all help in this respect, as long as the individuals concerned have enough time to study and digest the information they contain. Finally, the fronters should be educated in what to expect in the way of inspection protocols and regulatory practice. This aspect of training is likely to be tailored to the individual regulatory authorities; for instance, the FDA has a very different approach compared to many EU national

Table 5.4. Example Requirements Traceability Matrix

Specification									
URS Reference	**FS Reference**	**Design Specification Reference**	**DQ Reference**	**Source Code Review Reference**	**IQ Protocol Reference**	**OQ Protocol Reference**	**PQ Protocol Reference**	**SOP Reference**	**Comments**
1	Not Applicable	Not Applicable	Not Applicable	Not Applicable	Not Applicable	Not Applicable	Not Applicable	Not Applicable	
1.1	1.1	1.1, 1.7	1.1, 1.2	Not Applicable	IQ-ABC-001	Not Applicable	Not Applicable	Not Applicable	
1.2	1.2, 1.4	2.3	1.3	1.5	Not Applicable	OQ-ABC-001	Not Applicable	Not Applicable	
2	Not Applicable	Not Applicable	Not Applicable	Not Applicable	Not Applicable	Not Applicable	Not Applicable	Not Applicable	
2.1	2.1	2.1, 2.3, 2.7	2.1	2.2	Not Applicable	Not Applicable	Not Applicable	SOP-XXX-001	
2.2	2.2	2.1, 2.7	2.2, 2.3, 2.4	4.3	Not Applicable	OQ-ABC-002 OQ-ABC-003 OQ-ABC-004	PQ-ABC-001	SOP-XXX-001	
Not Applicable	2.2.1	2.2.1, 2.2.2	2.5	Not Applicable	Not Applicable	OQ-ABC-005 OQ-ABC-006	PQ-ABC-001	SOP-XXX-001	
2.3	2.3	2.1, 2.7	2.6	Not Applicable	IQ-ABC-002	Not Applicable	Not Applicable	Not Applicable	
		This table would be continued to include all URS and FS sections and subsections in an actual project.							
7	Not Applicable	Not Applicable	Not Applicable	Not Applicable	Not Applicable	Not Applicable	Not Applicable	Not Applicable	
7.1	7.1	Not Applicable	Not Applicable	Not Applicable	Not Applicable	Not Applicable	Not Applicable	SOP-XXX-002	
7.2	7.3, 7.6	Not Applicable	Not Applicable	Not Applicable	IQ-ABC-021	Not Applicable	Not Applicable	Not Applicable	
Not Applicable	7.2.1	7.2.1, 7.2.6	3.1	Not Applicable	Not Applicable	OQ-ABC-037	PQ-ABC-022	Not Applicable	
7.2	7.3, 7.6	Not Applicable	Not Applicable	Not Applicable	Not Applicable	OQ-ABC-037	PQ-ABC-023	SOP-XXX-002	
8	Not Applicable	Not Applicable	Not Applicable	Not Applicable	Not Applicable	Not Applicable	Not Applicable	Not Applicable	
8.1	8.6	Not Applicable	Not Applicable	Not Applicable	Not Applicable	OQ-ABC-038	PQ-ABC-024	Not Applicable	
8.2	8.7	Not Applicable	Not Applicable	5.6	Not Applicable	Not Applicable	Not Applicable	Not Applicable	

regulatory authorities such as the MCA. Those who front during an inspection need to be aware of these differences. Mutual recognition agreements (MRAs) should also be understood as information presented to one regulator in one context could be shared with another regulator out of context. Fronting an inspection can be a complex affair!

CONCLUSION: TRENDS IN GxP COMPUTER VALIDATION

Teri Stokes recently published an interesting analysis of increasing regulations affecting GxP compliance. Figure 5.1 is taken and updated from her book *The Survive and Thrive Guide to Computer Validation* [3]. Since 1991, international regulations and directives for the validation of computerized systems have increased 300 percent. In addition, 4 significant industry guides have been published since 1995 (see Chapter 3). The focus of these developments has been largely toward how to deal with modern IT technology and development methodologies. This is a significant learning curve for regulators and validation practitioners alike. It further highlights the need to monitor developments and take a GxP perspective.

The failure to validate to a regulator's satisfaction can have significant financial implications. Noncompliance incidents can lead to the withdrawal or delayed issue of a license to pharmaceutical manufacturers to distribute their product(s) in a given market such as an EU nation state or the United States. The U.S. Pharmaceutical Manufacturers Association (PMA) has estimated that a pharmaceutical company will spend between U.S.$50,000 and U.S.$100,000 for every day of delay during the pharmaceutical submission process. This is a small amount compared to the typical investment of U.S.$250,000,000 to bring a new drug to market. The real financial impact of GxP noncompliance, however, is the loss in sales revenue. For top-selling drugs in production, citations for noncompliance by GxP regulatory authorities can cost a manufacturer upwards of U.S.$2,000,000 per day in lost sales revenue. Some incidents concerning individual production units have cost the pharmaceutical manufacturer concerned over U.S.$100,000,000 in lost sales and revalidation. If a non-compliance was found with a corporate computer system, then the impact could be devastating. These systems by their nature are widely used within the pharmaceutical business. The declaration by a regulator that such a system is not fit for purpose and can no longer be used could disable an entire manufacturing site. If a satisfactory resolution cannot be quickly effected, then the pharmaceutical company concerned could be financially crippled, perhaps even leading to its liquidation.

The challenge is to conduct cost-effective, sufficient validation to ensure GxP compliance. As illustrated in Figure 5.2 [4], there is always debate over how much is sufficient to fulfill the regulator's expectations. Excessive validation may increase confidence in regulatory compliance, but it is expensive.

Figure 5.2. Good Practice for Validation

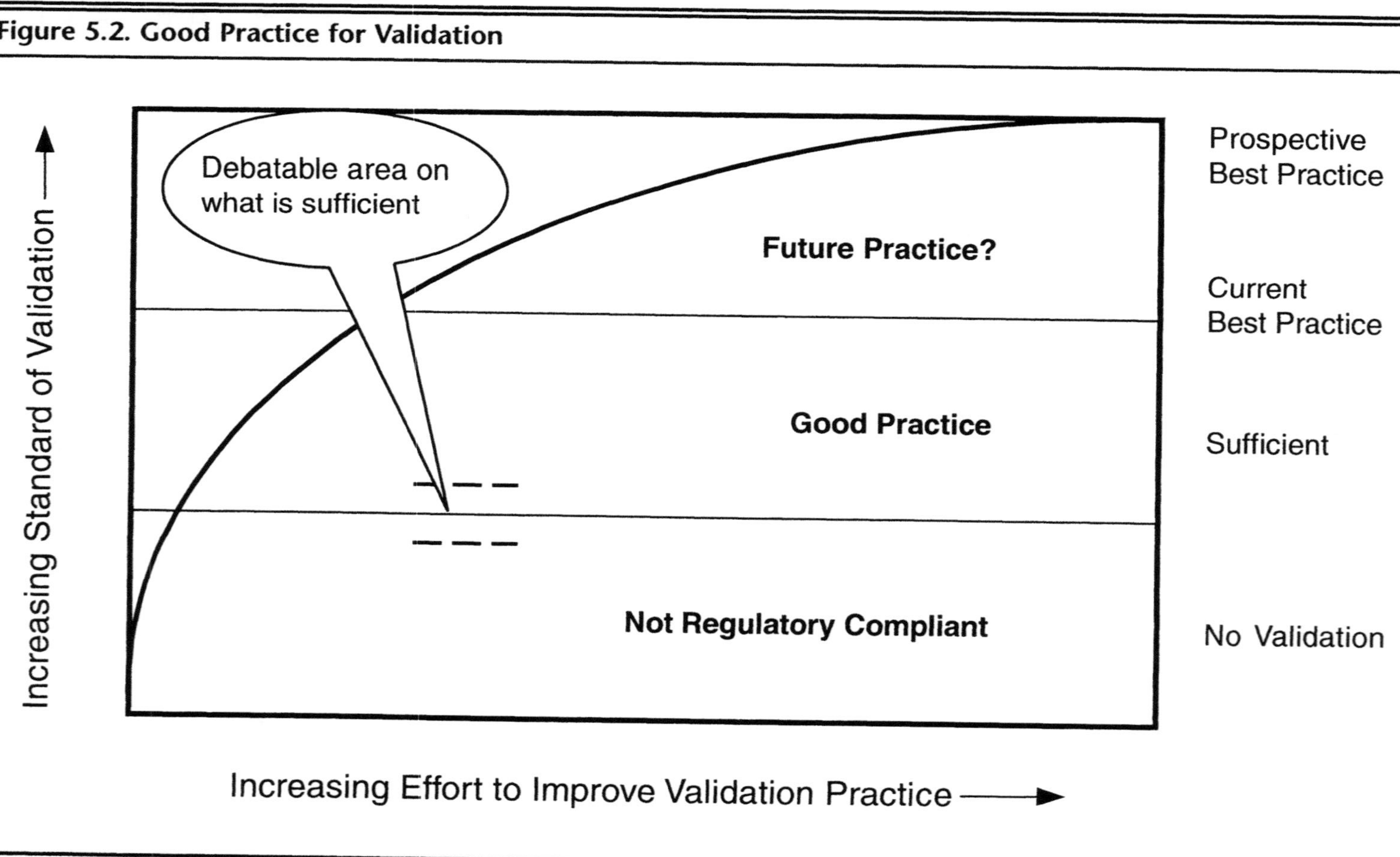

Inadequate validation may be cheaper, but, in the long term, the cost of regulatory noncompliance could be devastating. This book aims to clarify how much validation is sufficient for IT and network applications, suggest how it can be organized cost-effectively, and discuss areas of debate.

REFERENCES

1. FDA. 1999. U.S. Code of Federal Regulations, Title 21, Part 211: *Current Good Manufacturing Practice for Finished Pharmaceuticals.* Washington, D.C.: U.S. Government Printing Office.

2. European Union Directive 91/356/EEC. 1991. *European Commission Directive Laying Down the Principles of Good Manufacturing Practice for Medicinal Products for Human Use.*

3. Stokes, T. 1998. *The Survive and Thrive Guide to Computer Validation.* Buffalo Grove, Ill., USA: Interpharm Press, Inc., p. 148.

4. Wingate, G. 1997. *Validating Automated Manufacturing and Laboratory Applications: Putting Principles into Practice.* Buffalo Grove, Ill., USA: Interpharm Press, Inc.

6

Integrating Manufacturing Systems—A New Era in Production Control

Christopher Evans
Eutech Ltd,
Billingham, United Kingdom

David Stokes
Motherwell Information Systems
Sunderland, United Kingdom

Recent advances in the technology and integration of control systems and Manufacturing Execution Systems (MESs) means that project costs (including validation costs), capital costs and ongoing life-cycle costs (also including validation costs) may now be significantly reduced. This chapter explores the original driving factors leading toward such advances—the technology and the advantages it can bring. Specific focus is given to the associated validation requirements, long-term support issues and future trends.

WHAT ARE "OPEN APPLICATIONS"?

There has been a move toward so-called "open systems" and more specifically Open Control Systems (OCSs) since the early 1990s. There are many different definitions of "open", but members of the control and information technology (IT) industry would broadly define "openness" as *the ease with which different systems and sets of data can be integrated to provide a cohesive and encompassing solution.* This combination of systems and data can broadly be defined as "open applications", with no inherent implication on where the specific application is executed.

This generic definition is broader than that for open systems provided by the U.S. Food and Drug Administration (FDA) in 21 CFR 11: "Open systems means an environment in which system access is not controlled by persons who are responsible for the content of electronic records that are on the system" [11.3(1)].

In the sense of general control and the IT sector, a system could be open yet have sufficient security in place for it to be defined as a "closed" system under the definitions of 21 CFR 11. To avoid confusion hereafter, the term *open* will be used in the generic industry sense, and it should be assumed that such systems and applications may have such access restrictions in place to allow them to be defined as closed in accordance with 21 CFR 11 [1].

THE DEVELOPMENT OF OPEN APPLICATIONS

Open applications have been evolving over many years; indeed, the trend is probably as old as the IT sector itself. Early progress was limited and often provided no real benefit. Modern initiatives in the IT sector are overcoming these problems, and true open applications are starting to appear.

Early Open Systems

A typical pharmaceutical manufacturer will operate a range of computer systems, including the following:

- Intelligent instrumentation
- Barcode printers/readers
- Labeling systems
- Recipe Management Systems
- Laboratory Information Management Systems (LIMS)
- Management Information Systems
- Warehouse Management Systems
- Programmable Logic Controllers (PLCs)
- Distributed Control Systems (DCSs)
- SCADA (Supervisory Control and Data Acquisition) systems
- Batch record systems
- Building Management Systems (BMSs)
- Maintenance Management Systems (MMSs)
- Network controllers
- Weighing/dispensing systems
- Batch control systems

- MRP II (Manufacturing Resource Planning) systems
- Documentation systems
- Computerized analytical equipment
- Asset Management Systems
- Robotics

In the past, many of these would have been stand-alone systems (in part because they lacked the intelligence to be integrated). By the late 1980s/early 1990s, however, many pharmaceutical manufacturers had a strategic plan to integrate these islands of automation into a cohesive solution.

The cost of integrating systems from different manufacturers often proved to be very expensive, and, as a result, many automation projects failed to meet the required return on incremental investment. The main problem was that computer systems were not designed to be integrated in an open manner. Integration projects were typically dominated by the need to write bespoke software to link different systems together. The costs of validating a standard-proven solution increased by a factor of ten.

Today's Open Applications

The development of easily integrated (open) applications has been facilitated by the acceptance and deployment of standards initially used by IT applications now adapted or adopted for use by real-time control systems. Traditionally, the corporate world of IT (MRP II applications) was always seen as separate from the real-time problems associated with production. Using similar IT standards in both areas is now allowing complete integration of solutions, not only on single plants and sites but across sites and corporations as well. One such example of this has been the development of Object Linking and Embedding (OLE) for process control. This took an existing IT standard and enhanced its scope, speed and robustness so that it is now suitable for use in a real-time environment.

The challenge posed to pharmaceutical manufacturers is to maintain the security, data integrity and validation status of open applications that may now run across the corporate Intranet or even across the public Internet.

Integration Issues

There is a general realization in the pharmaceutical industry that the manufacturing process cannot be regarded as separate from the physical plant, nor the plant from the operating environment, and that the attributes of a particular product are closely bound to the process and plant at the time of manufacture. The validation process recognizes this as a requirement to qualify the complete manufacturing process/plant/environment, including associated production data.

Open applications must be able to fully integrate all aspects of production. This must be designed in such a way as to

- allow integration in a manner that is easy and relatively inexpensive to engineer and validate;
- overcome the data interchange problems associated with traditional interoperable (but not fully integrated) solutions; and
- support production throughout the operational life cycle of the manufacturing facility, not just the initial design and installation phases.

A hierarchy of systems can be considered to consist of OCS, MES, and Enterprise Resource Planning (ERP) systems (Figure 6.1). The issue of where individual systems implement particular functions in the hierarchy becomes blurred. As applications become more distributed, the ability to interchange data in a simple, low cost and validatable manner will become even more important. Open applications will therefore cover applications ranging across these three layers.

The open systems of the early part of the new millennium will act as a central focus for much of this, forming a hub which has spokes leading to multiple plant devices and multiple business system applications. Open applications will not only be able to interface to a multitude of devices in order to provide true integration but also will overcome the problems of data interchange.

The Process Control Environment

Depending on the particular product, process and plant, the OCS will need to communicate to an extremely wide variety of devices. Experience in the 1990s suggests that these will include traditional PLC systems, intelligent instrumentation and other control systems (e.g., barcode readers, barcode printers, weigh and dispense systems, etc.).

General Plant-Level Interfaces

There is a plethora of different interfaces available to communicate between different control systems and devices, such as barcode readers and printers, PLCs and so on. In the past, these have required a great deal of configuration, even when the interface was based on a fairly universal standard. This is because interpretation and representation of the data need to be coordinated at both ends of the link, and the description of the data was rarely contained in the information exchanged.

With the move toward such standards, many of these problems are finally being overcome, and interfacing to plant-level devices is being made much easier (whether real-time control systems or data-based systems).

Figure 6.1. The Three Layers Across Which Open Applications May Be Distributed

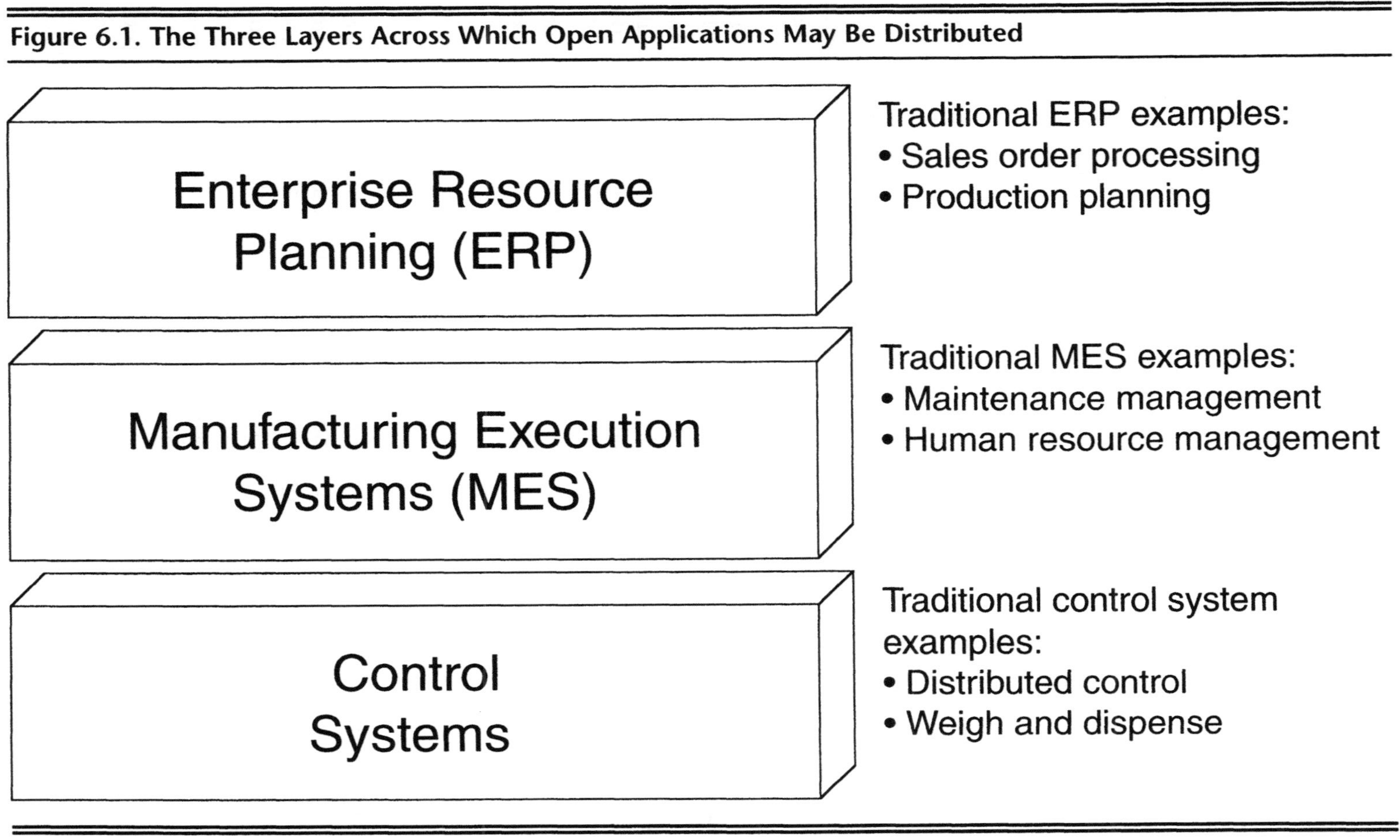

Fieldbus

Much has been written about the use of Fieldbus in process industries over the years, and it is true to say that the use of so-called intelligent instrumentation and digital communications has provided some benefit to users.

The major benefit to be realized from Fieldbus and intelligent instruments is primarily due to the ability to obtain on-line diagnostic information. This provides information on the performance of a control valve, for instance, or on the expected lifetime of a seal material based on the number of hours installed and the duty. These features lead to better planned maintenance (rather than preventative or run-to-breakdown maintenance), which in turn leads to lower overall maintenance costs and fewer unplanned losses of production. Validation of the on-line diagnostics will assist in the ongoing review of the validation status of the OCS.

The ability to configure certain parameters on-line is also useful, and certain instruments may be reranged remotely. However, the ability to rerange and remotely calibrate instruments in the field is not as attractive as it may first appear. Accurate calibration often needs to be validated for the product and process to be viable, and usually this can be performed only on a bench test or during a plant shutdown. In addition, the question of security of access to changing parameters must also be validated.

Many control systems fail to integrate Fieldbus instruments properly and cannot support the required data structures at all levels in the system. The modern OCS must be able to communicate with the intelligent instruments using whatever physical medium and protocol that the standard dictates, and the structure of the OCS must be able to pass the secondary attributes of the instrument to wherever the relevant software resides. This must be achieved without disrupting the real-time communications of the system and without corrupting the data structure which both the instrument and the associated software understands.

Trends

With the increasing trend to reduce even further the number of operators, some pharmaceutical companies are dispensing with the "traditional" control room (traditional in the sense of standard practice in the 1960s through to the present day). In many plants, operations are moving back out to the shop floor. This eliminates the need for two sets of operators (one in the control room and one in the production area), and often means a completely new method of operating the plant and interfacing with the control system.

The OCS may therefore need to have its primary interface in the production area (which may be a hazardous or zoned area) and also need to interface to additional production area displays and alarm/message paging systems. This will require open communications to a wide variety of operator interface equipment, including traditional or customized operator workstations that are increasingly PC (personal computer) based. The need to provide "mobile" interfaces also means that there will be a requirement for interfacing to such devices as radio paging systems and Personal Data Assistants (PDAs). Much of this technology is again being driven by the IT world, and rapidly evolving IT standards will again be the basis for the development of these interfaces.

Regulatory requirements, as detailed in 21 CFR 11 also allow for the OCS to be designed to incorporate or at least interface to equipment capable of biometric measurement (the measurement of a uniquely identifiable physical attribute that the operator possesses, such as fingerprints, retinal patterns, etc., or a set of repeatable actions unique to the individual).

These requirements are currently met by issuing each individual a unique log-in password, and this must be carefully managed. Password control is a procedural protection approach that relies upon the individual password holders to maintain the integrity of the passwords. If desired, the password approach may be enhanced by recording actual physical attributes as described above; this may become a standard requirement for OCS in the future.

The Production Environment

Whilst these cover the basic interfaces to the process, i.e., the process equipment and the operator, they do not cover the actual production environment itself. Many products require specific environmental conditions to be maintained, such as temperature, sterility, humidity and so on. These are often controlled by external systems, and the need to validate these conditions means that a separate, validated recording system is often provided for the air-conditioning system, which is known as a BMS.

It has been possible to interface such systems to the main production OCS for awhile, but this has been achieved via de facto standards that have made it difficult or even impossible to exchange additional information beyond the basic controlled parameters. For example, although the humidity and temperature of the production area ambient air could be measured and recorded within the OCS (via serial interface or hardwired transmitters, for instance), it has not been possible to monitor the load, temperature or efficiency of the main air-conditioning unit fans.

Whilst these are not important from a regulatory point of view, they may well be important from a business point of view. If the air-conditioning system fails unexpectedly, it may well necessitate a shutdown of production, and this should be avoided at all costs. Likewise, by monitoring the motor control centers and the electrical distribution, the complete process, plant and environment can be monitored. Whilst few of these may be absolutely required from a product point of view, they may all be important from a production, maintenance and business perspective.

The Corporate/Enterprise Environment

At higher levels, similar flexibility will be required to interface not just to manufacturing and business systems but also to individual MES and ERP applications. In a corporate environment that sees businesses choosing and integrating leading technologies from different suppliers for a number of applications, it is important that the applications can easily interface to any and all such applications.

These open applications may be distributed across numerous platforms within the corporate environment or may reside on any of the system platforms. Depending on the requirements of the individual business, a complete applications solution may well include finance, human resource, production planning, maintenance, health, safety and environmental issues. This is supported by the move toward more modularization of applications (with systems such as SAP R/3) and by the definition of standards for the interchange of information.

This information interchange is complicated by two major factors:

1. Different applications at different layers deal with data/information on different timescales (see Figure 6.2). OCSs typically work on the order of milliseconds and minutes, but ERP systems are designed to "think" in terms of months or even years.
2. Different techniques have traditionally been used to transfer data within the different layers. For example, OCSs typically may be interrupt driven or will work on the basis of polling for data, with each response being made before the next request is made. MES and ERP systems typically use transactional-based information interchange, where many requests can be outstanding at any time (due to the time taken to obtain the data and format the response).

These issues can be overcome through the use of specific tools, many of which are configurable instead of programmable, which therefore eases the task of validation.

An Example of an Open Applications Solution

Given that there are a wide variety of potential applications to integrate into a complete pharmaceutical "solution", what would a typical open applications solution look like?

Figure 6.2. The Different Timescales of OCS, MES, and ERP

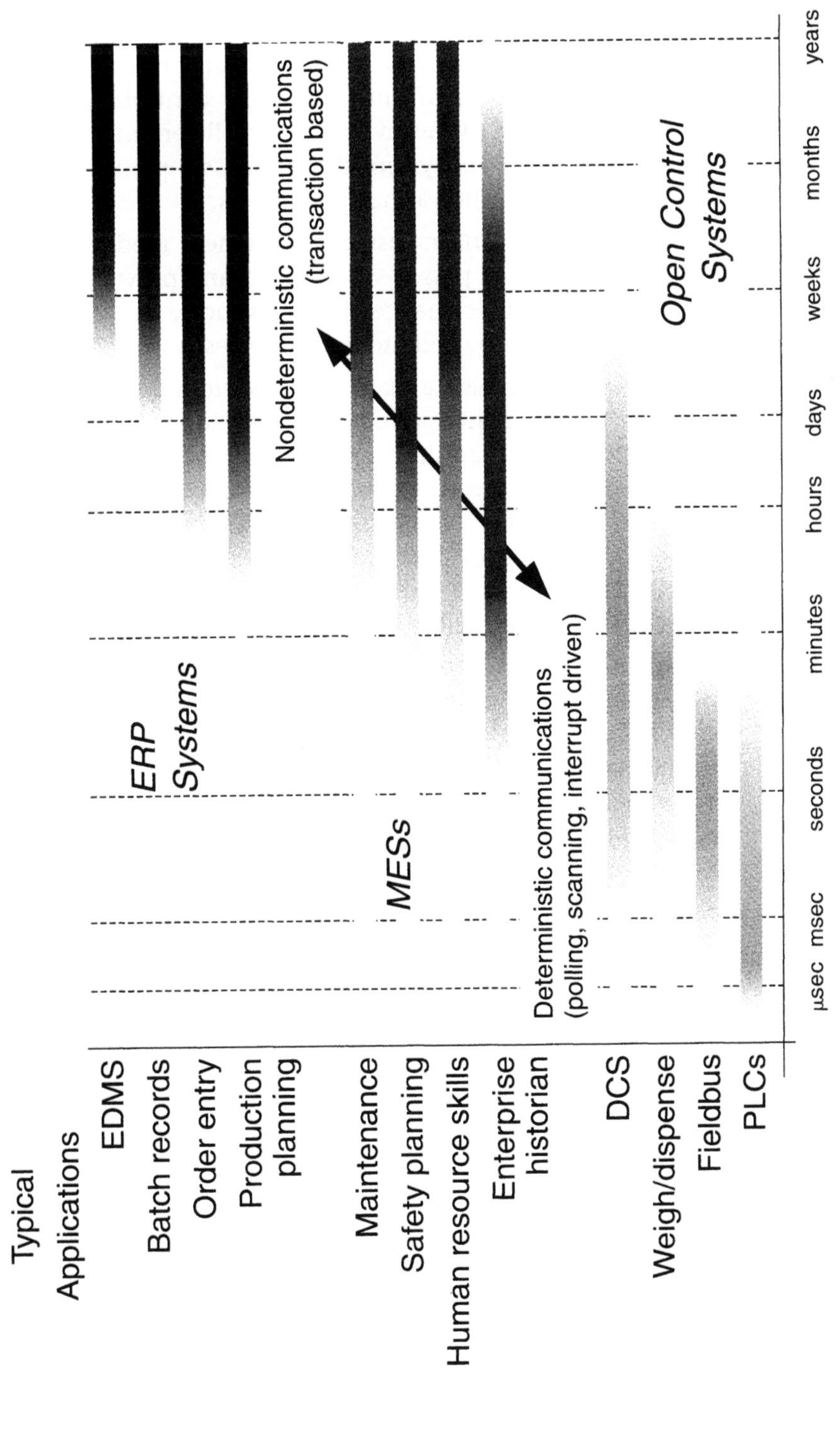

Figure 6.3 shows a typical solution, complete with all the necessary interfaces and the actual applications distributed in an open manner. It includes the following:

- An interface to the Fieldbus instruments, providing the OCS with the necessary data to perform supervisory control. Note that some control is performed at the level of the intelligent instruments themselves, and the control system downloads set points, monitors the control and records the various parameters.
- An interface to the on-line asset management module in the MMS. This uses data that is passed via the OCS and provides real-time information on the performance of devices such as transmitters, valves and so on in order to calibrate these devices.
- A MES level maintenance planning module, which uses data from the asset management module to schedule routine maintenance. The maintenance planning module can use this information to monitor long-term trends and adjust maintenance schedules to suit. This provides a full calibration system for the plant and schedules calibration of the plant instrumentation via the asset management module. This facilitates initial validation and provides long-term support throughout the life cycle of the plant.
- The OCS also has an interface to the PLC that is controlling the centrifuges and dryer. Supervisory commands are passed from the OCS to the PLC, and all data are recorded as part of batch logs. Note that the PLC has no user interface of its own except for local indicators on the machines themselves. These are "embedded" PLCs supplied by the machine manufacturers, networked to the OCS to provide the operator interface.
- The OCS interfaces to the weigh and dispense system and the Warehouse Management System in order to ensure the availability of raw ingredients prior to the start of a product run and also to schedule the pickup of finished goods by automated guided vehicles (AGVs). Note that although the OCS controller is connected to the Warehouse Management System via a dedicated link, the link to the weigh and dispense system is a "virtual" link across the corporate network.
- The weigh and dispense system and the Warehouse Management System are linked to the MRP II inventory management module to order to provide updates on inventory status. This is in turn linked to the purchasing module to ensure that new raw material stock is ordered on time and to the customer order entry module to advise availability for delivery.

Figure 6.3. A Typical Open Applications Solution

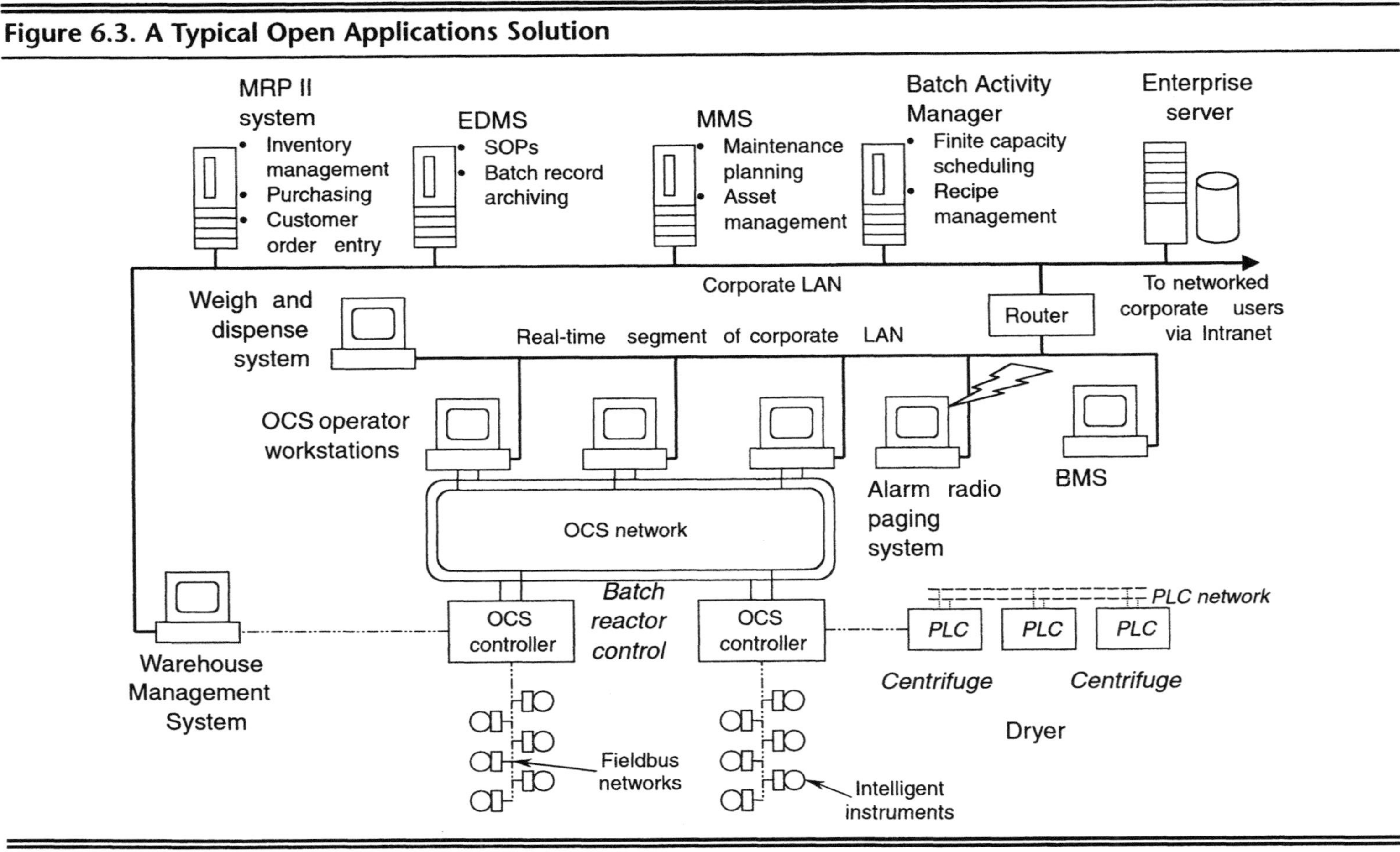

- The OCS also has an interface to the BMS, which controls the temperature and humidity of the production hall. These key parameters are also collected by the Batch Activity Manager and are recorded on batch logs on the enterprise server. Information on the BMS equipment (such as motor stop/starts, runtime, temperatures, etc.) are also passed to the Asset Management System and the MES maintenance module.
- The OCS controller provides the actual control of the batch reactor, following the recipes downloaded from the Batch Activity Manager. All data are recorded on batch logs.
- All key data are provided to the operators via the standard OCS operator workstations, which are able to access information from any of the systems described. These are all integrated onto the same displays, despite the fact that the data come from many distributed applications and databases.
- All real-time and historical information is stored on or available from the enterprise server. This can integrate information from multiple distributed applications and databases and consolidate the data into a single history system. This can then be made available as standard reports via networked printers or standard displays using Web browsers across the corporate Intranet. Copies of batch reports are written to local optical disks and also copied to the site Electronic Document Management System (EDMS) as a secondary backup and for long-term storage.
- Copies of Standard Operating Procedures (SOPs) are available to any process operator at any screen. Although these are managed and stored within the EDMS, they are called to the operator screen using standard document display formats.
- Operators are informed of alarms through the use of personal radio pagers. The OCS sorts alarms according to severity and plant area and then writes a suitable message into the database of the alarm paging PC which is installed on the site local area network (LAN).
- Overall production schedules are downloaded from the site MRP II system to the Batch Activity Manager, which breaks down the schedules using finite capacity scheduling in order to optimize the utilization of plant equipment and cut down on the number of product changeovers. The Batch Activity Manager also manages the recipes needed to produce the required product (recipe development, recipe release to production, recipe downloads, etc.). The detailed schedule is then passed back to the MRP II system along with updates on the actual production status. The production scheduling module then

updates the customer order module so that the company's salesmen can access the system and update their customers on the latest anticipated delivery.

Is this all science fiction? All of these open applications have been successfully implemented, and all are taken from actual projects. Although it is unlikely that all of these would be required on a single project, the ease with which this integration occurs becomes easier each year—a practical prospect.

EASIER VALIDATION THROUGH THE USE OF OPEN APPLICATIONS

It can be seen, therefore, that modern systems must be able to provide interfaces to a wide variety of devices. Key to all of these interfaces is the ability to receive and transmit, transfer, store, recall and display information in a wide variety of formats.

In a project such as the one described above, there are obviously a large number of systems to be integrated and interfaces to be considered. A method is required to allow such a complex solution to be validated. A logical and hierarchical approach is required that will ensure that the requirements of validation are addressed throughout the life cycle of the system.

Given the greater scope of today's (and future) open systems, it is important that all aspects of validation are fully reviewed and understood with respect to such systems.

The Validation Plan

The Validation Plan is the overall controlling validation document for the OCS and will be the first document created as part of the validation life cycle (Figure 6.4). The overall purpose of the Validation Plan is to define the approach that is to be taken to validate the system and to establish controls to ensure that satisfactory validation of the system is achieved. The Validation Plan is a live document for the full life cycle of the system (i.e., from concept to decommissioning).

Where a Validation Plan covers more than one distinct system, it may be appropriate to produce a Validation Master Plan (VMP), with subservient Validation Plans for the individual systems to provide a better focus on requirements.

The Validation Plan or VMP is required to cover a number of areas, for example,

- the scope of validation including system boundaries,
- the responsibilities of key personnel,

Figure 6.4. The Validation Life Cycle under Control of a Validation Master Plan

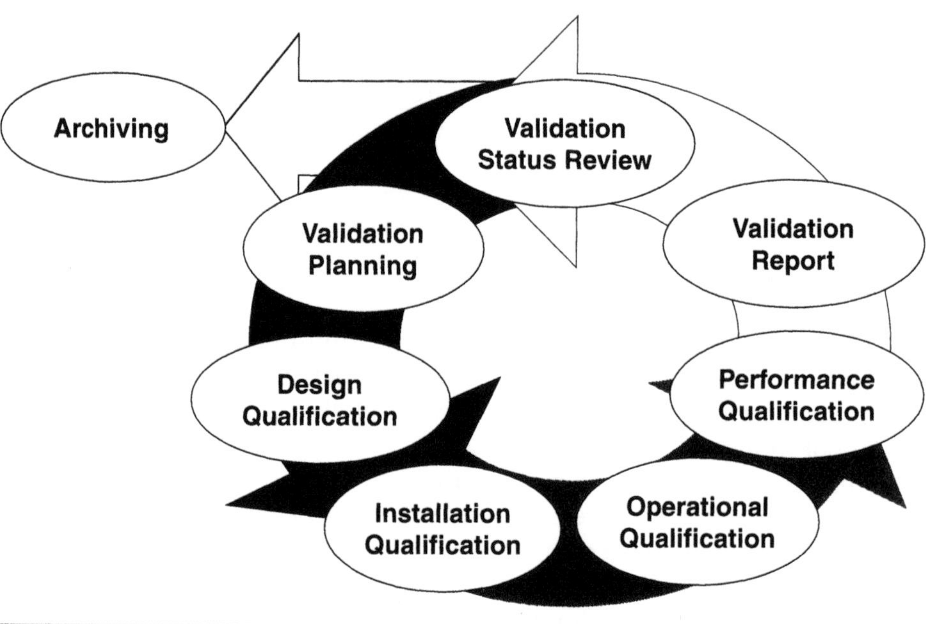

- references to related documentation and standards/regulations and
- Proposed timescales.

It is important to understand which validation tasks are performed when. Table 6.1 describes the links between the computer system project life cycle and the validation life cycle.

Design Qualification

Where open systems are concerned, there is a definite requirement to ensure that the design of the application is thoroughly reviewed and agreed on prior to moving into the qualification testing phase of the project. The potential complexity of the final system and its many interfaces is likely to make it virtually impossible to achieve a validated system if testing is the main focus. The intention of validating the system is to ensure that not only is there documented evidence that the system has been fully tested but also to provide evidence of control in the design process.

Table 6.1. Project and Validation Life Cycles

Project Phase	Validation Phase
Project concept phase	Validation planning phase
User Requirements Specification (URS) phase	Validation planning phase Supplier audit phase
Functional Design Specification (FDS) phase	Design Qualification (DQ) phase (GxP Assessment)
Software and hardware design phase	DQ phase (design documentation review)
Software and hardware implementation phase	DQ (Source Code Review) DQ and reporting phase
Integration phase	Installation Qualification (IQ) phase
Supplier acceptance testing phase	IQ phase
Site installation phase	IQ and reporting phase
User acceptance testing phase	Operational Qualification (OQ) and reporting phase Performance Qualification (PQ) and reporting phase
Site integration testing phase	OQ and reporting phase PQ and reporting phase
System handover	Validation reporting phase
Post-handover phase	Performance monitoring and reporting phase
Maintenance phase	Validation Review phase
Decommissioning phase	Archiving phase

The general approach required for open systems is to ensure that the individual parts of the system are designed in accordance with current best practice and then to ensure that the integration of the various systems does not affect the integrity of the system. A key requirement is that the team responsible for the integration of the open system have open personal communication links with the individual system suppliers. The validation of an open system will be very difficult if the culture is for each supplier to insist that any problems in interfacing are due to the other supplier's system.

The DQ process should cover a number of areas:

- *Supplier audit:* The overall goal of a Supplier Audit is to provide the pharmaceutical manufacturer with confidence in the computer system supplier and also to determine where validation efforts need to concentrated during the project. The results of this review should be documented in a Supplier Audit Report.
- *Review system's FDS against the URS:* This will determine if the design concept for each system within the open system being offered is in accordance with the original user requirements. The review will identify where functionality is not to be implemented (due to technology or cost limitations) and where extra functionality is offered (due to enhancements to operation recommended by the system supplier or because the functionality is standard on the system being offered). The results of this review should be documented in either a URS/FDS Review Report or as part of a DQ Report.
- *Review of design specifications against current regulations and guidelines:* This will assess the offered hardware and software along with the proposed functionality against current Good Automated Manufacturing Practice (GAMP). It is typically expected that the development of computer systems within the pharmaceutical industry should follow the guidance given in the GAMP guidelines [2] as well as 21 CFR 211 [3] and 21 CFR 11 [1] and the UK Medicines Control Agency (MCA) Orange Guide Appendix 11[4]. The results of this review should be documented as either a Regulatory Compliance Review Report or as part of a DQ Report.
- *GxP Assessment:* This will determine which parts of the system may have an influence on GxP critical system functionality. This will be achieved by assessing the functionality detailed in the FDS and assessing the impact these functions may have on product quality. With open systems, these functions may span across software packages and systems. The results of this review will determine the areas that should be given the highest degree of review and testing. The results of this review should be documented as either a GxP Assessment Report or as part of a DQ Report.
- *Source code review:* This will document a review of the source code produced in response to the agreed design. The Source Code Reviews will be based in particular on assessing the code responsible for the functionality identified in the GxP Assessment. This review will confirm that any bespoke code has been written and documented in accordance with good programming practice as discussed in the GAMP guidelines. The use of software objects and standard system functionality (e.g., configurable software) will dramatically reduce the effort that will be required in this area and is therefore to be

encouraged (see below). Once an object/module has be on reviewed, it may be utilized many times within the system to provide the required functionality. If the object/module has been reviewed for an application on another system, there is no requirement to review it again. The result of this Source Code Review will be confirmation that the requested and implemented functionality is the same and will also provide a good focus for the testing requirements for the system. The results of this review should be documented as a Source Code Review Report or as part of a DQ Protocol.

- *Failure Modes and Effects Analysis (FMEA):* This will document a review of the effects of failure of the component parts of the system. This review is mainly aimed at assessing the system hardware, interfaces and environment. This review should be performed on two levels: The first level reviews the possible failure modes of each individual system and the second level assesses the possible failure modes of the combined system. The overall objective of the FMEA is to identify the potential weak points and then to identify how these weak points may be designed out of the system. This may be achieved by installing redundancy, redesigning parts of the system, recommending procedural controls and so on. The results of this review should be documented as a FMEA Review Report or as part of a DQ Report.
- *Contingency planning:* This will document a review of the actions that should be taken in the event of failures as identified in the FMEA review and how the system will recover from these failures. Contingency planning may therefore not only cover the failure of the computer system but also the failure of associated services (e.g., electrical supplies and environmental control).

The DQ phase of the project is the time when quality may be built into the open system and is therefore the quality assurance (QA) part of the project. QA is the key to success, as it not possible to "test in" quality to the final system no matter how comprehensive the qualification testing may be.

System and Software Testing

The final documented evidence that the installed integrated open system operates in accordance with the agreed design is the testing documentation. For this documentation to be satisfactory, it is necessary to ensure that it is of the appropriate quality and provides sufficient rigorous testing of the system. At this stage of the project, we have now moved into the quality control (QC) phase. The ability to achieve the required result is based on the quality of documentation produced in the DQ phase of the project. This will include the range of documentation from the URS through the FDS, Hardware Design Specification (HDS), and Software Design Specification (SDS) to the System Integration Specification.

Testing Bespoke Applications

Some applications are project specific and will always require a high level of testing in accordance with GAMP category 5 [2]. Any such applications (whether within the OCS, MES or ERP layer) need to be rigorously and proactively tested. Such tests should not only focus upon the "normal" functioning of the application but also test the application's response to "abnormal" conditions.

For instance, what will the application do in the following conditions?

- The expected database does not exist or is corrupt.
- The data in the expected database are outside a specified range.
- The operator enters an illegal value (out of range, or a "garbage" string of characters).
- An interface to another device or application times out.
- Another application does not respond to a request for data or sends a response that cannot be interpreted.

This so-called stress testing of the application is important in order to be assured that nothing unexpected will result from a problem elsewhere in the system or the wider solution.

Testing Standard Applications

There are many applications which can be considered standard and will require less rigorous testing by meeting GAMP category 1/2/3 [2]. For this to be the case, the following conditions must be met:

- The same application (the same revision/release running on the same hardware platform and operating system) will have been used in a number of similar projects, usually within the pharmaceutical industry.
- There will be sufficient documentary evidence existing that the same application has previously been stress tested as above, either as part of its development or as part of another project.
- The supply of the product should be covered as part of an approved QA system (such as ISO 9000), which provides sufficient documentary evidence that the product supplied and installed is the correct one.

Many standard applications have a high degree of configurability that may be used depending on the details of project and the exact functional requirements (GAMP category 4). Where this is the case, it is still acceptable to treat this as a standard application so long as reference can be made in the validation documentation that the particular configuration has been covered in the stress testing referred to above.

However, because every overall solution is unique, the interface between various applications also tends to be unique. Many large corporate users also tend to have standard applications modified for their own use, and these modified versions should not be confused with the standard product—they are moving toward GAMP category 5. It is therefore important that particular emphasis is placed on which versions of the applications are being used and which version of other applications they are interfacing with. It may be that two applications have been interfaced on previous projects, but they cannot be considered to be a standard 'solution' unless the exact same version of the applications and interface are used.

It is often the case that the functionality of the particular application remains the same, but the interface needs to be developed or modified in order to integrate a solution. Where this is the case, so long as the supplier can provide documentary evidence that the base functionality has not changed and is not affected by changes to the interface mechanism, it may be possible to focus the stress testing on the interface alone. This is obviously aided by those suppliers whose base functionality and interfaces are separated in some way and can be separately tested.

Qualification Testing

Following the completion of application testing, the formal qualification testing process begins.

Qualification Testing Documentation

The qualification testing documents normally quoted by the pharmaceutical manufacturer are the DQ, IQ, OQ and PQ Protocols. This is, however, only part of the test documentation required to support an integrated system. There is a requirement for testing which is performed as part of the development of the open system to be documented. The main issue is that integration testing and internal software testing is performed by the company supplying the system and is not normally seen by the pharmaceuticals manufacturer. This may mean that the documentation supporting the testing is not to the appropriate standard; there is also the possibility that independence from the software development team is not provided. Software development personnel will often not appreciate the particular requirements of pharmaceutical regulations and guidelines. This therefore may be the weak link in the chain in the qualification process. It must be stated at this point that testing in accordance with validation requirements should be no more or less rigorous than should occur as part of Good Engineering Practice (GEP).

Qualification testing has two main parts: The first part is associated with the individual systems, and the second part is for interfacing the individual systems to form the OCS. Testing of the individual components will follow the normal V model approach as shown in Figure 6.5.

OCS Testing

In order to provide a logical approach to the testing and inspection of the OCS, we must clearly distinguish the separate applications that provide the system functionality and identify to which system each application belongs. As discussed previously, where the application resides in a modern distributed solution is perhaps less important that identifying its functional significance.

Each application needs to be allocated to a logical system in order to ensure that nothing is overlooked in the qualification process. Figure 6.6 shows how each of the various applications in our example has been allocated to either the OCS, the MES, or ERP system.

Figure 6.5. The V Model for System Testing

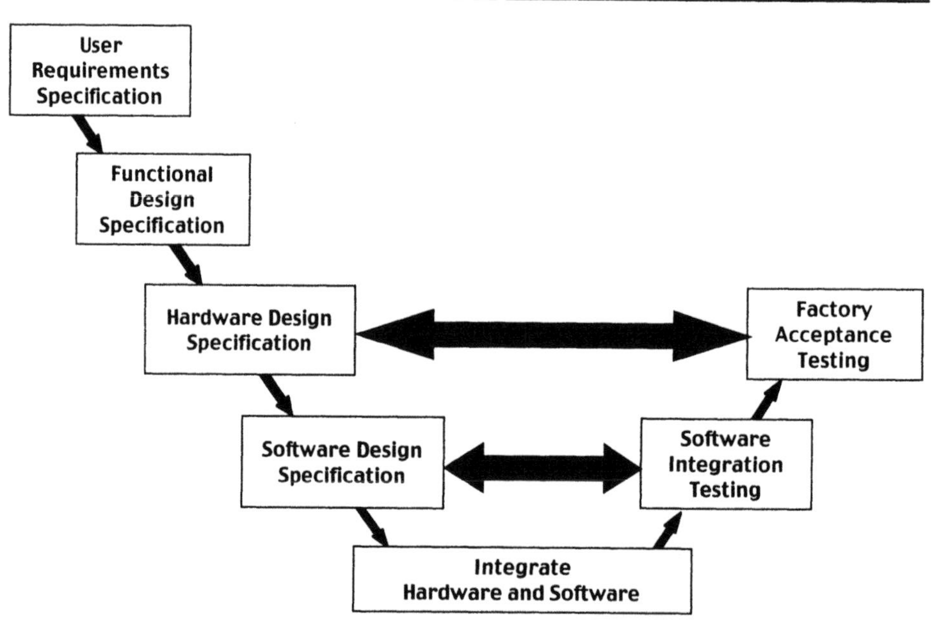

Figure 6.6. A Typical Open Applications Solution Mapped into Logical Systems

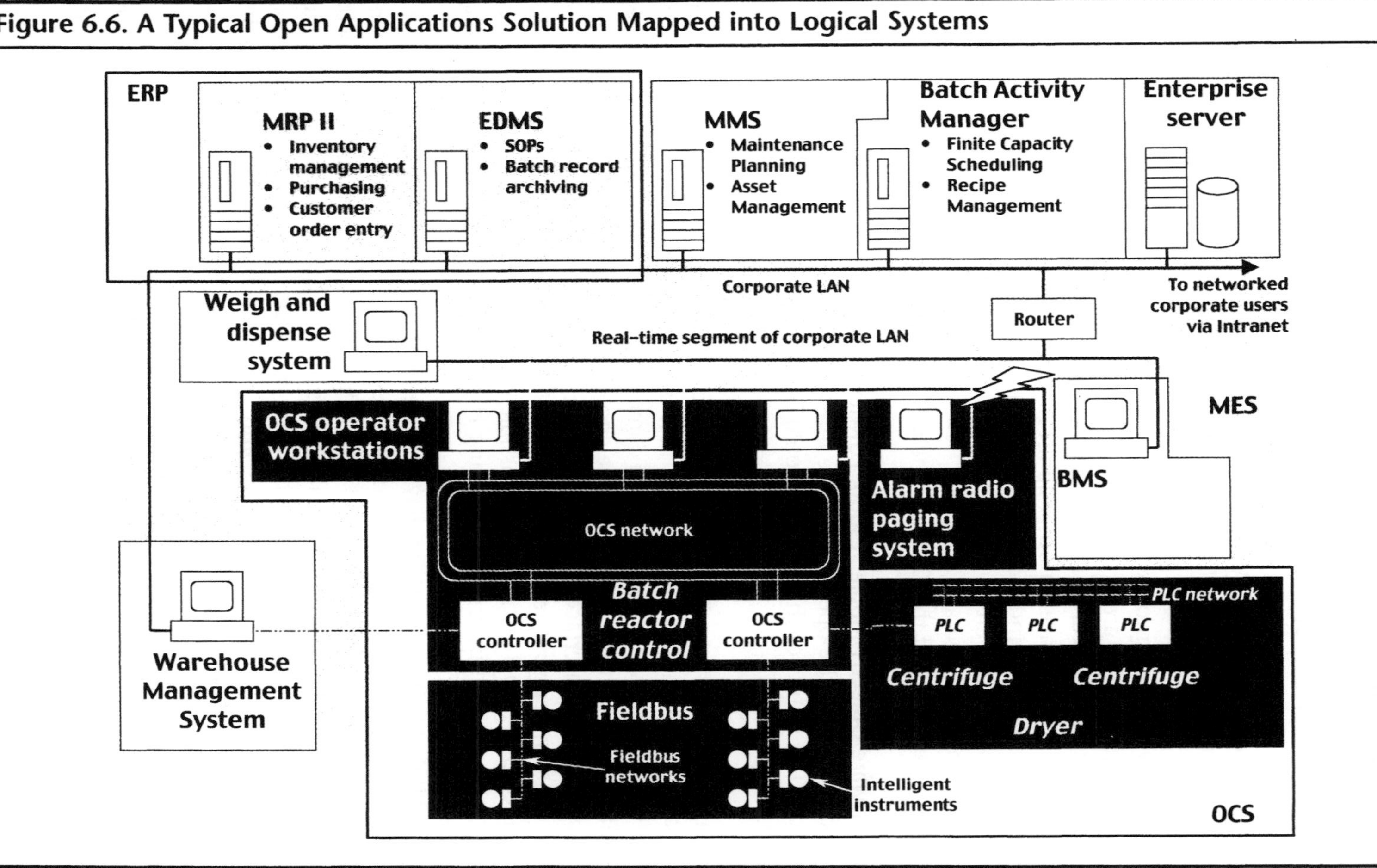

When assigning applications to logical systems, it is important to note that there will not always be clear distinctions between layers, and interfaces between applications may be within the same system or between systems. Where a group of applications only has interfaces between the applications, this is a good indication that the applications should form their own logical system (or subsystem). Although there will almost certainly be exceptions to any logical system of allocation, this systems approach will be identified in the Validation Plan and will tend to reduce the perceived complexity of the overall validation process.

Where there are a large number of interfaces between applications, it may not be practical to test the application in total isolation because of the applications' reliance on other applications for prerequisite data. The testing and validation of individual applications should be completed before wider systems are tested, and wider systems should be tested before the complete solution can be tested and validated.

The testing of the OCS will include the following:

Software Object/Module Testing. Low-level testing of individual software objects confirms the operation of the individual functions provided by the objects. The results of this testing should be documented as a Software Module Test Report.

Supplier Acceptance Testing. Supplier acceptance testing provides the pharmaceutical manufacturer with the first demonstration of the total functionality of the individual system and a demonstration of the interfaces associated with the OCS. It is, of course, to be encouraged that, wherever possible, true testing of the interfaces should be performed. This means actual connection to other systems rather than using simulation techniques. It is normally accepted that actual testing is not possible in all cases due to the complexity of the open system interfaces, so testing is performed using a validated simulation software or test harness.

Individual applications may be tested in isolation if they have simple or no interfaces to other applications. However, most applications will have interfaces to other applications and even other systems. These interfaces may be complex, and the major functionality of some applications may be tied up in the transfer of data across the interfaces, with very little in the way of actual data processing. These interfaces and information may be simulated by preparing suitable databases or by writing specific test or support routines. In this manner, the individual application may be tested prior to moving on to complete systems qualification testing.

There is always a question concerning the degree of testing that needs to be carried out in complex solutions. To a large extent, this depends on the application's "pedigree", criticality and amount of bespoke software.

Each test within the testing process should specifically relate to a function of the system as documented in the design specifications. These tests will cover the hardware, firmware and software for the system. At the end of supplier acceptance testing, there may be a number of outstanding issues as a result of test failures or items that have been tested only in a simulation environment and will therefore need to be verified following installation.

It is important that the documentation for supplier acceptance testing is utilized to support the qualification process, therefore, the data provided should be suitable for referencing directly from the IQ and/or OQ Protocol. This is most appropriate where the functionality of the system may not be fully demonstrated following the final installation (typically some alarm/interlock and failure mode tests may be difficult to reproduce without risk of damage to the installation external to the computer system).

At the completion of supplier acceptance testing, a report will be produced, which must be accepted by the pharmaceutical manufacturer prior to shipping the system to the site. There should also be a list of actions that must be performed prior to the commencement of IQ and/or Site Acceptance Testing.

IQ Testing. A record of the physical installation of the individual computer systems and their interconnections and interfaces is the result of IQ testing. The purpose is to uniquely identify the component parts of the individual systems, hardware, firmware and software and to verify that they were the items utilized in the supplier acceptance testing. It is often the case, however, that due to supplier acceptance failures, there may have been modifications to the software or replacement hardware performed under change control.

It should be standard practice to begin the IQ process with a backward looking review. This review should assess the outcome of the DQ activities as identified above and the supplier acceptance testing. Where issues have been identified and/or failures recorded, they should be assessed and appropriate action taken to ensure that the IQ process may continue. Any decisions made by the IQ tester and witness regarding these actions should be recorded in the IQ protocol. Table 6.2 represents the information typically recorded in IQ tests/inspections.

User Acceptance Testing (UAT). UAT provides the pharmaceutical manufacturer with the final demonstration of computer system functionality. In the OCS, there will be confirmation that the individual systems operate as expected; this will normally be a repeat of some or all of the tests in the supplier acceptance test specification. Following satisfactory demonstration of the individual systems, testing of the interfaces will be performed by the system integrator. Clearly the demonstration of the functionality in UAT is the same as is required for OQ testing, and the two activities should therefore run in parallel. The possible requirements for the on-site OQ are given in Table 6.3.

Clearly the complexity of the open system will require a phased approach to site testing. There is a definite requirement to implement the systems in a phased and logical manner, testing each system as it is installed and

Table 6.2. IQ Testing

Hardware	Documentation
• Computer main components: unique identification details • Peripherals: unique identification details (MMI, printers, keyboards, mouse, barcode readers etc.) and storage capacities (RAM, hard disk, floppy disk, DAT, CD-ROM, etc.) • Review of wiring in interface cabinets • Interfaces (I/O cards, dedicated/ network, cabling, transfer rates) • Settings (switch settings, firmware configuration) • Software versions (including third party) • Visual inspection of major hardware components • Interface card addressing • Wiring checks • I/O continuity • Electrical supplies • Diagnostic checks	• As-built computer system installation identification details drawings • Calibration certification • Computer system maintenance documentation • Record of settings (switch settings, firmware configuration) • Approved functional description • Approved design documentation • Approved GxP Assessment Report • Approved Source Code Review Report • Approved FMEA Review Report • Approved Supplier Acceptance Test Report • Draft maintenance procedures (e.g., backup and restoration, security handling, Contingency Plan) • Draft user manuals • Spare parts list
Software	**Environment**
• Version of operating system	• Earthing
• Versions of application software	• Temperature and humidity control of computer room
• Versions of communication software	• Temperature and humidity control of room containing interface equipment and peripherals
• Database structures	• RFI, UV, EMI
• File structures	• Power supplies
• Configuration files	• Backup power supplies (e.g., UPS)

MMI = man machine interface; RAM = random access memory; DAT = digital audiotape; CD-ROM = compact disk, read-only memory; I/O = input/output; RFI = radio-frequency interference; UV = ultraviolet; EMI = electromagnetic interference; UPS = uninterruptible power supply then again as each system is connected to the open system. The ability of the system integrator to provide a successfully integrated solution is key to the future acceptance of the system. This is the time when confidence in the system must be built up. Confidence is a like quality; if a system is not built with it, it is difficult to get it into the system at a later stage.

Managing Rapid Application Developments

With the tendency to move toward faster project timescales, there is less and less time to follow traditional routes for solution development. This is also being changed because of the tendency for users to work more closely with their suppliers.

Table 6.3. On-site OQ Testing

Hardware	Documentation
• Computer performance (CPU [central processing unit], bus, cache, clock, etc.) • Operation of storage media • Operation of peripherals • Operation of interface equipment	• Final versions of maintenance procedures • Final version of system documentation
Software	**Environment**
• Functional/operational tests versus user requirements and design specifications • Challenge testing against operating ranges (e.g., data entry, performance, maximum and minimum operating ranges) • Communication interfaces • Start-up and shutdown • Security and access • Backup and restoration • Data storage • Recovery from system failures • Alarm and event handling • Reporting and historical data	• Continuous measurement of environmental control

The process of developing software has traditionally been a time-consuming process, but this is also changing with the availability of more powerful software development tools. These enable developers to build applications much more quickly at a much higher level, often using very visual techniques, natural language structures and very intuitive tools. These tools provide full documentation and powerful debugging facilities. All this allows much more rapid application development (RAD), reducing the cost of developing the actual source code. The result of this is that the traditional method of defining the URS, the FDS, detailed design specifications and so on is becoming more of an overhead and often prevents the rapid development of applications and solutions.

Given the need to work faster and in closer cooperation with other parties, the traditional method of working is often condensed, with "hybrid" documents being developed and moved quickly forward to produce applications. As an example, the URS may now be completed in conjunction with the supplier and may include detail that would previously have been included in a separate FDS. A single, detailed SDS may then be produced, and the actual code may be produced from there, using the powerful development tools that are now available.

This does not follow the traditional models outlined in Good Manufacturing Practice (GMP) guidelines such as GAMP, but this is acceptable so long as the

- Validation Plans continue to detail how the development process and the application will be validated (and the process followed complies with the approach detailed) and
- Project Quality Plan details what documents are to be provided, how the various documents are related and how change control and testing will take place.

When the RAD route is taken, it is essential that all parties understand that the Validation Plan and Quality Plan are still key documents. Special emphasis must be given to qualification of the system because of the unusual route taken.

THE USE OF SOFTWARE OBJECTS

In an effort to reduce the complexity of engineering and software designs, many systems now use the concept of so-called "objects". These allow even the most complex engineering designs or software constructs to be built up from smaller building blocks known as objects. Most open applications are now based on object-oriented software, since this technique lends itself to efficient software development and easier software integration between different applications.

An object basically consists of inputs, outputs, a function (or functions) and parameters. When the object is processed (executed), the outputs are calculated based on the function of the object, the input values and the settings of the internal parameters. Some objects have a fixed function (such as an addition function, or a PID [proportional, integral and derivative] control loop), and some may have their object functions modified or even created from scratch.

Objects may be processed individually or connected together to perform more complex functions. These are sometimes called "function blocks" or "macros". The use of objects greatly simplifies engineering design, since standard objects are used wherever possible. This obviously has great advantages when it comes to the validation of software implemented using objects. Type tests may be performed on the basic objects, thus each subsequent instance of the object (each configured copy of the object) will not need individual testing.

Collections of objects can also be type tested to whatever level is required, so long as the internal functions are not changed and as long as the initial testing of the macro was sufficiently rigorous (using wide variations of input conditions and internal parameters). Integrated testing of unique (application-specific) collections of objects and macros is always required.

The use of object-oriented software has been estimated to save literally millions of dollars on new plants, and these savings only estimate the savings in actual engineering hours. These savings will obviously be reflected in the reduced cost associated with validating standard objects as opposed to application-specific or bespoke software.

The Applicability of Objects

As well as control systems being configured using objects, many other devices and systems are configured using this object-oriented approach. Fieldbus device descriptions are examples of objects that contain inputs, internal parameters, a function and an output. A Fieldbus device may typically have multiple functions, a dozen or more inputs and potentially hundreds of parameters and outputs that may be accessed.

The problem with objects is that they can vary considerably, from a simple addition function (two inputs, one function, no parameters and one output) to something as complex as an object used to describe a complete centrifuge.

Other objects may be used to describe the function of a motor or drive, a medium voltage circuit breaker, or a heat exchanger; a complete reactor may be described as a collection of objects. Emerging standards such as STEP (the Standard for the Exchange of Product Model Data, ISO 10303 [5]) are defining models for the interchange of electronic data describing all types of devices and equipment. These models can also be described as objects and may have a high degree of complexity.

In theory, objects can be applied to even the most complex of data. Biometric measurements were discussed earlier, and there is nothing to prevent such information being stored as an object. This may be a simple object such as graphical image of a fingerprint or retinal pattern, but the problem with such a simple object-based solution is that transferring, storing and processing of such information is not best suited to the world of computers.

It is much more likely that unique patterns in the graphical data will be mathematically represented and stored as one large object or a collection of smaller objects. Fingerprint analysis can now be carried out with mathematical rigor, identifying the topographical connection between definitive features such as forks and dead ends within the ridge patterns of the fingerprint. This can be reduced to a smaller data set and modeled using objects.

The use of objects can therefore be extended to many applications within the pharmaceutical systems arena, and support for such objects is an increasing requirement of modern open applications.

The Advantages of Software Objects

Using software based on objects has two key advantages: The first is related to the project stage of the life cycle, and the second is related to the later stages of the life cycle.

Greater Engineering Efficiency and Easier Validation

One big advantage of software objects is that they are modular and repeatable. This has great advantages when it comes to engineering efficiency. Once an object has been defined, it can be copied as many times as required ("instanced"). Such versatility makes it considerably more efficient to develop large applications with a high degree of repeatability.

However, because the objects themselves are designed as standard parts (starting at the lowest level in any solution), any project that is based on the use of objects should be easier to validate. Testing can start with the simplest objects, which are then built upon to form more complex objects. Because the lower level objects have already been stress tested, their basic functionality can be assumed to be acceptable when they are used at a higher level. This means that the testing of a more complex object can focus on testing the overall functionality rather than retesting each individual part.

This hierarchy of standard objects means that overall testing is reduced. There should be an inherent documentation trail that will facilitate of the requirements of validation.

Operational Advantages from Using Object-Based Software

The use of objects is also advantageous during the later stages of the life cycle—ongoing operations, system security, change control, periodic reviews and maintenance are all made easier through the use of object-oriented software.

Should any changes be required, a modular and hierarchical structure makes it easy to identify the impact of any changes. Change control can also be implemented at the object level, and it is therefore easier to set up software change control systems. Software security can be implemented at the object level, thus making it easier to control access to parts of the system and individual applications. This means that audit trails can become a standard part of the overall solution, again easing the ongoing task of maintaining the system's validation status.

If for any reason a low-level object needs changing or redefining, it is easy to see which higher level objects may need to be revalidated. This can significantly reduce the amount of time needed to identify the impact of any change and to implement, test and revalidate the changes. In a similar way, should any higher-level object need changing (such as the data to be included in a production schedule), the amount of retesting that will be required will be minimal, so long as the lower-level objects (on which this object is based) have not changed.

The structure that object-oriented software imposes also means that it is considerably simpler to obtain and understand information required for maintenance purposes, and it is considerably easier to conduct periodic reviews.

The use of object-oriented software therefore supports many aspects of the complete life cycle, providing positive benefit from the earliest days of the project until final decommissioning.

Coding Standards and Risk Assessments

Objects, and software generally, need to be coded or programmed using some form of language or tool. Increasingly, these tools are becoming easier to use and are extremely powerful and flexible. Although this has its advantages in terms of the speed of application development, there is also a danger that this speed and flexibility will lead to an increasing number of mistakes.

When the speed, power and flexibility of software was limited, there was generally plenty of time for Source Code Reviews, and the code could generally be independently reviewed by a colleague with appropriate training and experience. However, software code can now be generated very quickly, and tools exist to produce powerful code from graphical representations, plain text, software "wizards" and so on. There is generally less control over the resultant code, and the greater flexibility of the software increases the chances of a mistake being made, simply because the software can now do more than it could in previous generations.

It is therefore necessary to carefully review the software tools that will be employed on a project and to analyze the inherent risk. Some languages and tools enforce a fairly rigid structure (at the cost of less flexibility) but are easier to review.

Some software tools allow "rules" to be established which provide software conventions that will be enforced on the project. These will limit the variations in the software, help enforce consistency and help with the task of software reviewing (the "code walk-through") and validation. Other languages, such as C++, provide a great deal of flexibility for the programmer, and reviewing this code may be extremely difficult.

The careful selection of programming and software tools, the establishment of software conventions and standards and the enforcement of such standards should therefore be a part of every project and may be included in the Project Quality Plan. A review of these standards and the resulting software should also be part of the overall risk assessment and Hazard Analysis.

CURRENT TRENDS

In the rapidly changing world of IT, it is difficult to keep up with current trends, let alone predict future developments. However, current limitations in the use of open applications and object-oriented software have already been identified and are being overcome.

The Problem with Objects

Even with simple objects the lack of standards for object models has proven to be a problem in the past. The object used to describe a PID function in one control system will be different from the object used in a different system because of variations in the number of alarm settings, in the way the derivative action is calculated and so on. This variation in object design is one of the reasons why true integration between modern systems is such a problem.

To overcome these problems in the IT world, the industry has developed a component object model (COM) which basically defines what an object is and how it should relate to other objects. COM therefore allows software from different manufacturers to work together and is one of the things that makes such everyday functions such as "drag and drop" so easy to use. (The application you are dropping into needs to understand the format of what you are dropping, and the application you are dragging from needs to know how to how to place the data or object onto the "clipboard").

COM has been further enhanced to operate across multiple platforms in a network environment and in a distributed manner. This is known a D-COM (distributed component object model), but even D-COM fails to address some of the specific aspects of process and production control. Amongst these issues is the fast transfer and response times that are sometimes required and the so-called "high granularity" of the objects used (i.e., big objects are used by many of these applications).

Support for Multiple Object Formats in the Production Environment

It has been seen that to be really useful in the pharmaceutical industry, systems need to allow integration between all manner of devices and applications and also need to provide an environment that allows a wide variety of objects to be used.

The move toward common operating systems (i.e., Microsoft) means that for the vast majority of applications, the storage, recall and display of various object models is becoming less and less of a problem. Applications written to run under these operating systems are capable of storing, recalling and displaying their own object data and also transferring suitable objects between themselves using D-COM.

The real issue within the production environment is the transfer of various object data between applications. What is required is some sort of "object broker" or object management mechanism (OMM). This mechanism effectively handles multiple object models; it converts them into a common format that can be successfully transferred between various applications and then be reconstructed at the other end.

Practical Applications of Object Brokering

In order to enable the use of open applications, full support must be provided to make the transfer of object data as standard as possible. This will in turn reduce the time spent on validating such solutions and delivering ongoing advantages throughout the life cycle of the solution.

Many existing de facto standards do not explicitly make use of object-oriented techniques, yet use a standard data structure and format which can easily be mapped and transferred using an OMM. Examples of this would be ControlNet®, Modbus® or Data Highway Plus® protocols that use a structured addressing technique to transfer data between nodes on their highways. Such links could easily be connected to the OCS controller using standard physical media and protocols, and data are made available to the rest of the system by the OMM layer in the controller. Examples of this are shown in Figure 6.7.

Support for various Fieldbus standards can be easily incorporated into such a system, since the device descriptions can be passed between the field devices and the asset management software as another object class. This can be managed regardless of where the Fieldbus software resides.

Figure 6.7. Using Object Management Mechanisms to Transfer Nonobject-Oriented Data

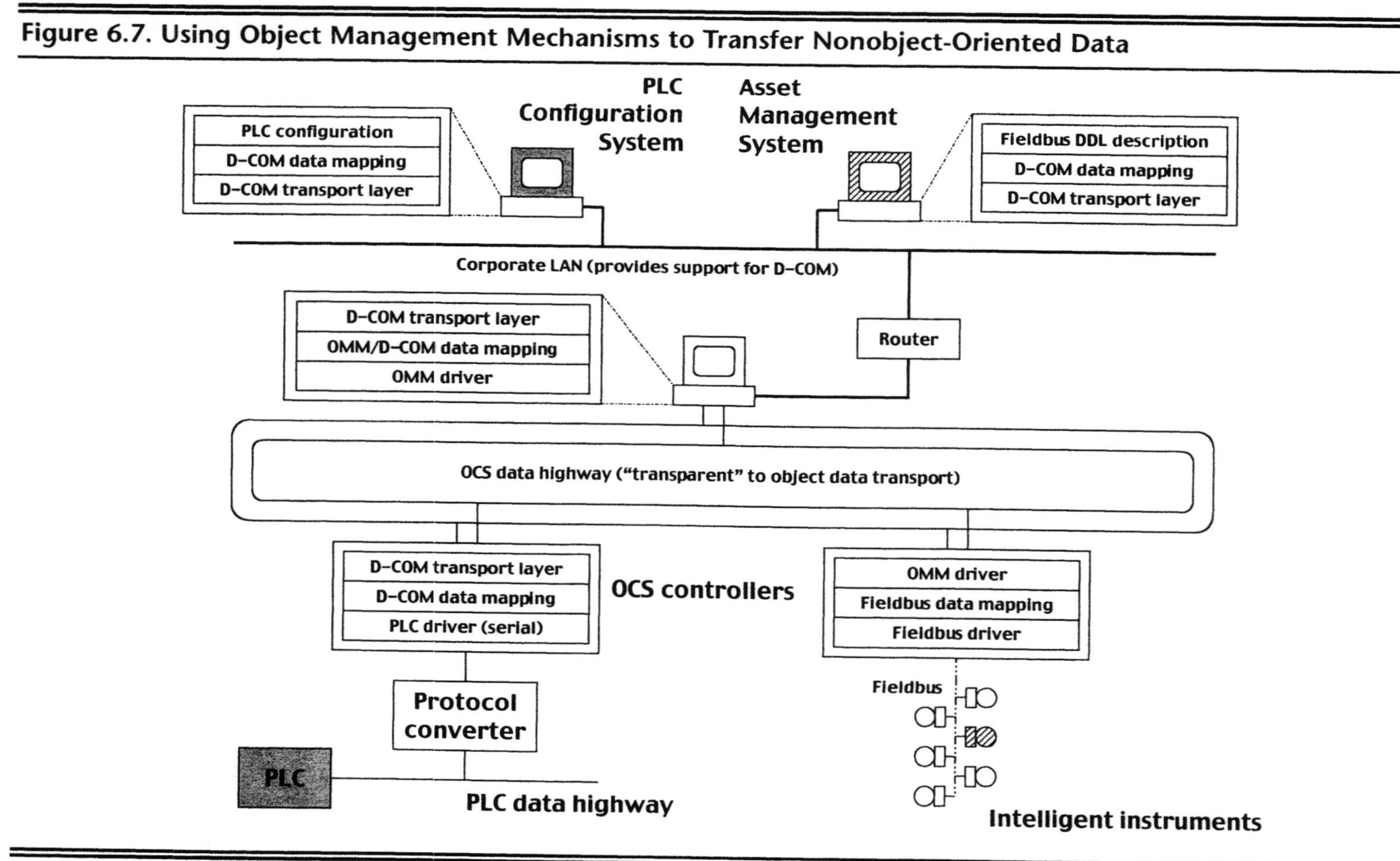

In the PLC example given, the PLC configuration data would be converted into object data using D-COM in the PC where the PLC configuration software resides. This would be transferred to the OCS via the corporate LAN, which provides inherent support for D-COM via the TCP/IP (Transmission Control Protocol/Internet Protocol). The OCS then passes the D-COM object across its own data highway, which is "transparent" to data which is "through routed" (i.e., the OCS data highway transport layer passes the data regardless of its content or format and adds error checking only to ensure data integrity). The OCS controller then converts the data from the D-COM object and breaks it down. As part of this process, it interprets the embedded addressing information and passes the configuration data to the PLC via the serial link, using the appropriate serial driver. This is then passed via the protocol converter and the PLC data highway to the PLC to be programmed.

In the Fieldbus example, the Asset Management System converts the standard Fieldbus descriptor (in device descriptor language [DLL]) to a D-COM object. This is then passed to OCS where it is converted from D-COM into the OCS's own object model. This is so that the relevant parts of the Fieldbus device descriptor can be distributed to the various OCS nodes for display, trending, alarm purposes and so on. The OCS then uses its OMM to transfer the data to the OCS controller, where it is again converted to DLL format prior to sending across the Fieldbus link to the intelligent instrument.

Although this sounds simple in principle, the use of multiple standards on the shop floor still requires that specific mappings be developed for individual standards and protocols. However, the existence of the OMM layer in an overall solution makes it possible for such interfaces to be developed, thus the resultant data can be made available anywhere else in the system.

This is the main difference between emerging solutions and the standard solutions of the 1990s. Although some form of bespoke interface may still be required to translate a particular data format and protocol to an object model, once this is developed, the resultant data are then available to all other users via the "object broker".

The real benefit comes when interfacing to the IT world, where COMs already exist and are already supported. Interfaces to third-party OCS, MES, or ERP applications can be handled so long as the system is connected to the appropriate corporate networks, and most transactions can be handled using standard IT technology.

A maintenance package can therefore request information on a transmitter, and the complete device details would be available. The object-brokering layer would convert from the appropriate device description and make this available to the maintenance package via D-COM. If the same maintenance package was looking for the total number of stop/starts and running hours on a motor, it would no longer be required to maintain separate counters and timers in the PLC or control system; the data could be obtained directly from the motor control center (MCC). The object-brokering layer would broker the request; the conversion to the MCC protocol would be performed automatically by the object-brokering layer in the controller and passed across to the MCC. The resultant data would then be converted to a suitable object model for transmission across the control network and converted back to D-COM for the maintenance package to understand.

In a similar vein, imagine a fuzzy logic controller running in a serially connected single loop controller. The controller requires the results of a laboratory analysis in order to operate optimally, and the analysis is performed on an spreadsheet in the laboratory. Traditionally, the operator would have to download these figures manually, or the analysis package running in the PC would need a driver developing to the single loop controller, with the link between the spreadsheet and the driver being programmed or set up using something such as dynamic data exchange (DDE). Such a traditional solution would require extensive module testing and validation. Using the object broker, a spreadsheet would pass the results to the OCS using D-COM. This would then be transferred via the object-brokering layer to the OCS controller, where it would convert the values contained in the object to the protocol required by the single loop controller.

Such systems are now available and are again based on some of the work coming from the IT sector. D-COM has again been extended to make it more applicable to the production environment and to allow it to operate in real time.

The use of a common object broker extends the advantages of object-oriented software and makes the practical integration of open applications much easier to engineering, test, validate and maintain.

Validating Object Brokering as a Standard Solution

As well as the operational advantages gained by achieving simpler integration of plant devices, there is also a significant savings to be made in the validation of such applications. Although the normal rules of validation will still apply, the advantage is derived from the fact that fewer and fewer bespoke interfaces will need to be thoroughly tested. Where the number of such interfaces installed in the field are such that they may be deemed as standard product features, a less proactive approach to validation may be adopted.

Where the OCS has a standard interface to a particular Fieldbus and the number of installations is such that this can be deemed as standard feature, there is little need for software module testing. This is true of any Fieldbus interface to any control system. In a traditional scenario, where data need to be transferred via the OCS to a maintenance application or a laboratory analysis package, the actual transfer of the data needs to be fully tested as a bespoke application.

Where object brokering is used as a standard part of the system and the interfaces between the Fieldbus and the laboratory analysis system and the object broker have been both previously validated, there should be no need to fully validate the application as a bespoke one-off solution. This may be the case even if these two sets of devices have never been connected in such a manner before.

Therefore, so long as any given interface to the object broker has either been fully validated by testing or by acceptance of the installed base, applications may be linked via the object broker and treated as standard product. This can significantly reduce the need for the full testing of interfaces and allows new combinations of interfaces to be made without application testing at the module level.

FUTURE DEVELOPMENTS

As ever, future developments can be predicted by looking at current requirements that are still not being met. The interfacing of open applications still needs to be made more efficient and easier to validate; this is where future developments will focused.

Toolkits for Developing Standard Interfaces

Systems already exist that provide true plant-level integration through the use of a object brokers. The ability of these systems to provide full integration in the future relies upon the development of suitable standard object broker interfaces to the large number of existing de facto standards that currently exist. It is, however, unlikely that specific interfaces will be written for many devices and systems that use less popular protocols.

What is more likely (and is already happening in some systems) is that generic, configurable interfaces will be developed. These will provide toolkits which allow transactional-based interfaces to be developed in a configurable manner.

Modern communications are much more transaction based (a trend developed in the commercial and business sector), and these are less suited to traditional control system interfaces. Control systems have traditionally operated on a real-time basis, polling for data on a routine basis or being interrupt driven by alarms and events. These traditional systems are not well suited to transactional-based communications, and the use of object broker interfaces will overcome some of these problems. The new object-oriented interfaces will again be easier to validate since the object broker interfaces will be configured rather than programmed and will use standard software blocks to build the interfaces.

As well as support for existing de facto standards, the future will see development focus on the provision of interfaces for true standards which should finally emerge within the next few years. The ability to develop interfaces for new object models quickly will be especially important given the current paces of new developments. As such standards emerge, the OCS, the MES, and the ERP market will largely be split into those systems which fully support such object-oriented interfaces and those that do not. Needless to say, those systems that do not will find it difficult to develop complete solutions and will lack functionality.

Standardization of Network Components

The bandwidth of the traditional systems will become more and more of a problem, given the need to pass more and more data and to transfer more complex object models. This will be especially true where control is distributed down to the device level. With the availability of deterministic Ethernet (using protocols such as MMS and running at speeds of 100 Mb or higher), previous objections to using the Ethernet in a real-time control environment will largely disappear.

Arguments that the collision detection and avoidance nature of the Ethernet causes networks to slow down will no longer be valid, and both traditional control networks (manufacturer's specific token-passing deterministic networks) and fast Fieldbus will almost certainly adopt some form of fast deterministic Ethernet (or a descendant thereof).

This will mean that standard network physical media can be employed, which will reduce costs. Standard bridges will become available between deterministic field/control networks and nondeterministic corporate networks, and this will mean that communications between field, control and corporate networks will be made even easier. Security will still remain an issue, with access to the control of the process being secured at a number of levels in the system. However, because this will again use standard solutions, the task of validation will again become simpler.

Further Distribution of Open Applications

The provision of secure, high-speed communication networks means that the trend toward distributed open applications will continue. Users will need to focus more on the logical makeup of their systems rather than relying on the definition previously forced upon them by which applications ran in which "box".

Validation of applications and solutions will have to be planned and executed according to their functional rather than their physical interfaces. Although initially confusing, this should give the user greater flexibility and control over their Validation Plan and make the process more straightforward.

CONCLUSIONS

Modern systems have come a long way in the last decade. From being tools to automate and control the process and regulate production, they are now tools to help design, engineer, support, maintain and validate the process, the plant and its environment.

When combined with standards from the commercial IT sector, tools such as object brokering allow the complete integration of the whole installation, not just the process and its associated control. Integration of motors, drives, switch gear, air-handling plant and BMSs, robotics and Warehouse Management Systems means that an overview of the complete production process can be gained. The merging of the real-time and IT worlds now means that these advantages of integration can be provided across the traditional layers of OCS, MES, and ERP.

Whilst this was theoretically possible in the past, it may now be achieved in a manner that uses standard products and tools and is therefore easier and less expensive to validate. The use of common databases and engineering tools also means that project costs may actually be reduced when an integrated approach is taken. Emerging communications standards such as Fieldbus and deterministic Ethernet will be quickly and fully integrated through the use of object brokering.

The support for multiple object models also allows an integrated approach to be taken toward engineering. This has advantages not just in the early stages of the development and validation life cycles but throughout the lengthy operational and maintenance phase. Current developments will broaden these advantages into the MES and ERP arena, providing a global plant database from which the whole business will gain competitive advantage.

Although existing approaches to validation will not change in the short to medium term, the amount of application-specific software will be drastically reduced. This will allow a less time-consuming (but nevertheless rigorous) approach to be adopted with respect to validation from the pharmaceutical manufacturer's perspective (relying on supplier quality). This will play a significant part in reducing project and operational and maintenance costs.

REFERENCES

1. U.S. Code of Federal Regulations, Title 21, Part 11: Electronic Records; Electronic Signatures (revised 1 April 1998). Washington, D.C.: U.S. Government Printing Office.

2. GAMP Forum. 1998.. *GAMP Supplier Guide for Validation of Automated Systems in Pharmaceutical Manufacture,* Version 3.0. Available from ISPE (the International Society for Pharmaceutical Engineering).

3. U.S. Code of Federal Regulations, Title 21 CFR Part 211: *Current Good Manufacturing Practice for Finished Pharmaceuticals* (revised 1 April 1998). Washington, D.C.: U.S. Government Printing Office.

4. MCA. 1997. *Orange Guide: Rules and Guidance for Pharmaceutical Manufacturers and Distributors*. London: UK Medicines Control Agency, ISBN 0 11 321995 4.

5. ISO. 1994. Standard for Industrial Automation Systems and Integration—Product Data Representation and Exchange (ISO 10303) Available in numerous parts. Geneva, Switzerland: International Organization for standardization.

7

Validating Enterprise Asset Management Systems

Chris Reid
Integrity Solution Limited
Middlesborough, United Kingdom

Tony Richards
AstraZeneca
Laughborough, United Kingdom

ENTERPRISE ASSET MANAGEMENT

Effective and efficient utilization of assets by pharmaceutical research or manufacturing enterprises is fundamental to the early delivery of new products to market and to satisfying customer demand once those products have been approved for release by the relevant regulatory authorities. Effectiveness and efficiency must be established from the outset of asset specification and design ("built-in quality") and cannot be delivered by fine tuning and testing ("testing in quality") when the asset is handed over from the development environment to the operational environment. Assets must be reliable, consistent and capable, that is, they must

- be available when needed and not fail during use;
- function consistently to predefined performance criteria; and
- meet performance criteria without undue stress, risk of failure or reduced asset life.

The continuous improvement of asset reliability, consistency and capability, either mutually or simultaneously, is the basic objective of the Enterprise Asset Management (EAM) strategy in order to reduce operation and maintenance costs and increase regulatory compliance. The foundation for continuous improvement is "information", without which it is impossible to establish a rationale for change. This foundation must be established at the start of the project, with the definition of the business need in measurable terms, i.e., performance criteria, without which there is no basis for design, testing, operation, maintenance, compliance and, consequently, continuous improvement.

Figure 7.1 defines a simple asset development and operational life cycle depicting the creation of critical asset management information at each phase. The information generated must be managed in order to facilitate structured access and controlled maintenance. Considering the volume of information supporting even modest-sized enterprises, it is essential that an information system (IS) strategy is developed to manage EAM information.

Business Need

Assets are only required to meet a business need, i.e., to develop or manufacture pharmaceutical products that can be marketed in order to realize a profit for reinvestment and/or remunerate shareholders. Business need must therefore be clearly defined and understood before initiating an asset development project that could require major financial investment and commitment of valuable resources. A business case must be developed that defines the strategic fit of the development within current and future business plans; the short-, medium- and long-term benefits of the development; and the payback on investment.

Map Processes

Once the business case has been accepted and investment received, it is necessary to define the process and functional requirements of the development. At this stage, we are not primarily concerned with the assets required to deliver the business objectives, although experience may suggest what form they may take.

A coordinated team of users, engineers and safety and quality representatives will map the operations required to meet the business need—the scientific research and development operations, the production process or the goods in process. The processes are often presented in a flow diagram supported by descriptive narratives to expand the process definition where required. Interaction between processes must be clearly defined.

Once processes have been established, the functions required to implement the processes should be defined, i.e., equipment sterilization and environmental controls such as temperature, differential pressure, particulate control and so on. It is at this stage that we must define the functional performance criteria that will provide the basis for design, testing, operation, maintenance and, ultimately, continuous improvement.

Figure 7.1. Asset Life Cycle

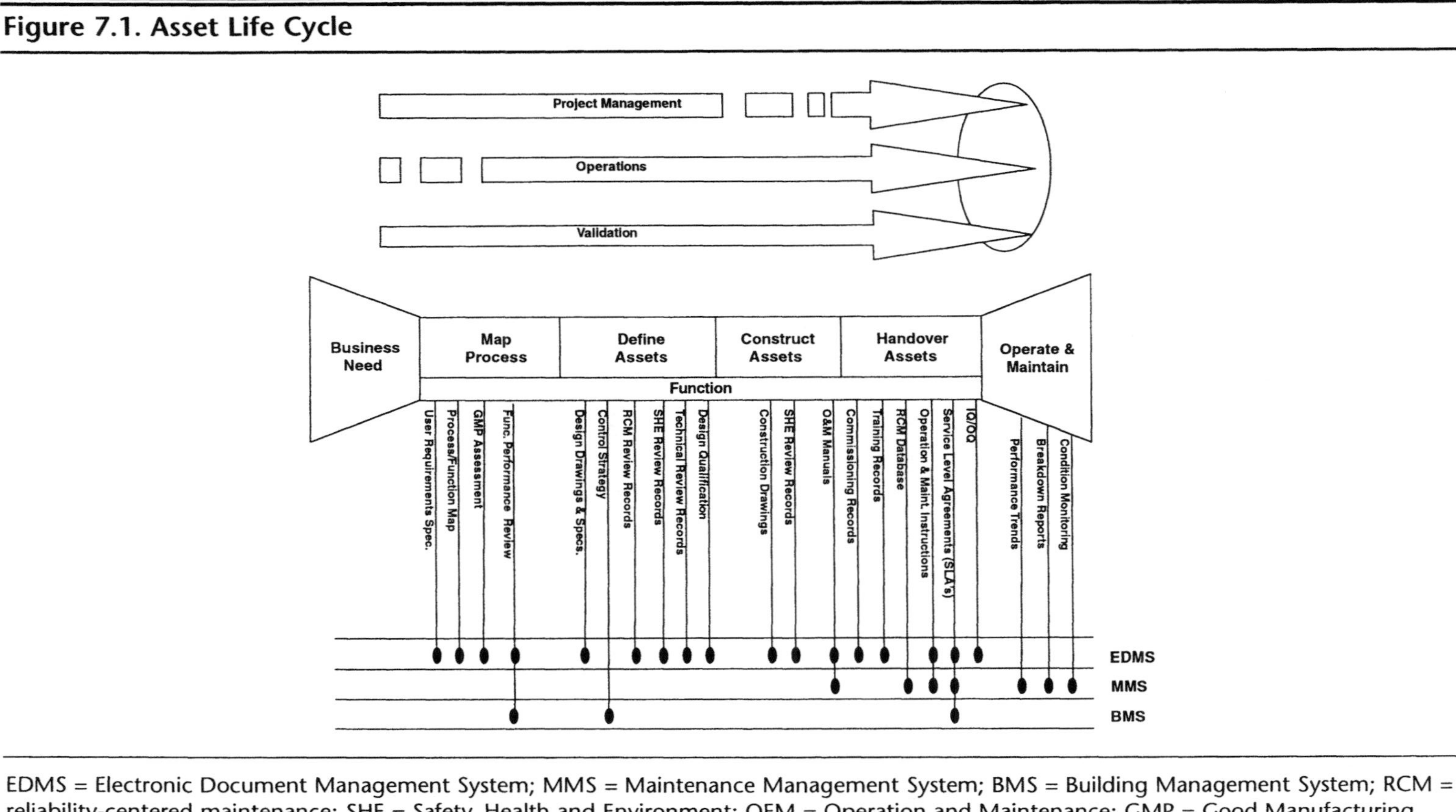

EDMS = Electronic Document Management System; MMS = Maintenance Management System; BMS = Building Management System; RCM = reliability-centered maintenance; SHE = Safety, Health and Environment; OEM = Operation and Maintenance; GMP = Good Manufacturing Practice; IQ = Installation Qualification; OQ = Operational Qualification

Functional performance criteria must not be simply stated as discrete values that do not provide any degree of tolerance or express the consequence of performance loss or interruption. Performance criteria stated as "Maintain room temperature at 18°C" are loose and ambiguous and should be more accurately specified as follows:

- Temperature range: 18°C–22°C
- Control accuracy: Set point ± 1°C
- Acceptable excursions: <5°C for less than 10 minutes

Processes and functional requirements must be reviewed before they are issued to the design consultant. The objective of the review, or in some cases multiple reviews, should be to ensure that processes and functions have been completely and accurately defined and that performance criteria are unambiguous. Reviews must also determine the consequence of function failure, that is, the risk to the research study, manufacturing process, safety and regulatory compliance. These consequences must be documented so that the delivered solution is appropriate to the business risk, i.e., the design must be relevant to the operating context of the asset.

The output of this activity will be the User Requirement Specification (URS) or Project Definition that will be issued to the design consultant and evolved into a detailed design defining the assets required meet the specifications.

Define Assets

Asset definition is a phased activity involving scheme, concept and detailed design, which delivers the specifications, engineering drawings, databases and so on that define the operational strategy and construction of the assets required to deliver the processes and functions defined within the URS. Typically, the assets of a pharmaceutical enterprise are managed as a hierarchy involving (see Figure 7.2)

- sites,
- buildings,
- rooms/Areas/Zones and
- systems (utilities, building services, process systems).

This hierarchical structure provides the foundation for information access—the information search capabilities that enable rapid access of, say, the calibration records for the heating, ventilation and air-conditioning (HVAC) system controlling zone 1 within the biochemistry building.

This chapter will focus on "systems", as they are the most diverse and complex asset in the asset hierarchy and the primary focus of continuous improvement strategies to improve system reliability, consistency and capability, leading to operation and maintenance cost reduction and increased regulatory compliance.

Figure 7.2. Asset Hierarchy

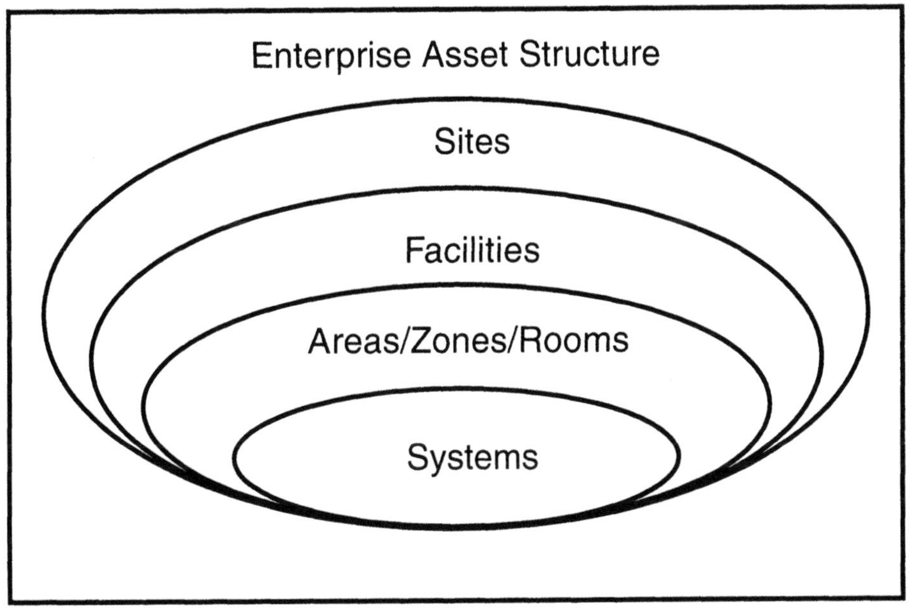

System Concept

Engineers and users frequently refer to vessels, pumps and valves. Although these items are critical components, they do not in isolation deliver the functionality required by the research or manufacturing process. It is the integration of such components into a "system" that enables the designed system performance to be delivered and maintained. The HVAC system delivers air to Class 10,000, temperature to 20°C ± 2°C and relative humidity to 50 percent ± 5 percent. The failure of a component, although important, becomes critical only if performance is lost. The design process must take into account the potential risk to the research or manufacturing process arising from the loss of performance and take remedial action to minimize such risk. Figure 7.3, differentiates between performance loss and system failure. GxP compliance is lost once the performance deviates from the predefined operating range, which is long before the system totally fails.

Essential information required for the design, operation and maintenance of the system must therefore be specific to the system. For example, the provision of a master valve schedule listing all valves within a facility will provide the necessary information to maintain all valves for all systems operating within the facility. However, retrieval of the information specific to a system that may have recently failed and that may be process or product critical will be cumbersome. The provision of system-specific valve schedules will

Figure 7.3. Performance Monitoring

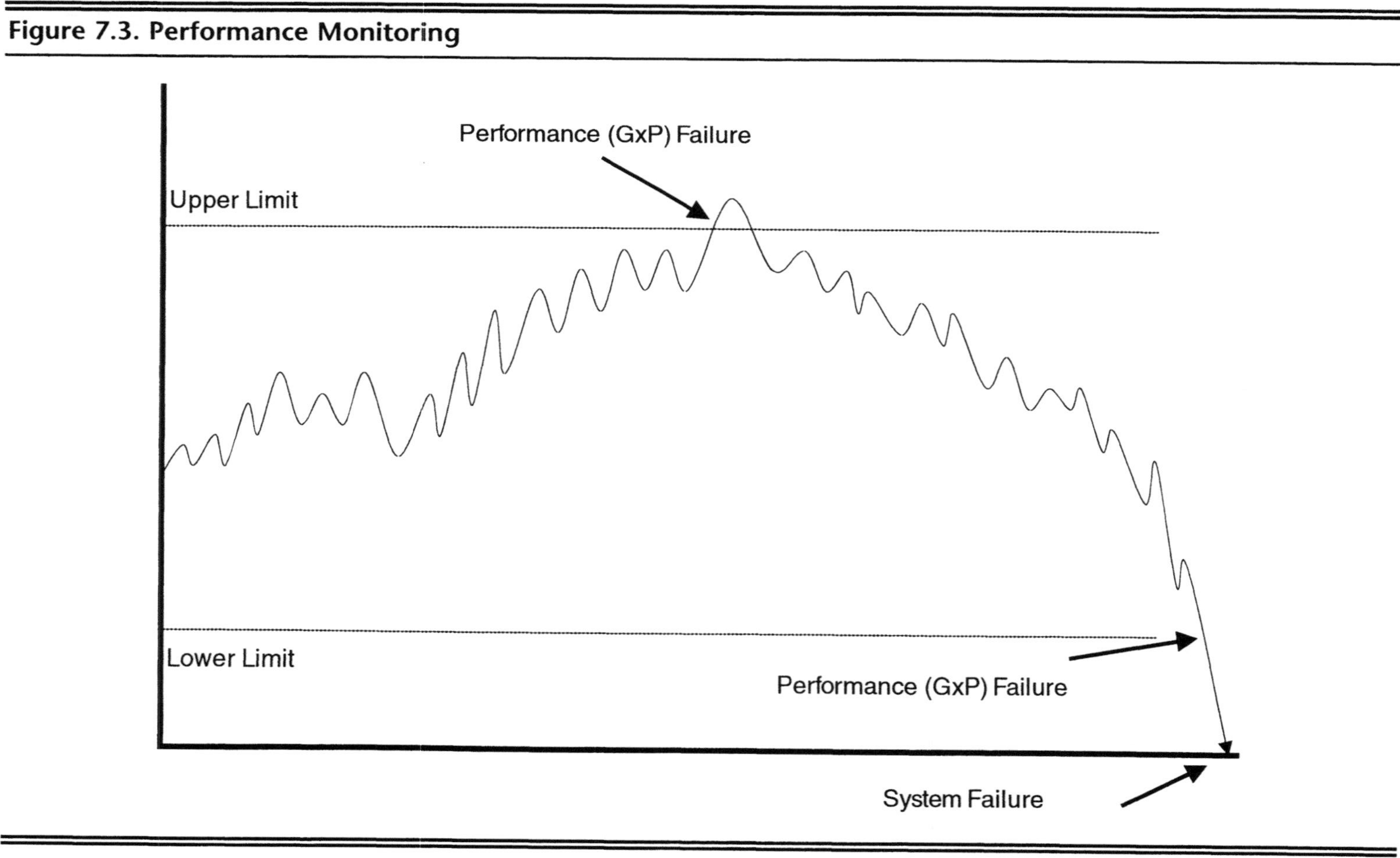

enable more efficient information retrieval. A further advantage of focusing on systems is that information can be more easily provided that is relevant to the operating context of the system. For example, two systems may be providing similar functions; however, one may operate within a GxP environment and the other not. The level of information required to support the system operating within a GxP environment is significantly higher than is required for systems operating within a non-GxP critical environment, e.g., materials specification for product contact parts, filter certificates and so on.

The relationship between the system and its component parts is synonymous with the relationship between information and data. Data in isolation are largely meaningless; however, when associated with other key data to create, say, asset failure reports that identify system function, function failure, failure mode and consequence of failure, then a powerful basis for continuous improvement is established.

Systems and Functional Performance Criteria

A measurement of system performance is essential to asset management. If loose and ambiguous performance criteria are defined, it follows that the basis for design is poor and that the system is unlikely to meet the business need.

Design reviews will ensure that the proposed system can meet the performance criteria in a reliable, consistent and capable manner. In addition to establishing a robust design that can be qualified against predefined performance criteria, it is essential to establish an asset management strategy that will effectively and efficiently maintain system performance, reduce maintenance costs and improve regulatory compliance.

Failure Modes and Effects Analysis (FMEA) is one tool that can be applied to challenge the design against the stated performance criteria and further provide the foundation of the asset management strategy to ensure that system performance is maintained (Figure 7.4). The FMEA process defines

- system functions (process requirements/objectives),
- function failures (failure scenarios),
- failure modes (reason for failure) and
- consequence of failures (impact on the business).

The output of the FMEA analysis is paramount to determining of the business risk (consequence of functional failure) presented by a system. This risk is used to determine the level of rigor applied to the validation, operational control, maintenance and documentation/information needed to verify and maintain system performance as indicated by Figure 7.5. It follows that the documentation/information supporting system function is as critical to the pharmaceutical enterprise as the system function itself.

Figure 7.4. Asset Management Strategy

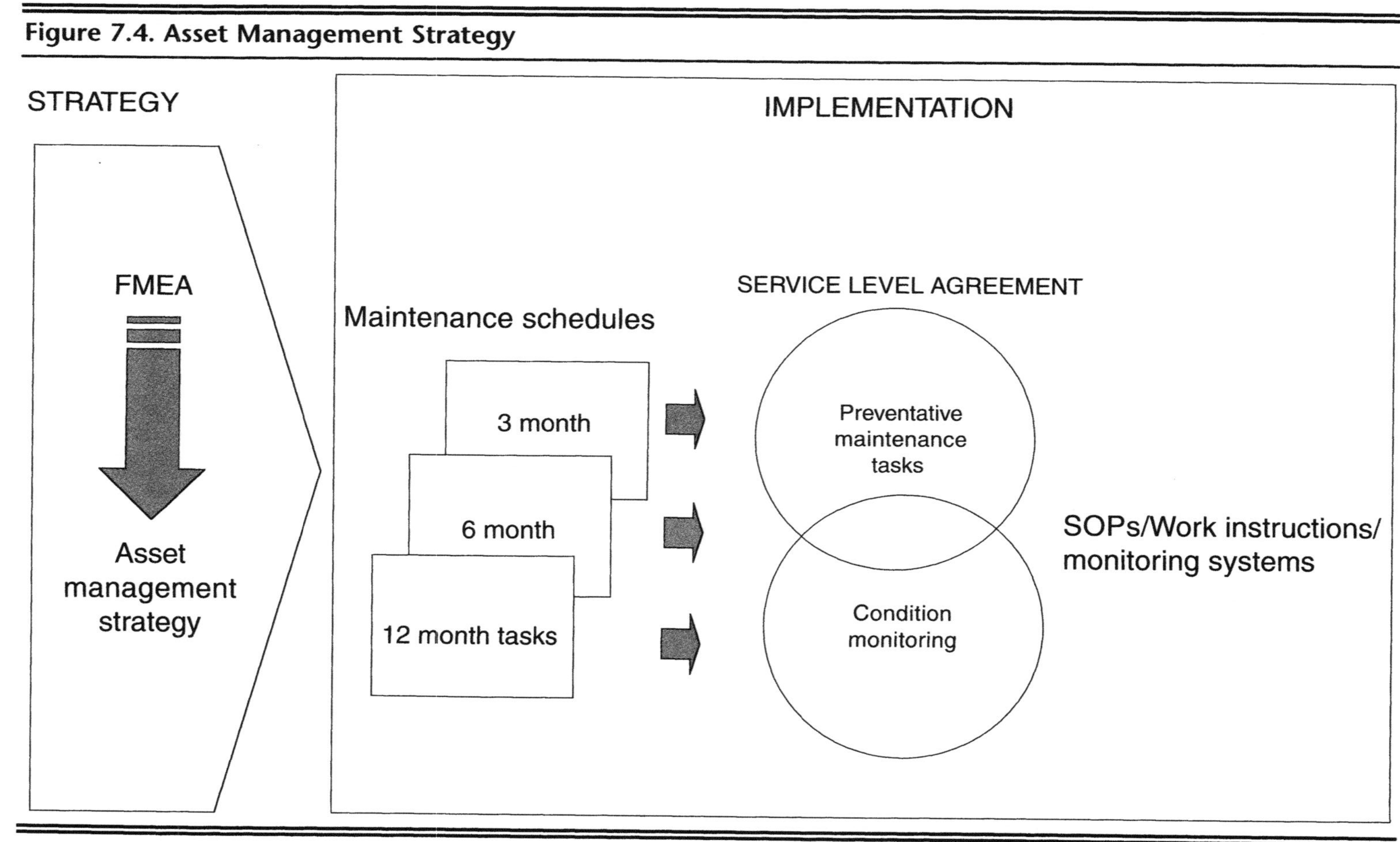

Figure 7.5. Functional Criticality

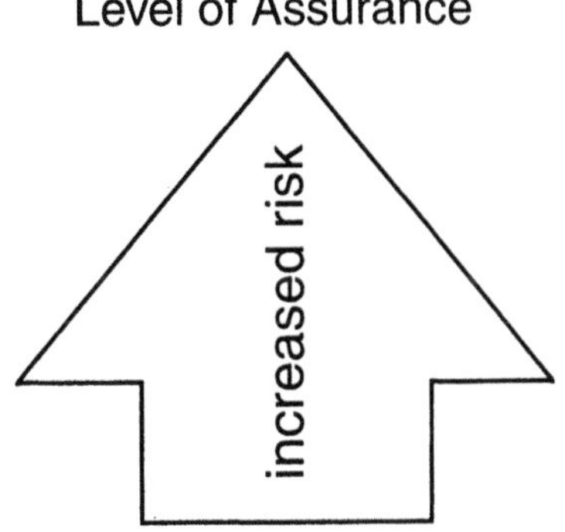

Functional Criticality

A	Operational safety
B	GxP critical
C	High business impact
D	Minimal business impact
E	None

The FMEA process is applied during the Functional Design and Detailed Design phases. The principles of FMEA may have already been used to review the processes and functions documented in the initial URS. At each phase, the outcome of the previous FMEA is refined. Each function of the system is challenged and assigned a criticality based on the consequence of functional failure. Table 7.1 provides a generic view of FMEA objectives at each stage of application.

The objective of this chapter is not to provide a detailed description of FMEA; however, it is clear that FMEA is a key ally of the validation process. Figure 7.6, identifies some of the key outputs of the FMEA process that support validation.

Construct Assets

Assets are constructed in accordance with the Detailed Design that provides a definitive description of the systems and components required to build facilities and assemble systems in a manner that will meet the business need. During this period, construction documentation is established to provide an accurate basis for system commissioning and validation. Further, operation and maintenance plans, instructions, SOPs and SLAs are developed in readiness for the handover of the assets to the operational environment. The time required to prepare, review and obtain approval of these documents must not be underestimated and should commence as soon as the FMEA process has defined the asset management strategy.

Table 7.1. Generic View of FMEA Objectives

Stage	System Evolution	Objectives
User Requirements	Functional performance criteria are known. System options are understood.	Ensure that the platform for system design is clearly understood and defined by users.
Functional Design	Tailored design is evolving. Performance criteria are clear. System construction is evolving, i.e., specific components are not known. System relationships/interfaces are known. Engineering line diagrams available.	Confirm system performance. Ensure the future maintainability of the system. Ensure that the evolving design addresses the consequences of potential functional failure.
Detailed Design	System construction details are largely defined.	The consequence of functional failure and the potential causes of functional failure are known. Maintenance tasks are defined. Service Level Agreements (SLAs) and Standard Operating Procedures (SOPs) can be developed.

Handover

Handover activities comprise commissioning and validation, operational and maintenance takeup and user takeup. Collectively, these activities ensure that

- facilities and systems are qualified against design and meet their predefined performance criteria;
- engineering and user training has been successfully delivered;
- operation and maintenance strategies, plans, SOPs and SLAs have been developed and issued; and
- documentation and information supporting the EAM strategy has been imported into the information systems.

Operation and Maintenance

Operation and maintenance of assets delivered by the development project are managed in accordance with the asset management strategy defined by the FMEA process discussed earlier. Figure 7.7 shows a typical operation and maintenance process that is a key component of EAM strategy. The process is

Figure 7.6. Relationships Between FMEA and Asset Validation

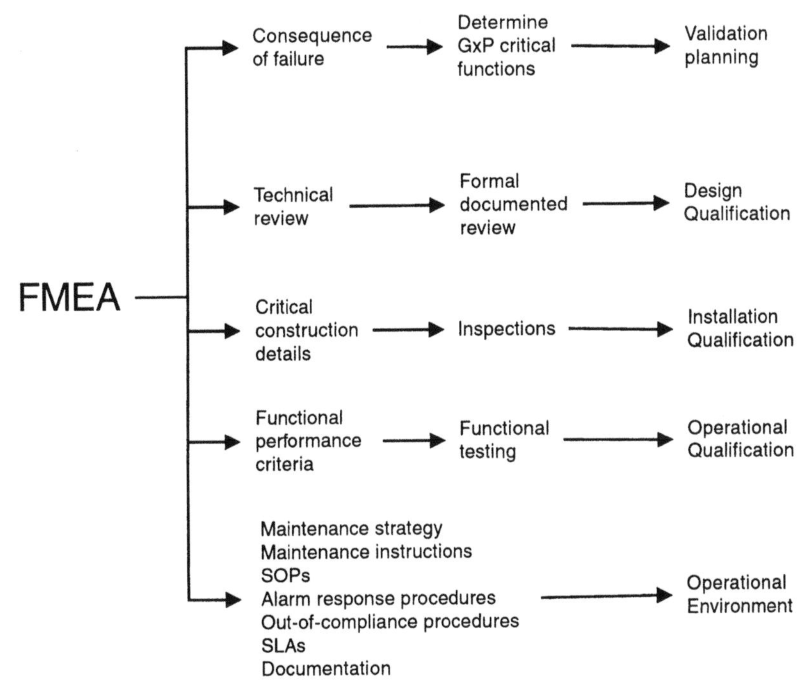

composed of three primary phases: "work order generation", "work environment" and "reporting and feedback". Work order generation controls the creation of work requests and the identification of work instructions, SOPs, documentation and SLAs required to conduct the work. Work environment controls the application of the work instructions and SOPs in order to confirm current system performance, carry out the defined maintenance tasks and reestablish performance prior to releasing the system back into the operational environment. It is essential to confirm system performance prior to carrying out the maintenance tasks in order to detect performance deviations requiring investigation and to assess the effectiveness of the EAM strategy in preventing such performance deviations.

The reporting and feedback phase of the process establishes information such as

- maintenance records,
- performance deviations,
- condition monitoring,

Figure 7.7. Operation and Maintenance Process

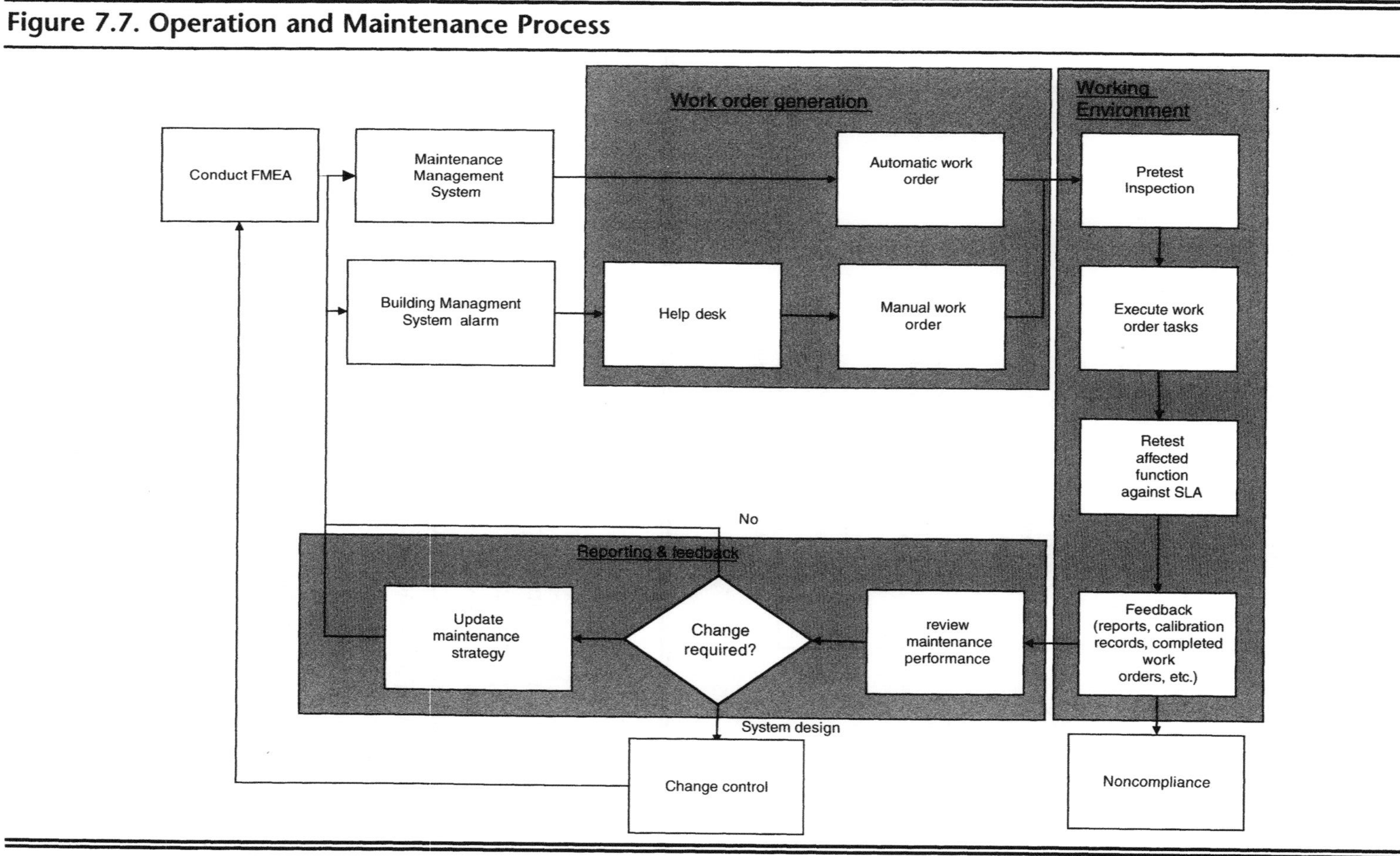

- failure reporting,
- out-of-compliance records and
- maintenance costs.

This information is used to establish trends that provide the basis for continuous improvement of the EAM strategy, including

- improved system design,
- system replacement,
- improved operation and maintenance strategy,
- improved operation and maintenance instructions and SOPs,
- improved technical documentation/information and
- improved maintenance reporting and feedback.

It is clear that a failure to define system function and performance criteria adequately would make such continuous improvement difficult if not impossible.

DEFINING INFORMATION NEEDS TO SUPPORT THE EAM STRATEGY

Asset management information provides the foundation for all other GxP documentation used in the research and manufacture of drugs (see Figure 7.8). The results of a scientific study or a manufacturing campaign are worthless if the systems used to control drug stability testing or the manufacturing process were not operated and maintained in accordance with their design and predetermined performance criteria.

Defining the information requirements to support asset management in a consistent manner that is understood by both the pharmaceutical enterprise and their suppliers is a considerable task. The pharmaceutical company must establish internal standards that define the following:

- System/asset numbering
- Asset management documentation needs
- Engineering drawing requirements
- IS information structures
- Engineering database structure (information templates for system types/system component types)
- Record requirements

Before establishing such standards, the pharmaceutical enterprise must assess the value of the information to the asset management process and consider the different requirements for information against the varied operating contexts of the business. The standards must also guard against information

Figure 7.8. Documentation Pyramid

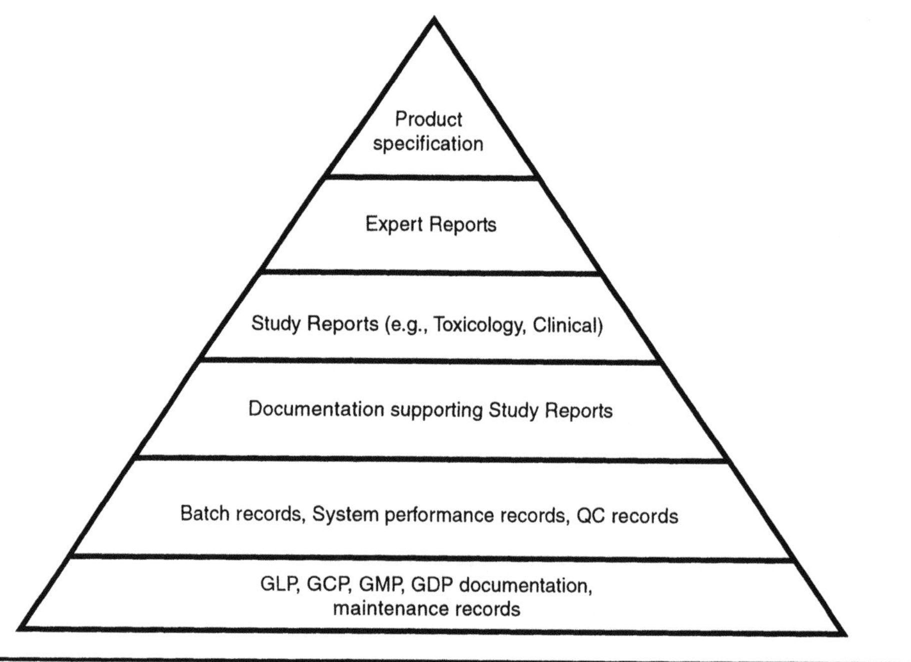

and documentation overload, which can be as detrimental as insufficient or unstructured information. The key objective of the standards is to ensure that information delivered by suppliers and maintained by the pharmaceutical enterprise is

- of the correct scope and depth,
- system specific,
- relevant,
- accurate,
- single instantiation (or as few instantiations as possible),
- easily retrievable and maintainable and
- controlled in accordance with risk.

Ensuring the above must not be underestimated; delivering information to the required technical standard in a consistent format and in a timely manner is a considerable project management task, especially when multiple suppliers are involved and when those suppliers have traditionally delivered information to their own and often unsatisfactory standard. Further, the import of such information into the pharmaceutical enterprise's EAM infrastructure is a demanding process. In some instances, there may be value in adopting the standards used by principal suppliers in order to minimize information translation and the associated risks.

Initiatives such as STEP (ISO 10303) attempt to define a standard to enable sharing and exchange of technical engineering data, independent of applications and organizations. Prominent industries involved in the development of STEP include aerospace, automotive, shipbuilding and process industries, and there is an increasing awareness of such standards within the pharmaceutical industry.

ROLE OF INFORMATION SYSTEMS IN THE EAM STRATEGY

It is clear that only the smallest of enterprises can effectively implement EAM strategies by means of manual processes. Pharmaceutical enterprises are increasingly looking to information systems to implement EAM strategies and to maintain the vast volumes of information that supports modern research and manufacturing operations.

Physical Architecture of the IS Infrastructure

Figure 7.9 presents a high-level representation of a typical IS architecture supporting the implementation of the EAM strategy. The engineering database is the hub of the environment, providing a repository for information related to each asset (site, building, room/area/zone, system, etc.). The nuclear power, oil and gas industries have taken the lead in the development and utilization of "intelligent" databases. Such databases provide hypertext links to the Electronic Document Management System (EDMS) and Computer-Aided Design (CAD) system in order to provide single-point access to data. For example, a Functional Design Specification (FDS), CAD drawing and SLA will contain hypertext links to the temperature performance criteria for an aseptic suite. If the engineer corrects an error on the drawing, the database information is automatically updated and propagates through the hypertext links to the FDS and the SLA. The information is therefore consistent (although not always correct) within all specifications, procedures, drawings and so on. The use of intelligent databases is increasing within the pharmaceutical industry, following the successes gleaned by other industries such as the process industries. The general principle that documents and information must be structured to minimize the number of occurrences of data is paramount to reducing the burden of information maintenance and the risk of regulatory noncompliance.

Access to information is as equally important as the controlled maintenance of information. For example, when a critical alarm is reported by the Building Management System (BMS), the engineer needs to promptly respond to instructions, SLAs, SOPs and engineering drawings. The integration of the

Figure 7.9. EAM Architecture

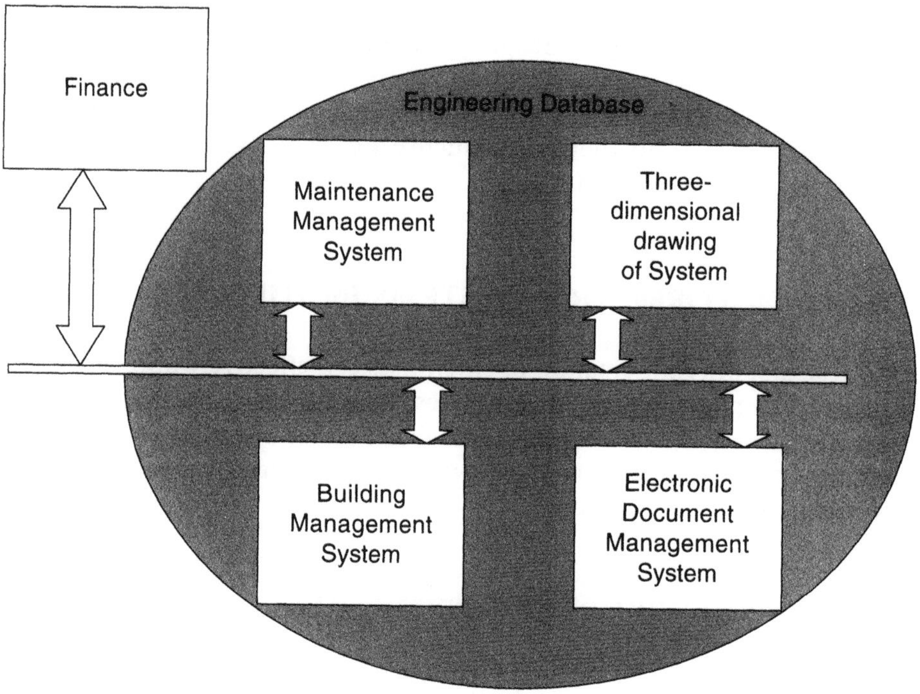

Finance

Engineering Database

Maintenance Management System

Three-dimensional drawing of System

Building Management System

Electronic Document Management System

EDMS, CAD and Maintenance Management System (MMS) with the BMS enables the documents and databases providing this essential information to be automatically available when the alarm is triggered.

Functional Architecture of the IS Infrastructure

There is obvious functional overlap between the systems described in Figure 7.9, for example, the MMS will provide some degree of document control that obviously overlaps with the EDMS. The following describes the generic functionality associated with the IS architecture.

Maintenance Planning

Maintenance tasks are either implemented proactively in order to "prevent" or minimize the chance of failure or reactively to "correct" a situation following failure. Preventive maintenance plans are derived from the FMEA process that will define the tasks and task frequencies required in order to maintain system reliability, consistency and capability. Having applied the

FMEA process in order to determine the maintenance strategy, it is essential that the system

- schedules planned maintenance at defined intervals;
- identifies relevant maintenance task schedules (consistent with FMEA requirements);
- references the SLA defining system performance criteria and system criticality;
- references documents, drawings and databases supporting the maintenance tasks; and
- references instructions and SOPs to ensure controlled and consistent execution of maintenance tasks.

Corrective maintenance, conducted following a failure, must be carried out in a similar controlled manner; however, in this instance, the engineer must manually construct the maintenance task schedule following investigation of the problem. The risk to GxP compliance, therefore, increases due to the manual intervention. The structured organization of the EDMS and MMS in particular, to ensure that SOPs, documentation, drawings and information are easily accessible through the relationship to a specific system, is essential.

Operation and Maintenance Implementation

Work instructions and SOPs define the operations required to start up, operate, monitor, shutdown and maintain systems. These work instructions and SOPs must be controlled in order to prevent inadvertent and unauthorized modification and to ensure access to only the latest revision of the document. Work instructions and SOPs must therefore be held in accessible but secure areas that are periodically backed up and archived. IS access must be controlled by a hierarchical security system that constrains system operations in accordance with the role, responsibilities and competency of the user. Access to and modification of the information supporting such work instructions and SOPs, for example, engineering drawings and specifications must be equally controlled.

System Control and Performance Monitoring

The BMS and similar systems integrated into the IS architecture provide control and performance monitoring functionality to ensure that performance criteria are met and performance deviations are detected. Functional performance deviations will inevitably affect product quality and, consequently, GxP compliance. It is essential that the design process "builds in quality" to ensure that the system is reliable, consistent and capable of meeting the predetermined performance criteria. Monitoring functions, although GxP critical, should provide only the fail safe mechanism for detecting and reporting failures. Monitoring of process variables using available technologies such as BMS and ultrasonic and vibration analysis may often be deployed in order to predict pending failures, enabling corrective action to be taken before performance and hence GxP compliance is lost.

Maintenance Reporting

Maintenance history is an essential component of GxP compliance. Work instructions and SOPs controlling maintenance operations should ensure that the maintenance engineer records all performance measures, observations and maintenance tasks in a consistent manner, for example, calibration records containing calibration parameters, calibration procedure, reference to calibration equipment, name of engineer, date of calibration, next due date and so on. Where automated condition and performance monitoring are employed, the integrity of the recorded data is obviously a GxP issue.

Records generated by the asset management process are used to bring about continuous improvement in order to increase the effectiveness and efficiency of the EAM strategy. All changes arising from the review must be controlled and documented.

Technical Issues

The review of the asset management process can be broken down into the technical issues that could potentially impact GxP (see Table 7.2).

Table 7.2 is by no means an exhaustive list of issues that could potentially impact GxP compliance; however, it is a good indication of the criticality of the IS architecture.

REGULATORY IMPACT

Earlier discussions have provided a strong indication of the GxP criticality of the EAM architecture. However, the GxP impact can be determined only in the context of the operating environment. Two identical mechanical systems may provide similar functionality; however, the fact that one system operates within a GxP environment and the other operates within an office block is fundamental and inextricably linked to the consequence of system function failure. It therefore follows that the regulatory impact of information and the information systems that manage such information is inextricably linked to the functional criticality of the mechanical system.

Table 7.2. Technical Issues with Asset Management Functions

Asset Management Function	Technical Issue
Maintenance scheduling	Accuracy and consistency of interval between preventative maintenance work order issue
Assignment of maintenance tasks	Referential integrity between work order and maintenance task schedule/plan
Accuracy of instructions and procedures	Templates, print controls
Accuracy of information	Screen input formatting and input verification, data recovery, referential integrity, data transfers, print controls
Automated condition and performance monitoring	Scanning frequency, data communication integrity, records retention, time and date stamping
Manual condition and performance monitoring	Screen input formatting and input verification, data recovery, time and date stamping
Work order assignment	Robust relationship between tasks and trades, e.g., don't allow assignment of a mechanical installation to an electrician
Work order traceability	Event sequencing, e.g., accept, work in progress, wait for parts, approve, complete
Maintenance record traceability	Record "key" management and assignment of record to correct system
Archiving of maintenance records	Accurate retrieval
Change	Record locking to prevent parallel access, security to prevent inadvertent or malicious modification, disaster recovery, maintenance of referential integrity

GxP Assessment

As previously discussed, the FMEA process can be used to determine the consequence of functional failure. This process can in turn be applied to the functionality provided by the information systems supporting the EAM strategy. For example, if an information system fails to generate preventative maintenance plans that are required for the periodic calibration of critical temperature control loops, then the functionality of the information system responsible for the generation of these plans must be deemed GxP critical. Once again, we can see that the operating context must be considered, as it is the environment within which the plans are applied that determines GxP criticality.

The GxP Assessment is conducted in accordance with FMEA principles. Figure 7.10 provides a high-level representation of the GxP Assessment process. The flowchart is supported by standardized questions that challenge the impact of the EAM function on GxP compliance. For example, will the total or partial failure of the information system lead to

- loss of or interruption to process system performance?
- failure to conduct critical maintenance activities in accordance with a predetermined schedule?
- use of superseded or wrong maintenance procedures?
- incorrect maintenance/failure/performance reporting?
- incorrect chronological reporting of operation and maintenance tasks?
- loss or corruption of operational/maintenance data?
- loss of database referential integrity?
- system security violation?
- failure to recover following system failure?

Figure 7.10. GxP Assessment

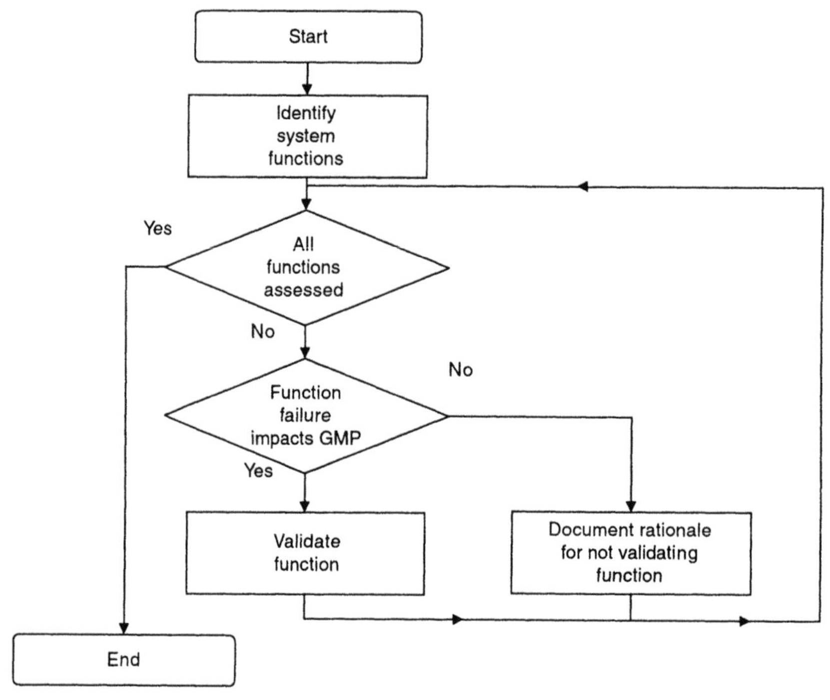

The above is not an exhaustive list but provides insight into the extent to which EAM functions can impact GxP.

EAM VALIDATION STRATEGY

Earlier discussions have referred to systems in terms of mechanical assets. The following discussion refers to the systems in terms of the information systems supporting the EAM strategy.

Principles of Criticality-Based Validation

The cost of validation is much publicized as is the debate regarding the extent to which information systems should be validated. *Validation* is essentially the term adopted by the pharmaceutical industry and its regulators to define the additional rigor required to confirm GxP–critical aspects of information systems throughout the development and operational life of those systems. Given that pharmaceutical regulators have the power to withhold, suspend or withdraw product licenses, it essential that pharmaceutical enterprises validate GxP–critical functions. In order to maximize business efficiency and minimize cost, it is also essential that pharmaceutical enterprises differentiate those functions of the information systems that are GxP critical and those that are not and, hence, focus valuable and limited resources where they are most warranted.

Processes used to determine functional criticality, such as FMEA, have already been discussed within this chapter. Similarly, FMEA and other risk assessment tools can be used to determine the scope of validation. The risk of failure increases as information systems supporting the EAM strategy deviate from a standardized solution, that is, the level of tailored development increases. In addition, the extent to which a product is utilized within an industry, in particular pharmaceuticals, must be taken into consideration when determining the scope of validation.

The GAMP guideline [1] provides guidance on the extent to which operating systems, third-party packages and applications utilized by the EAM System must be validated from the simple recording of the version number for extensively used, industry-standard operating systems to full life-cycle validation for tailored applications. There will, however, always be gray areas where the level of validation documentation required is under debate. A general principle is "if in doubt, err on the side of caution" as the cost of the additional effort may not always exceed the cost of the debate (Figure 7.11).

Figure 7.11. Documentation Levels

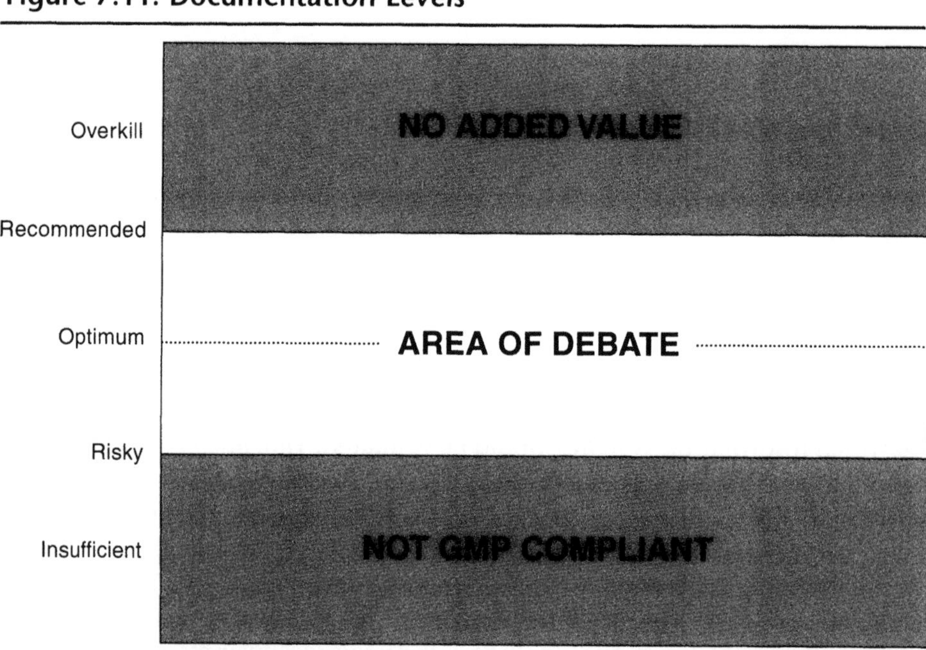

IS Validation Life Cycle

Figure 7.12 depicts a typical validation life cycle from validation planning to validation reporting and ongoing support. Responsibility for each phase of the validation life cycle will switch from the pharmaceutical organization to the supplier at certain key phases. It is, however, the fundamental responsibility of the pharmaceutical organization to assure themselves that all phases of the life cycle have been conducted in a quality manner, consistent with the expectations of pharmaceutical regulators [2,3,4]. A detailed discussion on validation life cycles can be found in *Validating Automated Manufacturing and Laboratory Applications: Putting Principles into Practice* [5].

Validation Master Plan

The pharmaceutical organization must develop a Validation Master Plan (VMP) to define the validation strategy for the implementation of the IS architecture. The VMP should address the process by which the pharmaceutical organization assures the quality of the products being procured and the strategy for validating the specific implementation of those products.

Regulators consider the VMP to be the first fundamental commitment of the pharmaceutical enterprise to the validation process. In particular, the VMP must recognize that different approaches may be required to meet the

Figure 7.12. Validation Life Cycle

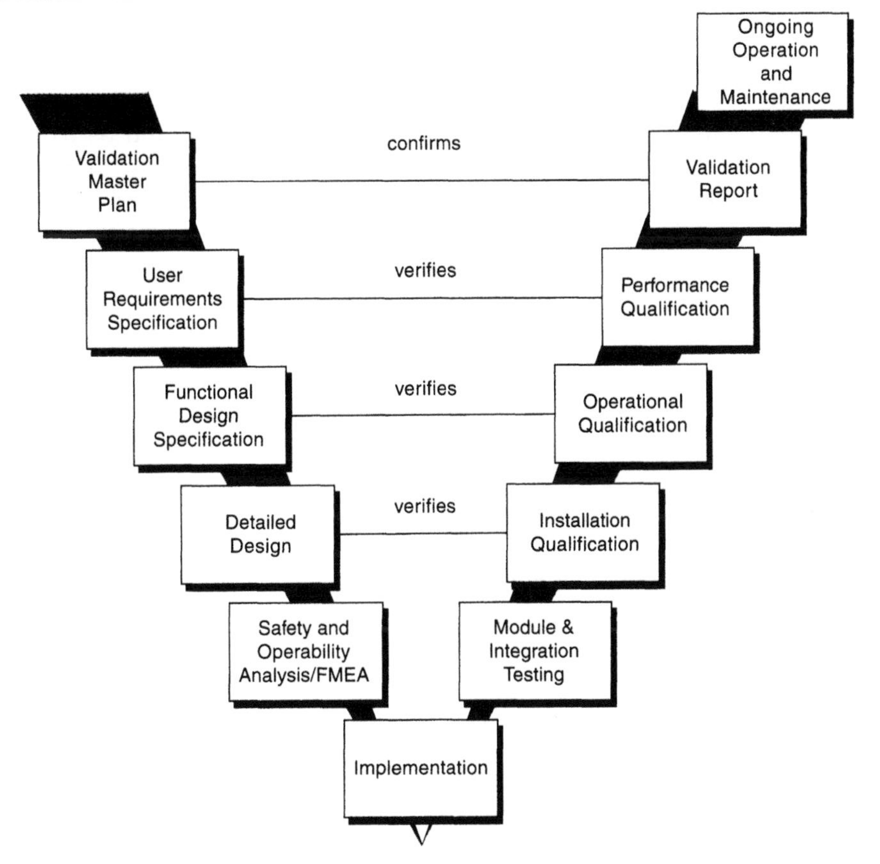

differing capabilities of the various suppliers contributing to the IS architecture. The VMP must be reviewed and approved by the project sponsor, GxP process owners and Quality Assurance (QA) representative. The typical contents of the VMP are as follows:

- Validation scope
- Validation strategy (core products and applications)
- Validation organization
- Roles and responsibilities
- Validation documentation requirements
- Document approval authorities
- Key phasing and milestones

User Requirements Specification

The URS is developed by the pharmaceutical organization and forms the foundation of the project. It should convey the asset management processes and functional requirements performance criteria to the supplier. The objective of the URS is to convey business needs rather than technical solutions (other than where corporate standards apply). The URS should be written in a clear, concise and unambiguous manner that facilitates traceability throughout the design and testing phases. The GAMP guideline [1] provides a guideline for developing the URS, which is supplemented by a further guideline from the Supplier Forum [6]. Below are the typical contents of the URS:

- Asset management processes
  - Maintenance strategy development (FMEA, performance monitoring, etc.)
  - Maintenance strategy implementation
  - Document management
  - Drawing management
  - Change control
- Functionality
  - Process step definition
  - Functional performance criteria
  - FMEA
  - Failure mode recovery requirements
- Informational requirements
  - Information structures
  - Legacy system interfaces
  - Data entry range
  - Data retention requirements
- Man machine interface (MMI) requirements
  - Screen specifications
  - Data entry modes
  - Refresh rates
- Data migration
  - Legacy system data structures
  - Manual process data
  - Data transformation requirements

  - Data cleansing
  - Data archive and restoration
- Security requirements
  - Access levels
  - Security mechanism
- Communication interfaces
  - Information transfers
  - Transfer frequencies
  - Legacy system protocols
- Client/server infrastructure
- Standards
  - Corporate hardware standards
  - Current installations
- Environmental conditions
  - Hazardous, static electricity, dust, etc.

Supplier Audit

All suppliers, system integrators and consultants contributing to the supply of systems and advice that may impact GxP regulations must be audited. Often, the supplier development organization is logistically separate from the support organization and will utilize different Quality Management Systems (QMSs) in the execution of their services, which further complicates the audit plan. Where the IS architecture is complex and comprises several systems from a variety of organizations, the cost of the audit phase can be considerable. The increasing use of postal audits by pharmaceutical organizations has, however, had a significant and positive impact on the efficiency and cost of auditing. Site audits are only conducted when there is considerable risk arising from the use of the application or where the results of the postal audit are dubious or indicate serious weaknesses that need to be investigated in greater detail.

The Supplier Audit establishes whether the controls applied to the development of the core product and application configuration are consistent with GxP requirements and whether the organization is technically, organisationally and commercially capable of supporting the application for its anticipated life. The Supplier Audit will collate information for review and, where required, corrective action and whether a follow-up audit is required. An example postal audit questionnaire is presented in Table 7.3. The questionnaire is not exhaustive; however, it clearly demonstrates the objective and scope of the postal audit.

Table 7.3. Sample Postal Audit

Question	Yes/No Comment	Evidence Attached? (Yes/No)
Organization		
Is the company organization documented?		
Does the organization include specific responsibilities for quality?		
Are roles and responsibilities of the organization documented?		
Is there a project management structure?		
Quality Systems		
Is there a documented QMS?		
Has accreditation/registration been achieved for a. BS 5750 Pt. 1 or 2 or ISO 9000? b. TickIT?		
Please detail any other quality accreditations/ registrations held.		
Is the QMS based on a life-cycle approach?		
Does the QMS cover postdelivery support and maintenance?		
Have procedures been established to control each phase of the development and operational life cycle?		
Are internal quality audits/inspections conducted on a regular basis?		
Planning		
Do you develop detailed project plans (e.g., Gantt charts)?		
Do project plans include task dependencies?		
Do project plans identify resources assigned to each task?		
Do project plans identify critical paths?		
Do you develop Project Quality Plans?		
Are plans reviewed and/or approved by your customers?		

Table 7.3 continued on the next page

Table 7.3 continued from the previous page

Question	Yes/No Comment	Evidence Attached? (Yes/No)
Specification		
Do you insist that your customers issue a URS?		
Do you develop an FDS in response to the URS?		
Do you develop a cross-reference matrix to relate ALL URS clauses to the FDS?		
Design		
Are detailed Software Design Specifications (SDSs) produced?		
Are Software Module Design Specifications (SMDSs) produced?		
Are detailed Hardware Design Specifications (HDSs) produced?		
Are customers invited to attend design review meetings?		
Are design reviews minuted?		
Are design changes a. Proposed? b. Approved? c. Implemented? d. Controlled?		
Do controls extend to subcontractors?		
Are subcontractors audited?		
Implementation		
Do software coding standards exist?		
Are Source Code Reviews (SCRs) conducted?		
Are SCRs documented?		
Are changes traceable from the code?		
Is there a procedure for data migration?		
Is there a procedure to control the inadvertent or malicious modification of software?		
Testing		
Is software module testing conducted?		
Is software integration testing conducted?		

Table 7.3 continued on the next page

Table 7.3 continued from the previous page

Question	Yes/No Comment	Evidence Attached? (Yes/No)
Is hardware testing conducted?		
Is Factory Acceptance Testing (FAT) conducted?		
Are test specifications produced to cover all testing?		
Are test acceptance criteria defined?		
Are test results recorded?		
Is test evidence retained (e.g., printouts, screen dumps, etc.)?		
Installation		
Is installation carried out by subcontractors?		
Is installation controlled by procedures?		
Are installation inspections conducted?		
Is Site Acceptance Testing (SAT) conducted?		
Are customers invited to witness tests?		
Are installation reports produced?		
Support and Maintenance		
Is an SLA established with the customer?		
Is a help desk facility provided?		
Are customers notified of faults/defects/anomalies within your product?		
Do you have procedures to control bug fixes?		
Do you have procedures to control hardware and software upgrades?		
How long do you support a. Hardware? b. Software?		
What is the minimum notice period before support is withdrawn for a hardware or software product?		
Is there a change control procedure?		
Are change control records maintained?		

Table 7.3 continued on the next page

Table 7.3 continued from the previous page

Question	Yes/No Comment	Evidence Attached? (Yes/No)
Personnel Development		
Is personnel development planned?		
Are records of personnel development maintained?		
Are records of personnel experience maintained?		
Document Management		
Are procedures and documents reviewed and approved prior to issue?		
Are procedure and document changes managed through formal change control?		
Are obsolete documents withdrawn?		
Are subcontract documentation standards audited?		

Where a site audit is required, the audit can focus on the main areas of concern raised by the postal audit, thus reducing the duration of the audit and enabling a more in-depth review of the main areas of risk.

Auditors must assure themselves that there is sufficient documentary evidence available to demonstrate that quality controls appropriate to the pharmaceutical industry are in place and are being routinely applied. The supplier should have produced Project Quality Plans for all projects, similar in objective to the VMP produced by the pharmaceutical organization.

The following sections define the typical areas to be challenged by the audit team and the scope and content of documents produced at each phase of the project life cycle.

Project Quality Plan. Where a Project Quality Plan exists, it will often be used as a guide to the audit. Specific activities and documentation stated in the Project Quality Plan should be reviewed against their controlling procedures. The Project Quality Plan should also reference the project plan, usually presented as a Gantt chart that defines

- project tasks;
- task start, end and duration;

- task dependencies;
- resource allocation; and
- critical paths.

The project plan should be reviewed to ensure that all tasks have been defined and that the estimates are realistic and will not compromise the delivery and quality of the information system. Ongoing monitoring of the project plan is essential, as project slippage and the associated commercial implications generally lead to shortcuts in key quality controls, such as the application of procedures, Detailed Design, reviews and testing. The earlier that project slippage can be detected, the earlier corrective action can be taken to minimize the risk to project delivery and quality. The typical content of the Project Quality Plan are as follows:

- Project background and scope
- Organization, roles and responsibilities (internal and external)
- Approach (life-cycle activities, tools and methodologies)
- Quality standards and procedures (internal and external)
- Review points
- Communication channels
- Key deliverables (including documentation)
- Milestones

Functional Design Specification. The supplier should be able to demonstrate that an FDS is produced in response to the pharmaceutical organisation's URS. Typically, the FDS will be a standard document for the core product application, with deviations from the standard offering documented in an addendum or separate project-based document. Deviations usually warrant some degree of tailored software development and should, therefore, be of prime consideration during the validation exercise. Table 7.4 provides the typical content of the FDS.

The FDS is the first critical, technical document produced by the supplier. It will demonstrate the supplier's understanding of user requirements and form the foundation of the final technical solution. If either of the above are inaccurate, inconsistent or ambiguous, the likelihood of project failure is high.

The FDS must be fully traceable to the URS, clearly demonstrating that all URS clauses have been met. Where the FDS deviates from the URS, a rationale for the potential impact of the deviation must be provided.

Hardware Design Specification. The HDS will define the hardware platform to support the EAM architecture. It is likely that the pharmaceutical organization will impose corporate standards to ensure compatibility with other installations on the site. In many instances, the supplier will simply state

Table 7.4. Typical Contents of the FDS

- System architecture
  - Diagrammatic representation of the major hardware components of the system
  - Client/server architecture
  - Hardware component interfaces
  - Geographical location of major hardware components
- Software module architecture
  - Hierarchical structure of software modules and packages
- Process definitions
  - Diagrammatic representation of business processes implemented within information system (e.g., maintenance management, change control, document management, etc.)
  - Work flows
- Functional definition (flowcharts, narratives)
  - Function inputs
  - Function objectives
  - Functional performance
  - Function outputs
  - Functional failure modes and reporting
  - Functional failure mode recovery
  - Function synchronization (events, relationships)
- Information structures
  - Database schema
- Data migration strategy
  - Migration of information from legacy systems
  - Import of information from manual systems
- Data storage
  - Folder/directory structures
  - File structures
  - File sizes
  - File properties
  - File access
  - Storage media
  - Storage capacity
  - Data retrieval rates
- User interfaces
  - Menu/display hierarchy
  - Display structure
  - Screen formats
  - Screen access security
  - Toolbar options
  - Message bars
  - Data entry field configuration
  - Window configuration
  - Refresh rates
  - Input devices (mouse, touch screen, keyboard)
- Communication interfaces
  - Wide area networks (WANs) to other sites
  - Local area networks (LANs)/Intranet
  - Serial interface protocols
  - Message packet formats
  - File Transfer Protocol (FTP)
  - Transfer frequency
  - Transfer rate
- System reliability and performance
  - Server performance
  - Network bandwidth
  - Serial communication performance
  - Printer performance
  - Backup and restoration
- Expansion/enhancement capability
  - Redundancy
  - Hardware upgrade paths

the hardware requirements and allow the pharmaceutical manufacturer to procure the hardware, especially when the EAM architecture comprises a number of systems from a variety of suppliers. The typical contents of the HDS are as follows:

- System architecture diagrams
- Layout and wiring diagrams and drawings
- Main component specifications
- System interface specifications
- Performance (CPU [central processing unit], bus, cache, clock etc.)
- Capacities (RAM [random access memory], hard disk, floppy disk, DAT [digital audio tape], CD-ROM [compact disk, read-only memory], etc.)
- Peripherals (MMI, printers, keyboards, mouse, bar codes, etc.)
- Interfaces (communication cards, network connections, cabling, speed)
- Settings (switch settings, firmware configuration)
- Environment (temperature, humidity, RFI [radio frequency interference], UV [ultraviolet], EMI [electromagnetic interference])
- Electrical supplies (UPS [uninterruptiple power supply], earth requirements, filters, etc.)
- Define relevant standards (safety, electrical, etc.)

The Installation Qualification (IQ) is developed to verify the major critical components stated in the HDS.

Software Design Specification. The SDS provides a detailed decomposition of the processes and functions defined in the FDS. The audit should establish that appropriate design methodologies have been applied, leading to a structured modular and logical design. The SDS should provide sufficient detail to enable unambiguous implementation of the software. The typical contents of the SDS are as follows:

- Module architecture
- Module descriptions
- Module interfaces
- Module relationships (events, timers, handshaking)
- Database schema
- File structures
- System interfaces

Specific module specifications should be produced for complex software. The SDMS has the following typical contents:

- Module input parameter definition (integer, real, character)
- Global data definitions
- Local data definitions
- Parameter passing mechanism (pass by value, pass by reference)
- Detailed Functional Definition
- Returned Values
- Programming Standards
- Test Harnesses

Good Programming Practice and Source Code Reviews. Suppliers should have established standards to govern the development of software. The objective of such standards is to ensure a consistent and structured approach to the development of software, thus minimizing the risk of software failure, enhancing software maintainability and avoiding personal style whilst trying not to suppress creativity.

Suppliers should conduct SCRs on all critical software modules in order to capture deviations from good programming standards, identify logic errors and ensure software modularity. Tailored software developed to satisfy user requirements not catered for within the standard product offering should be a particular focus of attention, as the risk of software failure increases for new software developments. SCRs should be documented in order to record observations raised against the software and resultant corrective actions. Further, documented evidence of the implementation of corrective actions should be available for inspection. Where software modules present a major risk to GxP compliance or evidence of internal SCRs is limited, the pharmaceutical organization should consider additional independent reviews. The scope and content of good programming standards are given below:

- Data scoping
- Module size
- Module layout
- Module cohesion and coupling
- Naming conventions
- Use of control blocks
- Module structure
- Commenting

Configuration Management. Configuration management ensures that hardware and software are controlled during the development and operational phases. Configuration management extends beyond the control of software modules during development to the control of the development environment (development system configuration, operating systems, command files, compilers, linkers, etc.) and documentation. The requirements of configuration management are as follows:

- Development is controlled within project-specific areas on the development system.
- The development area is organized into a meaningful folder structure to facilitate controlled access to files.
- Tested files are held in secure read-only areas.
- Files are "booked out" of secure areas before modification can take place.
- Files are "booked in" to secure areas once tested.
- Simultaneous file access is prohibited.
- File access is restricted to authorized users.
- Version control is applied to track file modification.
- Command files are controlled in the same manner as software and configuration files.
- Records of the development environment are maintained to enable reconstruction of the development environment for subsequent modification (e.g., compilers, linkers, assemblers, operating system versions).

The whole project development environment must be routinely backed up during the development phase and archived at the end of development to enable reconstruction of the development environment in order to facilitate subsequent maintenance activities and disaster recovery.

Configuration management must also consider hardware configuration, both of the development environment and the test environment, especially where the supplier maintains systems for a number of years.

Supplier audits must establish whether such configuration management requirements are implemented by the supplier organization. Experience shows that all controls are rarely applied; however, there is an increasing use of standard configuration management products that, as a minimum, control access and modification of software, configuration and documentation files.

Application Configuration Specification. The Application Configuration Specification (ACS) documents the application configuration required to meet the URS. The ACS records system setup parameters, process configuration, database configuration, file structures and so on required to implement the specific business implementation of the system. Where a standard implementation of the core product is adopted, this will be the only custom specification delivered to the pharmaceutical organization, accompanied by standard technical and user manuals.

Supplier Testing. Testing is conducted in several phases depending on the complexity of the software design. The supplier is responsible for developing Module, Integration, Factory and Site Acceptance Test Specifications to demonstrate that the design has been fully and accurately implemented.

Module tests are conducted against the SMDS in order to verify the discrete functionality of the module and simulate the input and output interfaces of the module. Module testing often requires the development of "test harnesses", simulation software specifically written to supply inputs to the module and interpret the returned result. The test harnesses themselves pose a further risk to GxP compliance, as they must be appropriately developed and controlled.

Integration tests are conducted to challenge the integration of new modules into the system. Table 7.5 provides the typical scope of the Integration Test Specification.

One of the greatest challenges to system suppliers is demonstrating that the integration of new software modules has not had a detrimental impact on the existing software. Regression tests must therefore be conducted in order to provide reasonable assurance that existing modules have not been affected. In extreme circumstances, it may be necessary to compare the image of the new software release with the image of the previous version in order to determine which modules have been affected and then provide a documented rationale for the potential impact on each module.

FAT is conducted within a simulated environment to demonstrate that the system meets the URS and FDS. FAT will only be conducted if the software

Table 7.5. Integration Testing

• Module interfaces – Number of parameters compared – Parameter types compared – Parameter passing mechanism (e.g., by name, by value, by reference) • Module synchronization – Handshakes (e.g., events, interlocks) – Sequencing • System performance – Functional performance – Information storage and retrieval (e.g., database queries) – Screen refresh – Data entry response times – Serial communication interfaces	– Network performance – Data storage capacity – Multiple user access • File and data integrity – Shared file access – System failure during read/write operations – Data retention following system failure – Cyclical file management – Prevention of duplicate record creation – Referential integrity • Impact on existing modules – Regression testing

is a new development or major adaptation of the standard product. FAT is the first opportunity for the pharmaceutical manufacturer to test the system in its entirety. The scope of the testing should be wide enough to ensure that most problems can be identified and rectified within the development environment. Where tests are not dependent on the operating environment, for example, data entry validation, FAT may serve as the validation record for that function. However, in order to adopt this approach, formal test specifications must be developed and approved prior to executing the tests, and the results must be clearly recorded.

Supplier Audit Reports. A Supplier Audit Report will be produced for each supplier, documenting the positive and negative observations made during the assessment of the response to the postal audit questionnaire and/or the site audit. All corrective actions must be followed up, possibly requiring further site visits, in order to ensure that nonconformance issues have been appropriately addressed in a timely manner to minimize the impact on project success.

Audit reports must be factual and not subjective, clearly stating the basis of observations and the criticality of the observation (major or minor). Suppliers must be allowed to review the Supplier Audit Report prior to issue to enable observations to be agreed on or, where a misunderstanding has occurred, enable additional mitigating information to be provided.

Supplier Audit Reports should avoid the detailed specification of corrective actions. It is the supplier's responsibility to define how observations will be addressed within the context of the supplier organization and quality systems.

In certain circumstances, the supplier may consider corrective actions contrary to their business objectives. For example, a supplier who is dependent on the pharmaceutical industry for only 10 percent of its business may be reluctant to implement corrective actions specific to pharmaceutical regulatory requirements. The pharmaceutical organization must then determine whether to seek an alternative supplier or implement corrective actions within its own organization in order to address the issues.

Installation Qualification

IQ is the responsibility of the pharmaceutical enterprise. IQ defines methodical inspections that verify the hardware and software installation against the design. Each inspection is conducted in accordance with a detailed inspection method, and the outcome verified against unambiguous acceptance criteria. The results of the inspection must be recorded on a test result sheet, referenced to or contained within the IQ Protocol. An inspection will be deemed to have "passed" if all the acceptance criteria set forward have been satisfied. Table 7.6 provides the typical contents of an IQ inspection.

Table 7.6. Typical Contents of an IQ Inspection

• Installation plans/procedures – Satisfactory execution of installation procedures • Software installation – Correct executable images installed (including versions) – Correct third-party software packages installed (including versions) – Correct folder/directory structure created and files installed within folders/directories • Software configuration completed satisfactorily – Site and system identification – User access groups – Security configuration – Menu/display access configuration – Logical device connections • Inspection of critical hardware components – Servers in correct locations – Processor speed – Cache size – ROM bios – Memory capacity – Peripherals – Storage devices – Input devices – Network interface cards, addressing and connections	– Printers – Connection – Bit switch settings – Printer driver installation • Networks – Connection – Network addressing – Server synchronization (e.g., date and time) – Network conflicts • Electrical installation – Cable connections – Electrical testing – UPS • Input/output – Outstation connection and configuration – Field device connection and calibration – Diagnostic checks – System performance • Documentation – User manuals – Technical documentation – Availability of user SOPs – Availability of Disaster Recovery Plans • Training – Availability of Training Plans

Operational Qualification

Operational Qualification (OQ) should be integrated with the SAT normally conducted by the supplier. OQ verifies the functionality of the system within its normal operating environment. An OQ Protocol should be developed to clearly define the methodology by which the tests are conducted and the acceptance criteria that determine the success or failure of the test. Figure 7.13 provides an example of a typical OQ test script. OQ must reasonably challenge the operating boundaries of each function (although never to destruction). For example, data input functions should be challenged by entering

Figure 7.13. Sample OQ Test Script

OQ Reference:	001. Recipe Save and Retrieve
Prerequisites	Recipe to be created shall not exist.
Test Equipment/ Simulators/ Harness	None
Function/Purpose:	To ensure that recipes are correctly saved to file FDS Ref: 3.1.2
Method:	1. From Recipe Edit Screen select "Create New Recipe". 2. Enter a value in each field. 3. Print screen and verify each field against entered values. 4. Select "Save Recipe". 5. Exit Recipe Editor. 6. Reenter Recipe Editor. 7. From Recipe Edit Screen select "View Recipe". 8. Enter number of newly created recipe.
Acceptance Criteria:	1. Recipe Edit Screen is displayed. 2. New values are accepted, values out of range are not accepted. 3. Printed fields match enter fields. 4. Message "Recipe Save OK" is display. 5. Main menu is displayed. 6. Recipe Editor is displayed. 7. Prompts for recipe number. 8. Recipe is recalled and displayed. 9. Values match those on printout from Step 3.
Results/ Observations	
Acceptance Criteria Achieved (clearly write YES or NO)	☐
Tested by:.. Date:	Witnessed by:.. Date:

- nonnumeric characters in numeric fields,
- values on and slightly outside operating ranges,
- field lengths that exceed the permitted number of characters for the field,
- negative values in positive value fields and
- decimal values in integer-configured fields.

Table 7.7 provides the typical contents of the OQ test.

As with IQ, all protocols must be pre- and postapproved, and tests independently witnessed or reviewed. When writing OQ tests, it is important that the acceptance criteria are clear and not embedded in the test method where they may be overlooked. Tolerances on the acceptance criteria must be in line with design and should not be so wide as to guarantee success.

Performance Qualification

Performance Qualification (PQ) demonstrates the consistent operation of the system once released for operational use. The scope of PQ for an information system is described in Table 7.8. During the PQ period, it is preferable that legacy systems and/or manual systems are operated in parallel. This approach is not without difficulty, and the strategy for parallel operation must be carefully planned. PQ will require the collation of records, such as change control, operator logs and so on across many sites. The procedures and responsibilities for the collection of records in support of PQ must be established in advance of system cutover. Further, the PQ review team must be determined to ensure that personnel are available to conduct the review, considering the pressures that will be exerted by the business to divert resources onto the next project.

Table 7.7. Typical Contents of the OQ Test

• Process flows	• Security and access
• Functional operation and performance	• Data storage and retrieval
• Failure processing, reporting and recovery	• Error reporting
• Multisite challenge	• Backup and restoration SOPs
• Operating boundaries (e.g., data entry)	• User SOPs
• Network interfaces	• Contingency Plans
• Serial communication interfaces	• User and system administrator training (delivery)
• Start-up and shutdown	

Table 7.8. Typical Scope of PQ

• System stability – System failures – Design changes – Implementation changes – Operator observations – Anomalies – Expectations • Maintain business critical functions against their performance criteria	• Assess suitability of security levels and procedures • Effectiveness of security measures • Assess effectiveness of training • Verify backup and restoration procedures

The outcome of PQ may indicate that there are areas of weakness in the design or implementation of the application, in particular if high levels of design changes or functional failures are observed within a localized area. In such instances, it will be necessary to reconsider to adequacy of OQ, and further testing may be required to determine the root cause of the failures. The outcome of PQ should further define the need to extend the PQ duration.

Validation Report

The Validation Report responds to the VMP, providing a summary of the actual approach taken and the documentation produced. Any deviations from the approach prescribed by the VMP must be justified, including an assessment of the consequences of any deviations. Where a deviation is not acceptable, a corrective action plan must be formulated to address the issue. It may be possible to implement manual procedures to overcome the issue in the short term to enable the system to move into the operational environment whilst the issue is being addressed. Where there are no issues or they can be overcome by manual procedures, the system may be recommend for "operational use".

The Validation Report should clearly demonstrate that a suitable operating environment has been established, including procedures to control documents, changes, backup, archive and restoration and security and that appropriate service contracts have been established.

Operational Environment

The validated status of information systems must be maintained during the operational life of the system. It is therefore essential that procedures are developed to control operation and maintenance of the system after cutover. Such controls include

- system operation,
- change control,
- backup and restoration,
- system upgrade;
- Contingency Plans and
- SLAs.

All procedures must be developed, reviewed and approved prior to OQ. The effectiveness of the operation and maintenance procedures and plans should, where possible, be verified during the OQ and PQ phases.

ASSET MANAGEMENT ORGANIZATION

As a final note, the culture of the asset management organization is fundamental to establishing and maintaining GxP compliance of the process and information systems comprising the asset management infrastructure. Senior management must assign high priority to GxP and not abdicate responsibility to users because they believe they are in the front line for regulatory inspection. As was depicted by Figure 7.7, asset information and validation documentation are the foundation of regulatory submissions. If the foundation crumbles, so does the credibility of those submissions and, subsequently, the reputation and profitability of the pharmaceutical enterprise.

Effort must be expended to raise GxP awareness and to ensure that personnel view the rigors of producing GxP documentation as an integral part of the asset management process. Documentation should be seen as the definition and recording of essential processes, not as a paperwork exercise to satisfy regulators. Further, continuous improvement of EAM strategies, processes, information and information systems is paramount to the evolution and success of the pharmaceutical company and to satisfying increasing regulatory expectations. Establishing such a culture is not easy in an area that has traditionally not recognized the need for such control and documentation. Investment in change management programs may be needed to bring about such cultural change; ultimately, though, senior management must lead by example.

REFERENCES

1. UK GAMP Forum. 1998. *Suppliers Guide for Validation of Automated Systems in Pharmaceutical Manufacture,* Version 3. Available from the International Society for Pharmaceutical Engineering.

2. European Union Guide to Directive 91/356/EEC. 1991. *Directive Laying Down the Principles of Good Manufacturing Practice for Medicinal Products for Human Use, European Commission.*

3. MCA. 1997. *Rules and Guidance for Pharmaceutical Manufacturers.* London: Her Majesty's Standards Office.

4. U.S. Code of Federal Regulations, Title 21, Part 210: *Current Good Manufacturing Practice in Manufacturing, Processing, Packaging, or Holding of Drugs (General)*; Part 211: *Current Good Manufacturing Practice for Finished Pharmaceuticals.* Washington, D.C.: National Archives and Records Administration.

5. Wingate, G.A.S. 1997. *Validating Automated Manufacturing and Laboratory Applications: Putting Principles into Practice.* Buffalo Grove, Ill., USA: Interpharm Press, Inc.

6. Supplier Forum. 1999. *Guidance Notes on Supplier Audits Conducted by Customers,* Rev. 4. Available on the Web at www.eutech.com/forum.

8

Validating Enterprise/Manufacturing Resource Planning Systems

Owen Salvage
CSR
Chatswood, New South Wales
Australia

Melanie Snelham
Eutech
Northwich, United Kingdom

This chapter is concerned with the validation of Enterprise Resource Planning (ERP) systems in the pharmaceuticals industry and seeks to establish a project model whereby the objectives of validation become embodied in the objectives of project implementation.

The following passage is a typical statement from a senior manager looking for the company's benefits from an ERP system:

> *I want a system to replace all the stand-alone pockets of automation handling the warehouse, purchasing and materials management systems. The ERP system is to interface with my laboratory system and provide Internet links with my customers and suppliers. . . . I expect to be able to reduce my inventory, reduce lead times and IT [information technology] support costs, and install a system which flexible in the face of changing market conditions.*

These are common requirements for an ERP system, and as so often happens, there is no mention that the system needs to be validated. The pressures on an ERP implementation are massive, and any such system is vital to the continued operation of the company and is business critical.

There are examples where an ERP implementation has been attributed with the failure of a pharmaceutical company [1]. More often, a poor implementation can leave residual problems which severely strain a company's operations for a considerable period following "go live". Getting the implementation right is vital for the business, and getting the validation right is crucial to allow the business to operate.

Often, companies have a Computer System Validation (CSV) Manual that guides the validation engineer through the process of specifying, designing and installing a process control system, but the CSV Manual is frequently deficient in offering advice for a business system.

With the majority of ERP projects in the pharmaceutical industry, only relatively small changes in project direction are necessary to produce the required evidence for a validated system. The production of specifications, test plans and overview documents are all activities that appear on the standard ERP project plan. In many cases, the issue appears to be in organizing the documentation to ensure specifications are categorized into the V model categories, overview documents and summary reports are written, test evidence is collated and comprehensive Source Code Reviews are retained.

Many ERP projects are well managed, with the project team conducting a business process review and generating specifications and plans for the desired method of operation. The number of documents generated in a typical ERP implementation can be in the thousands. A major difficulty following implementation is finding a way through the large quantity of documents and being able to present a regulator with a coherent set of specification and test documents that approximate a standard validation life-cycle model, such as the V model.

BUSINESS CASE AND MANAGEMENT ISSUES

In justifying the potential installation of an ERP system, there are many organizational and data accuracy advantages to be highlighted. Any system able to assist the pharmaceutical manufacturer's business operations and help speed the dispatch of pharmaceutical products to market will have a strong case. An ERP system can contribute to this by reducing the lead time from customer order to product dispatch, reducing warehouse stocks and maximizing the business cash flow potential. With top selling products attracting sales revenues of up to U.S.$2,000,000 per day [2], the swift dispatch of batches from the warehouse can generate more cash for the company.

Data Integrity and Analysis

A single integrated, validated ERP system avoids the double entry of data and errors associated with data transcription between previously unrelated systems. All data are available to the users, and planning data can be traced from rough cut plans through to completion of work orders to dispatch of finished goods. The audit trail is established and easy to track. Once the system is established, components are assigned automatically and accurately to meet each specific order requirement. Communication errors between previously manual or unrelated functions are reduced (e.g., once plans are established on the ERP system, all users can gain access to the plan, and the data interchange gives rapid and consistent reporting of order and production status).

ERP Supplier Capability

A critical part of the successful introduction of an ERP system is in the early stages where Supplier Audits are vital to ensure confidence in the design of the software and control of software changes. It has been shown that acceptance of a software vendor simply on the basis of a quality accreditation, e.g., ISO 9000, is inadequate. The customer should carry out an audit of the supplier to review written proof of use of the claimed quality system. It is also essential to verify the quality system was used for the particular version of software to be implemented.

Consideration must be made as to the suitability of the ERP product, vendor and, increasingly, the third-party implementation partner. It is estimated that in the year 2000, over 30 percent of all ERP implementations will be carried out by third-party implementers, so a check on the level of their validation competence is essential.

Regulatory Climate

Computer validation has been and remains a very topical regulatory area. In a review of U.S. Food and Drug Administration (FDA) 483s, the number of 483s issued citing noncompliance of automation from 1984 to 1996 shows an increasing trend [2]. Typical inspection findings include but are not limited to the following:

- Qualification and testing
- Change control
- Validation approach, quality assurance and auditing
- Specification, design and Source Code Reviews
- Planning documents
- Operating procedures, including security practice
- Inadequate design documents
- Inadequate software version control

Examples of FDA 483s for Manufacturing Resource Planning (MRP II) systems illustrate the fact that, in many cases, the same types of deficiencies are being found in 1998 as found in the early 1990s. A 483 from a 1993 inspection cited inadequate detail in the software specifications. A 483 issued to a UK pharmaceutical company in 1998 cited no structural and functional design included in the validation. It is therefore not surprising to see that more and more inspections are including detailed inspections of computerized systems.

It has been suggested that half of all GxP regulatory inspections will include a detailed inspection of computer systems in the year 2000. This would seem to be a well-founded prediction with FDA, UK Medicines Control Agency (MCA) and other national GxP regulatory authorities actively training their investigators in the inspection of computerized systems.

Given this level of expectation and the known levels of citations, the case for ensuring that validation is a key objective of any pharmaceutical ERP project would seem conclusive. The various areas covered by regulatory requirements for computerized systems can be summarized in Table 8.1.

Validation Costs

When considering the choice between retrospectively validating an existing system or carrying out prospective validation as part of the initial ERP implementation, it should be remembered that the validation costs for validating an existing system are approximately 5 times that of prospectively validating a new system. Validation should be considered as a means of protecting the investment in the ERP system, not only from adverse regulatory findings but also to maximize value for money from its installation. Over half the cost of a system can be attributed to its use rather than development. Poor development can significantly increase the cost of support. The authors have experienced an ERP system where validation reduced support costs by 75 percent.

SYSTEM FUNCTIONALITY

The key functions of an ERP/MRP II system can be summarized in Figure 8.1 [2].

Planning/Forecasting

The planning functionality of ERP/MRP II systems allows for long-range sales projections and forecasts to be developed into firm plans to ensure that items with long lead times are available as required to fulfill orders. Having the planning facility on the same system as ordering, scheduling and manufacturing reduces the potential for errors in the transfer of information between systems.

Table 8.1. Regulatory Requirements for Computerized System

	U.S. Code of Federal Regulations, Title 21, Part 211.68	European Union Directive 91/356/EEC, Annex 11	Australian Code of GMP, Medicinal Products, Part 1, Section 9	European Union Guidelines on Good Distribution Practice 94/C 63/03
Personnel	✓	✓	✓	✓
Equipment siting		✓	✓	
Building/Facility	✓			
Facility management				✓
Equipment design	✓			
Validation life cycle		✓	✓	
Equipment maintenance	✓			✓
Contracted out services		✓		
Standard Operating Procedures	✓			✓
Contingency plan		✓	✓	
Detailed system description		✓	✓	
Transport conditions				✓
Authorized Supplier/ customer				✓
Record Retention	✓	✓	✓	✓
Management of counterfeit product				✓
Calibration	✓			✓
Software quality		✓	✓	
Change control	✓	✓	✓	
Product recall				✓
Testing	✓	✓	✓	✓
Backup files	✓			
Error recording		✓		
Data storage	✓	✓	✓	
Hard copy records	✓			
Ongoing evaluation	✓			
Security	✓	✓	✓	

Figure 8.1. ERP/MRP II Functionality

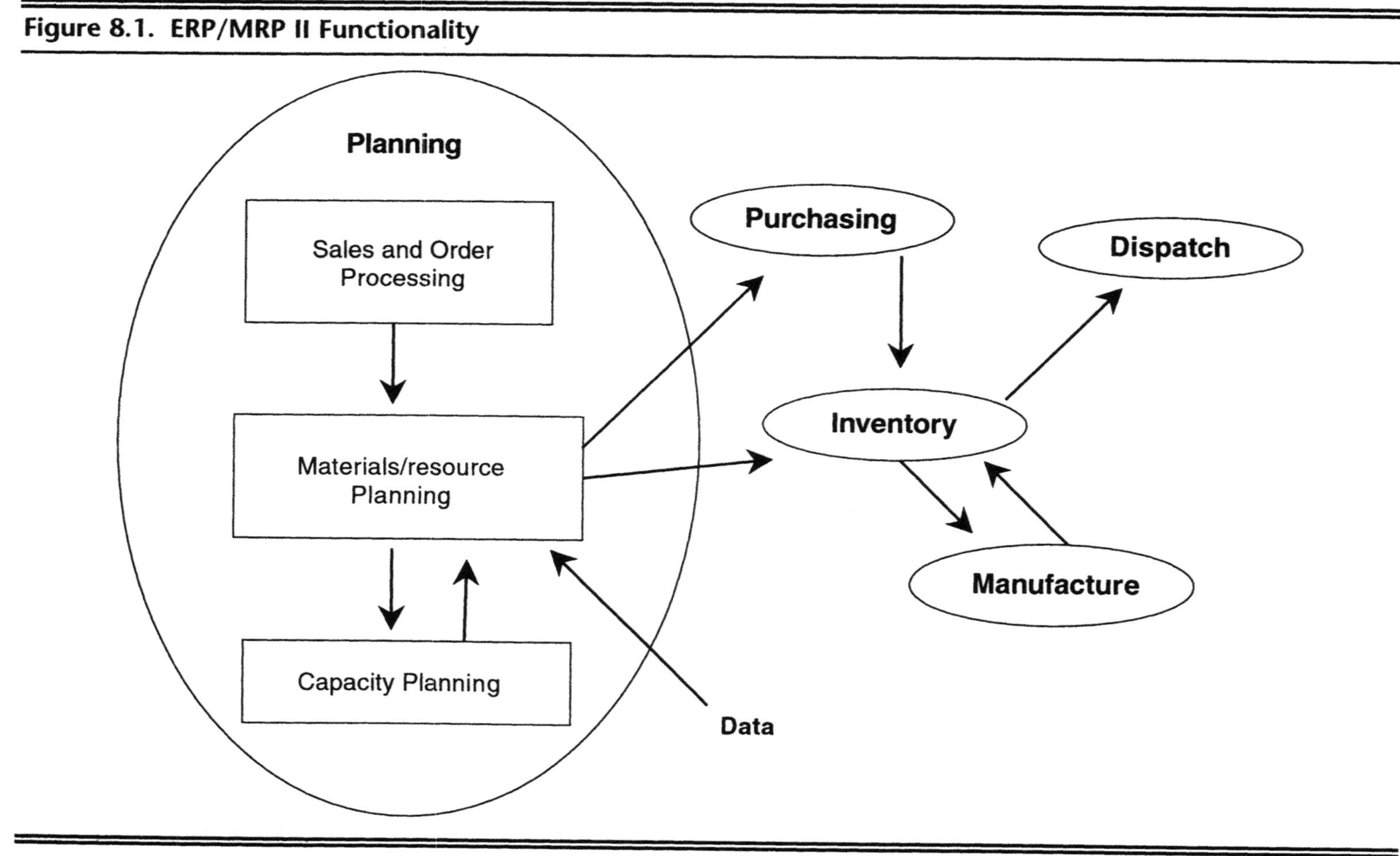

Purchasing

The capacity to have purchase orders produced by the same system as planning gives a greater level of control. There is also a clearly demonstrable system in place for purchasing only from approved vendors.

Warehousing

The level of control exerted over warehousing is often seen by the warehousing staff as very restrictive. Instead of being able to pick items from the nearest available batch for a given item, there is a requirement to pick a specific batch from a specific location and to confirm that this has been done. Whilst this may seem restrictive, it is vital to ensure full traceability and good stock rotation.

Quality Management

The control of authority levels for quality control (QC) and quality assurance (QA) staff is paramount to the control of materials. A restricted number of personnel would have the authority to change the status of materials, for example, from quarantine to available, and the system would not pick a material for manufacture with a status of quarantined.

Manufacturing

The issue of traceability is again key within manufacturing, where various batches of raw materials with their own lot/batch numbers are used in the manufacture of a material with a separate unique identity.

Goods Dispatch

The main advantages of having an ERP system in dispatch include the selection of goods available. On a manual or unrelated system, there is often the issue that the production plan for a given week may be altered due to unplanned occurrences, such as machine breakdown. There is then a complex and time-consuming activity for the dispatch staff to identify what items are actually available for dispatch. Dispatch control also ensures a full traceability on a single system, which would prove invaluable in the case of a product recall.

The authors have experienced the following issues arising at pharmaceutical/healthcare companies when implementing or using such systems:

- A UK healthcare company amended expiry dates of materials following testing, but it was later found that the system was allocating materials on a nearest expiry date first use basis, thus the items with extended expiry dates went to the end of the queue.

- Several companies who implemented new systems transferred data from previous versions of the system or other systems in a bid to reduce project overhead. On checking a sample of the data, it was realized that the data had not been transferred correctly and had resulted in data either populating the wrong database location or in corruption of the data. The transfer of open orders is particularly problematic for the project.

PROJECT MANAGEMENT

The classic project model of specification, design, build, install and test is too inflexible for an ERP implementation impacting the work of large numbers of people. Often, not enough is known of the technology by the client organization and of the business by the implementer. Only in very few cases does the ERP product closely match the organization's operations, or the organization is flexible enough to change to match ERP product structures.

A typical approach to implementation is as follows:

- Project initiation
- Install development system
- Define current business processes
- Define future business processes
- Conduct conference room pilot
- Review legacy data
- Data upload
- Readiness review
- Go live

At the same time as attempting to install a major information system (IS) the opportunity is often taken to reengineer the current business processes to take advantage of the technology. The existing business processes are defined, the desired processes are identified and these are prototyped during the conference room pilot (CRP) exercise. During the CRP, the requirements for modifications to the ERP functionality are noted, and bespoke code is written to adapt the basic ERP product to match the needs of the client organization. In reengineering the business processes, there are often resultant changes in working practice that lead to issues of change management in addition to the required "selling" of the system to the users.

Data to be transferred from existing systems are reviewed prior to transfer to the new system. Redundant data are removed from the transfer list, and a data upload mechanism is identified. Data upload is carried out, either manually or automatically, and a readiness review is conducted to check all items are in place before the big day and "go live" (see Figure 8.2).

Figure 8.2. ERP Implementation

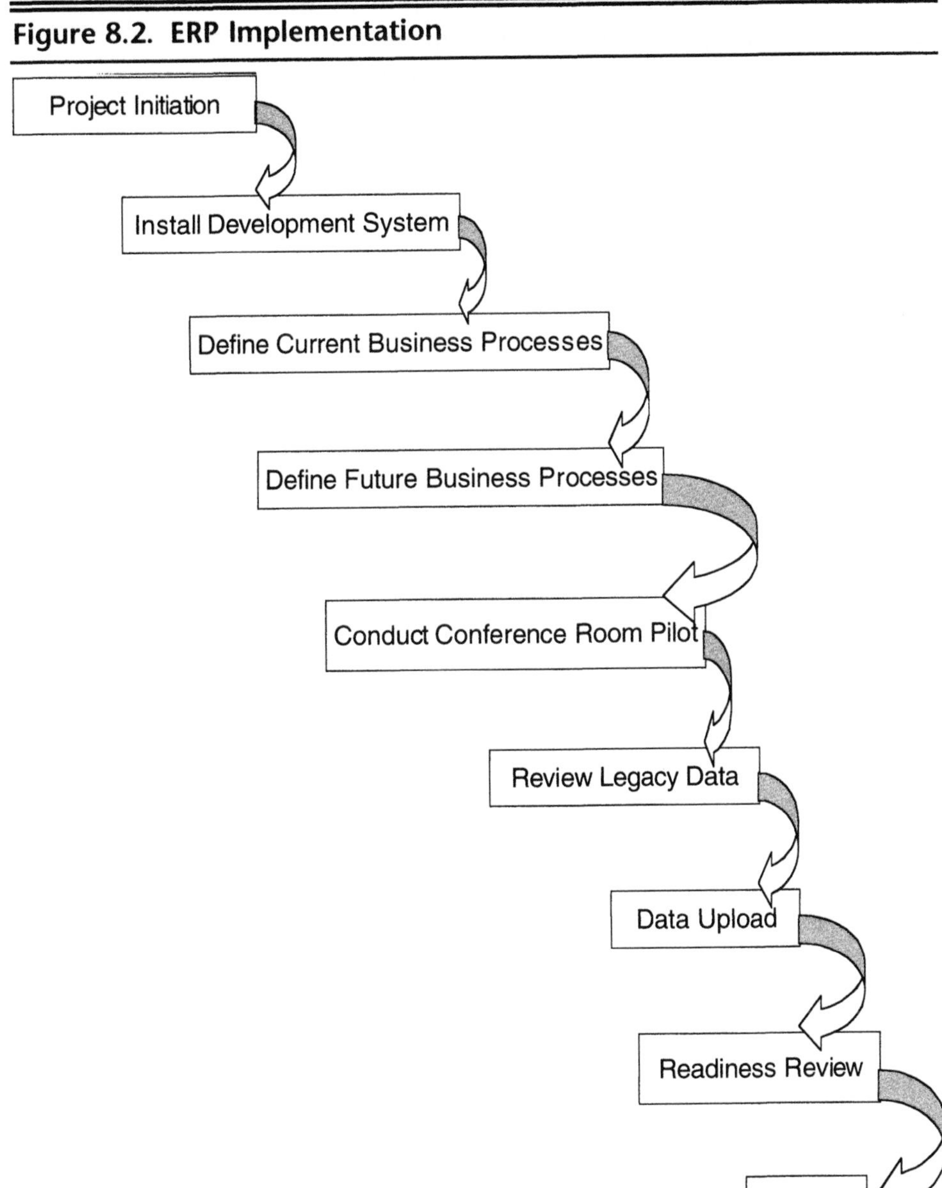

What is wrong with this model? Nothing in principle, but the question arises, "What about validation?" The validation life cycle includes specification, design, build, installation and qualification, which is not greatly different from the classic project management model that was rejected earlier in this section.

The problem lies in that the specification of the system is not fully known or defined until the CRP is all but complete, and this is typically immediately prior to go live. On a pressured-driven, business-critical project, where is the opportunity to produce an approved System Definition, Installation/Operational/Performance Qualification (IQ/OQ/PQ) test protocols when the information comes so late in the day?

If validation is not a key objective for the project, on a par with on-time delivery and within budget, the likelihood is that validation will not be adequately conducted. On far too many occasions, validation is crammed into the last part of the project, and it is only when notification of an inspection arrives that real attention is focused on validation. A rushed validation effort will often result in poor evidence, which will present a poor impression to a regulator. Thought should be given at various points throughout the project as to the level of quality of the validation documentation being produced. It is too late to resolve poor quality validation evidence immediately prior to inspection.

The recommended approach to project management for the implementation of an ERP/MRP II system would include the following key issues, which are often not included or are given only the minimum amount of thought and planning:

Education

The system to be introduced will often dramatically change the ways of working by the majority of staff, and there is a natural suspicion on the part of the users. It is vital that education be undertaken at the earliest possible time in the project life cycle and continued throughout the project. In companies where the education and training of the users has been a key consideration, the implementation has clearly benefited from their understanding and commitment.

Validation Objectives

There is a tendency with the validation of these projects to rush in at a late stage in the project life cycle, without a clear view of the validation objectives. In order to be able to educate and win the support of the project team for this vital part of the process, it is essential that the objectives of the validation be demonstrated.

Validation Model

Another weakness of many projects is the lack of planning as to the validation model to follow. It is all too easy to fall into the trap of assuming that definition documents will be produced by the vendor, and that validation for the project team will only involve some qualification testing. Unless the validation model is agreed on before commencing the project, it is highly likely that the resulting validation will fall prey to regulatory criticism, even at the most cursory examination.

Roles and Responsibilities

In order to conduct the agreed-on validation in accordance with the agreed-on model, it is essential that the roles and responsibilities for the validation activities are decided before commencing the validation. It is all too easy for a project member to assume that a particular part of the validation falls within the role of someone else in the team, and it is only at the end of the project that a gap is revealed in the process.

Supplier Audits

The vendor of the system can have a very positive or negative effect on the level of effort required for validation, depending on the levels of communication established at an early stage. The approach of the vendor to documentation and testing is vital to validation. If the vendor has a well-defined and documented approach to development, it is likely that the definitions required for user validation will be more easily produced. A good working relationship with the vendor also facilitates a smooth progression through validation testing. An example of an audit checklist for a Supplier Audit can be found in Appendix 7 of the GAMP guide [3]. It is important to include audits of any integrators or consultants, as well as the developer of the original software product.

Supplier Commitment

As detailed above, the commitment of the supplier should be a major consideration when selecting a supplier. It is important that the supplier understands the needs of the customer in terms of its validation requirements and that, wherever possible, the supplier provides the required level of support. A long-term relationship should be developed on both sides and should not be viewed simply as a short term means to buying/selling a system.

GxP Assessment

Because of the very complex nature of MRP II/ERP systems, it is important that GCP/GLP/GMP (GxP) assessment is carried out to identify the areas of the system that are GxP critical. This is the first step in establishing how to approach the testing of the system and the scope of validation testing. Of equal importance is the resulting report that would specifically detail which areas of the system are deemed non–GxP critical and are not to be included in the validation testing. Justification for these decisions should be documented in the report and could be used to demonstrate to a regulator that consideration has been given to the entire system, and that any parts of the system which have not been tested are in the report as specific exclusions. Failure to document this adequately would be seen by a regulator as an incomplete validation. A guide to the GxP–critical parts of an MRP II system are given below (GAMP Section 3.9) [3]:

- Information associated with lot
  - Lot number
  - Lot status
  - Dates associated with lot
  - Quantity/potency
  - Conversion factors
  - Lot notes
  - Quality
- Information associated with shop orders
  - Shop order number
  - Quantities
  - Receipt date
  - Transactions
  - Shelf days/retest days
- Information associated with purchase orders
  - Purchase order numbers
  - Vendor lot number
  - Dates
  - Transactions
- Information associated with items
  - Item number
  - Item notes

  - Bills of material
  - Location
  - Type
  - Quality
- Information associated with customer orders
  - Customer order number
- Information associated with the supplier
  - Quality approval
- Information associated with users
  - People (including security)

Project Members

Bearing in mind the complexity of MRP II/ERP systems, it is usual for implementation/validation projects to be carried out over a long period of time. During this time, it is likely that there will be changes in staff or movements of staff from one role to another. Whilst this can be an advantage for the project in the introduction of a new perspective, it often highlights weaknesses in the documentation of the project. It should be remembered that at any time a person with no prior knowledge of the project could be brought in and expected to follow the proceedings. Documenting the progress of the project in each phase will avoid issues in this area, and any new members of the team will be able to read a comprehensive account of the progress to date, enabling them to contribute more rapidly. Whilst it is not usually to the detriment of the project if some of the members change, it is important that not too many changes are made in rapid succession, or there is a danger of the knowledge and skills being diluted. There is also the added issue on some projects of a merger or takeover occurring during the life of the project. Any changes to the project which may arise from this should be documented.

Documenting Evidence

At all times throughout the project, it is important to remember that on completion of the project, the documentation is the only evidence that the project was carried out effectively. The standard of the resulting evidence will be the key means for a regulator to judge the level of quality of the implemented system. A poorly defined and documented set of evidence will raise immediate concerns as to the level of quality of the system and is more likely to lead to a thorough inspection of the system.

Overviews of Functions and Documentation

One aspect of validation documentation that is continually being cited for deficiency is the absence or inadequate evidence of overview documents. In order to assess the computerized system and the level of quality of its implementation and validation, a regulator will usually start by requesting overview documents. These may be specific overviews for the system being inspected or of the validation approach of the company. Either way, the inability to provide these does not get an inspection off to a good start. In the absence of these documents, more detailed documents will be examined; the more detail that is examined, the more likely that deficiencies will be identified. With the number of FDA 483s raised on the issue of inadequacy of overview documents, it is very surprising that these are not tackled as a matter of urgency by other companies. One way to test the adequacy of the documentation is to arrange an audit by a third party, either from another area of the company or an independent consultant with expertise in regulatory inspections. This person would assess the level of quality of overview documents, which could help the company avoid receiving citations in this area.

Validation Deliverables

The Validation Master Plan (VMP) or site plan will usually detail the deliverables to be produced from the validation. At the end of the validation exercise, it is important to check for the completion of this list. If there have been any changes of requirement during the project, and a deliverable is no longer required, it is essential that this be documented. The absence of such a note could lead to criticism from a regulator that the validation deliverables are incomplete.

THE VALIDATION LIFE CYCLE

The principles of Good IT Practice for the use of computer systems in business applications are now well established as the means by which quality assurance is built into the development of automated systems. ERP systems fall into the category of automated systems, and it is appropriate that a ERP installation be conducted in line with a defined life cycle.

The validation of an ERP system proceeds under the authority of a Validation Plan and follows the V model leading from specification, GxP Assessment, through to implementation and qualification. The phases of the validation life cycle for an ERP system and the sequence of activities are listed below:

- Validation Plan
- User Requirements Specification (URS)

- Supplier Audit and vendor selection
- Functional Design Specification (FDS)
- GxP Assessment
- Design Qualification (DQ)
- Source Code Review
- Installation Qualification (IQ)
- Operational Qualification (OQ) including data verification
- Performance Qualification (PQ)
- Validation Report
- Ongoing support and integrity checks (system management)

The ERP validation life cycle follows the classic V Model as defined in the following categories and shown in Figure 8.3.

Validation Plan

The Validation Plan defines the scope, organization and timetable of the validation program. Supplier organizations may not be known at this stage and may be generically identified within the Validation Plan. Change control procedures during and following the validation exercise are defined and any necessary Standard Operating Procedures (SOPs) are identified. The Validation Plan also outlines the general and specific acceptance criteria for validation, establishes the responsibilities for performing the validation and specifies the documentation requirements for defining and controlling test protocols, results recording and reporting.

Predicting deliverables and identifying staff can be difficult with a long-term project. It is possible that amendments or additional detail will need to be documented at a later stage.

User Requirements Specification

The URS is a statement of the system requirements in terms of how the ERP application is to operate in the intended environment. The URS should clearly identify the following:

- Integration of the ERP system with other systems
- List the number and types of relevant equipment
- System size including spare capacity (e.g., number of workstations)
- Requirements for data presentation, records and reports
- Any networking requirements to local/wide area networks (LANs/WANs)
- Communication links

Figure 8.3. Validation Life Cycle for an ERP System

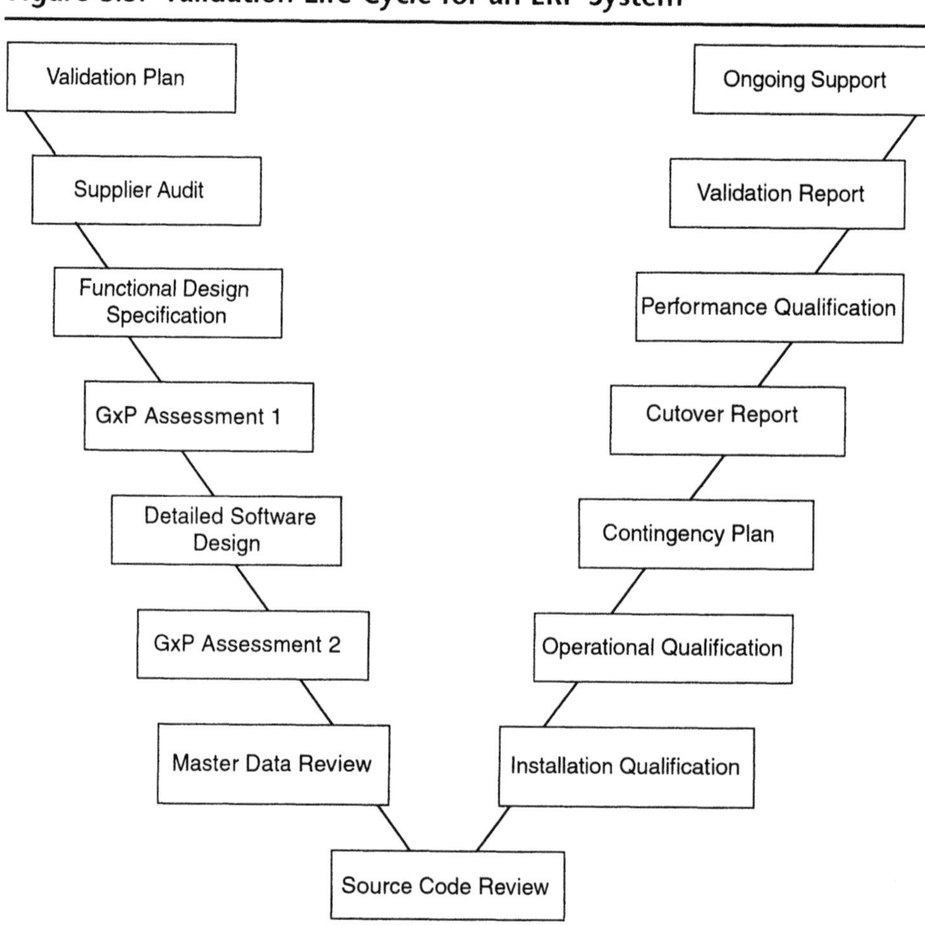

- System performance, capacity and availability targets
- Maintenance and requirements
- Documentation requirements

End user involvement in the development of the URS is strongly recommended to ensure good understanding of the operating environment of the ERP system and the functionality required of the installed system. Examples of reports required from the system, engineering drawings, graphics and location sketches greatly assist the ERP designer.

Supplier Audit and Vendor Selection

The pharmaceutical manufacturer is accountable to the regulatory authorities for any deficiencies in their chosen supplier's capability. The pharmaceutical manufacturer will be expected to bridge any gap in standards by assigning their own staff or employing a consultant to assist the supplier support the installation of a fully validated system. It is worth taking the supplier's validation credentials into account when selecting a vendor.

Selecting a supplier accredited to ISO 9000 or ISO 9000-3 (TickIT) gives confidence that a defined project life cycle will be followed. A Supplier Audit is recommended to confirm the supplier is capable of delivering the correct standard of software engineering and documentation required. A supplier Project Quality Plan identifies the organization, standards, software tools and timetables proposed by the supplier in the execution of the project. It should never be assumed that accreditation negates the need for an audit of the supplier, as the authors have experienced audits where the quality system in place was not actually being used and where the accreditation was only applicable to a small section of the company and did not include software development. During the audit, there should be confirmation that the specific version of the software being considered for implementation is covered by the quality system for its development and testing.

System Definition/Functional Design Specification

The System Definition captures the functionality of the ERP system, hardware platform, performance expected, business and maintenance procedures. It translates the requirements of the URS into the technical solution proposed to fulfill the project requirements. Operational and performance criteria for the ERP system are stated in such a way as to facilitate testing of the installed system. The System Definition should specify the following:

- Functional requirements to facilitate software development
- System performance (e.g., timing, memory storage, availability and spare capacity)
- Communications and network protocols (e.g., TCP/IP [Transmission Control Protocol/Internet Protocol], Ethernet, Token Ring)
- Method of storing and retrieving data
- Terminal emulation or client server requirements for personal computers (PCs)
- Site environmental conditions including power supplies
- Software configuration details and terminal emulation requirements
- Peripheral devices (e.g., printers and backup devices)
- Number of users and response time required from the ERP

- Functional descriptions of ERP operations
- Business operating procedures
- Contingency plans, maintenance and operating procedures
- Application macros, bespoke software and reports
- Security and access control methods

GxP Assessment of Critical Components

A GxP Assessment of the ERP system allows the validation effort to be directed toward those system elements with a direct GxP impact, such as critical analysis equipment and the components crucial for the accuracy of data. The GxP Assessment also identifies the key elements of the calibration regime.

On the data management side of ERP system, GxP data elements include item number, shelf life and retest days, sample date and number, sample specification version and status. Also included are sample method references, study references, purchase order information and ERP access security identifiers (user names).

Design Qualification

DQ checks that the design documentation matches the requirements stated in the FDS. Layout drawings, operating and maintenance procedures, parts lists and system descriptions are included in the DQ. Both hardware and software configuration are reviewed together with the supplier's product documentation. DQ offers the opportunity to monitor the progress of project validation as well as any operational and safety issues, such as the use of radioactive instruments or access to analysis equipment for maintenance.

Source Code Review

For any bespoke macros or configuration, a Source Code Review assesses the application coding against Good Programming Practice (GPP). The goals of a Source Code Review are to identify logic errors, assess the application of programming standards and comments, identify redundant code and check critical algorithms. A Source Code Review Report summarizes the review findings and may identify further actions as part of the OQ to test specific code modules. If the Supplier Audit resulted in a low degree of confidence in the supplier's own software development process for the ERP product, then it may be necessary to recommend that the supplier conducts a Source Code Review of their core product code. The quality of documentation of the Source Code Review is an area frequently criticized during inspections.

Installation Qualification

IQ consists of documented checks that all equipment, parts, services, and documents have been supplied and installed as agreed. The equipment supplier is responsible for all construction and installation activities. IQ activities include the following:

- Hardware and software versions and manuals
- Diagnostic self-tests for analysis equipment
- Network compatibility of analysis equipment and peripherals (printers)
- Calibration certificates for connected instruments
- Network connections to LAN/WAN
- Power-up/power-down tests
- Installation of services, e.g., power supplies and earthing
- Security access testing

Supplier testing carried out at the factory includes hardware and software tests. Successful completion of the tests allows the scope of the IQ to be reduced.

Operational Qualification

OQ is a series of tests to demonstrate that all functions of the equipment operate as designed. Typically, a number of functional test scripts test each of the functions identified in the FDS. OQ includes the following activities/tests:

- Communication driver tests
- Integration tests
- Data upload and migration checks
- Data integrity checks (e.g., range checks, "are you sure?", validation of inputs)
- Core functional checks of core ERP product
- Special configuration functions
- Testing of macros and bespoke software
- Signal diagnostics from linked analytical equipment
- Special calculations and algorithms
- Software backups and restoration of data
- Business operating procedures
- Archive and retrieval of documents and records
- Verify operating manuals including challenge testing

Calibration certificates must be available for the test instruments used. Also, Contingency Plans must be confirmed as operable. The degree of testing may be alleviated by the amount of predelivery testing conducted by the supplier in association with the confidence in the supplier's software development practices determined by the Supplier Audit. Predelivery tests should be witnessed or at least the results reviewed and a summary report produced.

Data Verification

The accuracy of data within the ERP system must be confirmed. This involves checking of data loaded manually or via automated upload from legacy systems.

Performance Qualification

PQ documents that the ERP analysis equipment performs effectively and reproducibly across all operating ranges, and the data management functions meets its specification and quality attributes specified in the URS. Following a period of operation, typically 4–6 weeks, PQ summarizes the ongoing monitoring of the ERP system in the immediate period after going live. Over this time, the ERP system is expected to have stabilized in terms of user support requests, system performance, change requests and network security integrity.

Validation Report

The Validation Report reviews the validation activities with reference to the Validation Plan. Any outstanding issues are summarized, and ongoing actions are given a forward audit trail using a change request reference.

It should be possible for a regulator to get a complete picture of the validation exercise from the Validation Plan, Validation Report and overview documents. The Validation Report should be reviewed to ensure that any changes not already documented during the project are detailed so as to give a consistent picture from planning to reporting.

ONGOING SUPPORT AND INTEGRITY CHECKS (SYSTEM MANAGEMENT)

Maintaining the validated status of the ERP system requires management of the network, servers, ERP functions and data. To ensure the validated status of the ERP system is retained during operation, change control must be in place, along with periodic reviews of system status. Implementing a series of procedures and maintaining physical access control to ERP servers supports the maintenance of validation over the life of the system.

Change Control

Change control of the ERP hardware, software and associated documentation (SOPs and operating manuals) is necessary and requires that changes be recorded, specified and assessed for GxP impact. Changes are to be approved by qualified personnel and tested, with records of testing retained.

Maintaining compliance through business change is a major issue and relies heavily on an effective change control mechanism. This is no straightforward task, since the business organization is almost certainly going to change as a result of installing the ERP system. Compliance of data, backups, specifications and testing evidence and maintaining the validation record of the system all require attention when an ERP system influences the business organization. As the business organization changes, it is vital to ensure that the group responsible for change control is identified and given the power and resources to enforce that control.

IT Management

Administration of the database and servers needs to be managed. Reviewing the ERP databases ensures data accuracy is monitored, and the system data are updated in line with data required by the laboratory operation. The ERP servers require management. Regular backups are vital to ensure data are secured and available in the event of a server failure (e.g., hardware failure or fire). Periodic restoration of data verifies that the data are intact and confirms the restore procedure is operable. Server administration includes resolving user problems, adding and deleting users, setting user privileges and managing upgrades.

Network Security

Effective network security is important to ensure ERP data are not corrupted or accessed by unauthorized personnel. Threats can come from viruses or external sources such as vandals via the Internet. More than 10,000 viruses exist, with an estimated 6 new viruses appearing each day. Network virus protection combats this threat but requires constant updates to detect new strains (e.g., the recent Laroux virus). Installing a firewall is one of the best ways to protect networks from attacks via the Internet. Whatever the protection strategy employed, close monitoring and auditing is important to ensure the protection is effective.

Upgrading the ERP System

The ease with which hardware and software may be implemented also implies that change is straightforward. Implicit with computer systems is the rate of change of the available hardware and software. Current computer hardware and software has a life cycle of approximately two years, and upgrading of systems is now a regular feature during the lifetime of a system. Changes should be developed on a separate development system and are subject to the normal validation testing and documentation updates of the ERP System. Changes to hardware are assessed for compatibility with the intent of the ERP System prior to implementation. The System Design Specification should be updated, and the new equipment tested prior to return to service.

As part of the ongoing relationship with the supplier, the approach to upgrades should be discussed. It is important to understand how the vendor approaches upgrades in terms of frequency and for how long the supplier will support previous versions.

Contingency/Disaster Plans

Contingency Plans (sometimes referred to as Business Continuity Plans) must be in place and tested regularly, including any standby systems. It is estimated that only 15 percent of organizations have a tested and effective plan. Such plans would include details of how the system would be recovered in the event of one of several disaster scenarios. Details of a backup system would be outlined, whether an alternative system or paper data. Companies who have such plans in place run periodic tests to ensure that the plan would actually work in the event of a disaster.

Training

Training records of users must be retained along with the IT support and development staff records. Service Level Agreements (SLAs) with suppliers provide a further level of support for the system in the event of failures. If SLAs are in place, it is very important to ensure that they are monitored and reviewed.

Periodic Reviews

Periodic reviews should include an assessment of system performance, changes, authorized user access privileges, network integrity and operating procedures. Analysis equipment calibration results can be reviewed using statistical process control techniques to indicate the health of equipment.

PRACTICAL ISSUES

Data Migration and Risks

It is important to fully investigate the potential risks associated with data migration from existing systems over to new systems when they go live. On several occasions, the authors have experienced assumptions being made as to the risks; only after migration has it been realized that the structure of the data has been changed or values have been amended. This has led to very time-consuming 100 percent manual checks of the data that could have been avoided. Discussions with vendors may be another area in which the vendor-customer relationship can be used to the benefit of the project.

Resourcing

The adequate resourcing of implementation and validation is key to the success of project timescales. Often, staff are put onto project teams in addition to carrying out their normal roles within a business. On a project of this complexity, it is unlikely that they will be able to fulfill both roles adequately.

Update of Plans

It is more than likely that during the course of an ERP project, there will be changes from the original plan. Bearing in mind that the plan is considered to be the evidence of the planned activities and will be used to compare to the evidence of activities carried out, it is obvious to see that any changes to the plan must be documented. If a regulator took the plan and compared it to the resulting report, he or she should have a full overview of the project.

Documentation

As mentioned earlier, the quantity of documents produced during this type of project can number in the thousands. The organization of a document numbering and management system is essential to controlling these documents. It is often the case that document management is further complicated on multisite implementations where documents travel between sites for review and signature. Responsibilities should be clearly established from the beginning of the project for the control of these documents.

Overview and summary documents are invaluable and continue to be an area of weakness, as many FDA 483s will verify. It is also important to have a consistent approach to the documentation and standard formats to make for easier reviewing. If information is easy to locate in a document, it will be looked upon more favorably than a complex, disorganized document that requires reading from cover to cover.

Change Control

The issue of change control can be responsible for the undoing of a long and complex validation project. It is little use having documented evidence of the validation effort if there is a lack of change control within the organization. Pharmaceutical companies will usually have strict controls over changes within the production environment, but weaknesses are often found when it comes to computerized systems. Unless a full history of any changes and the consideration of potential impact are documented, a system cannot be considered to be in a validated state.

Effect on the Organization

When a site moves to an ERP system, there are issues to be resolved regarding the boundaries of data ownership. Previously, systems came under the direct control of the department which required the system (e.g., purchasing controlled the purchasing system, warehousing controlled the warehousing system). Issues such as data management were handled within the department, and the control hierarchy was clear.

With ERP systems cutting across departmental boundaries, aspects of the ERP system which affect all departments can fall between the cracks, e.g., master data. Who owns master data such as item master data, bills of materials, routings and supplier details? Who has the authority to change these data and where do the skills reside?

Ownership and responsibility for the accuracy of the data is a real issue with an enterprise wide system, and departments are no longer in direct control of data uniquely used for their own purposes. The responsibility for the specification and authorization of changes must be clearly defined.

Upgrades and Hot Patches

Upgrades, including the installation of hot patches (bug fixes), must be controlled to ensure the ERP system is maintainable into the future. Examples exist of systems implemented only a few years ago for which replacement parts and software enhancements are no longer available, let alone the relevant software skills to make your own bug fixes or developments. Choices must be made whether to upgrade in an evolutionary way, taking into account the cost of buying upgrade versions and any revalidation requirements, or to upgrade in a "big-bang" approach. Management of the ERP upgrade path is a key skill in protecting the company from threats to the ERP system's validated status and serious maintenance problems in the future. As always, proactive management is better than reactive management. Documentation is also essential as evidence of the upgrades/patches.

Electronic Records and Electronic Signatures

Since August 1997, the FDA has been enforcing 21 CFR 11 [4] in order to tackle the long debated issue of electronic signatures and electronic records. The requirements for electronic records are to ensure that they are as secure and capable of audit as paper records. It must be possible to identify valid and invalid records, with any changes followed in the correct sequence with the appropriate authorization to progress. Also, in order to be as secure as paper data, it must not be possible to make a correction which obscures the original entry, nor should it be possible to amend the audit trail.

The key requirements for electronic signatures are to ensure that the signature is as secure and reliable as a written signature. Any electronic signatures must be unique only to a single individual and must be associated with a timestamp which takes into account different time zones. Protection must exist from unauthorized copying or amendment to signatures, and any electronic signatures must appear in full when electronic documents containing them are printed. Any electronic signatures used must be a legal equivalent to a written signature.

REFERENCES

1. *London Financial Times,* 11 August 1998.

2. Wingate, G. A. S. 1997. *Validating Automated Manufacturing and Laboratory Applications: Putting Principles into Practice.* Buffalo Grove, Ill., USA: Interpharm Press, Inc.

3. UK GAMP Forum. 1998. *Supplier's Guide for Validation of Automated Systems in Pharmaceutical Manufacture,* version 3. Available From the International Society for Pharmaceutical Engineering.

4. U.S. Code of Federal Regulation, Title 21, Part 11: Electronic Signatures; Electronic Records (revised 1 April 1999).

APPENDIX 8.1: CHECKLISTS FOR MRP II VALIDATION

The following checklists give examples of the content of various documents produced during the course of the validation of an MRP II/ERP system.

Functional Specification

- Overview
  - System description
  - User requirements
- Functionality
  - The MRP II/ERP system
  - Business model overview
  - Performance
  - Access and security
  - Environment
  - Configurable items
  - Traceability
  - Data
  - Interfaces
  - Nonfunctional attributes
  - Glossary
  - References

GxP Assessments

Assessment of Business Processes

- Sales and operational planning
- Purchasing
- Inventory/warehousing
- Quality control
- Quality assurance
- Shop floor control
- Master data
- Manufacturing

Assessment of Business Procedures

- Sales and operational planning
  - Master production schedule
  - Capacity planning
  - Routings
  - Shop orders
  - Bill of materials
  - MRP
- Purchasing
  - Purchase of materials
  - Planned orders
  - Order amendments
  - Repetitive supplier scheduling
  - Schedule batch release
- Manufacturing
  - Issue of materials
  - Manual allocation of materials
  - Shop order release
  - Shop order close
- Warehouse/inventory management
  - Goods receipt
  - QC inspection
  - Movement of raw materials
  - Location creation
  - Movement of work in progress
  - Movement of finished goods
  - Returns to supplier
  - Scrapping of materials
- Quality Control/Quality Assurance
  - Release of materials to production
  - Quarantine of materials
  - Scrapping of materials
  - Testing of materials

  - Retesting of materials
  - Release of material (by qualified person)

IQ Protocol

- Methodology
- Hardware
  - Client
  - Server
  - Network
- Software
- Vendor-supplied manuals
- Operating manuals
- SOPs
- Environmental checks
  - Power supplies
  - Backup power supplies
  - Temperature
  - Humidity
- Hardware Security

OQ Protocol

- Receipt, approval and release of materials
- Release of raw materials from quarantine
- Movement of raw materials to scrap from quarantine and release status
- Movement of materials to quarantine
- Control of amendment to retest dates
- Booking in of materials against purchase orders
- Control of amendments to purchase order quantities
- Issue and control of stock items
- Movement of stock
- Return of materials to supplier
- Raw material issue
- Transfer of goods to warehouses (on-site and off-site)
- Disposal of items from scrap

- Confirm shop order correction on location transfer
- Release of product from quarantine
- Rejection of product
- Item master preparation
- Item master amendment
- Control and use of effectivity dates
- Location creation
- Location amendment
- Creation of bill of materials
- Amendment to bill of materials
- Creation and release of shop order
- Amendment of MRP
- Creation of product routing
- Amendment to product routing
- Mass replace of bill of material
- Explosion of MPS (master production schedule)
- Control of discontinue dates as the result of effectivities
- Item master maintenance
- Shop order closure
- Shop order amendment
- Shop order deletion
- Manual allocation against shop orders
- Confirm rejection of incorrect item number
- Confirm rejection of incorrect lot number
- Confirm that expired goods cannot be issued against shop orders
- Confirm that nonreleased items cannot be issued
- Confirm control of authority levels
- Confirm control of approved/unapproved vendors
- Confirm security controls, e.g., user profiles

PQ Protocol

- Change control
- User request log
- Fault/problem reporting and follow-up
- Review of procedures
- System performance

- Access control
- Backup control
- Ongoing maintenance
- Master data
- User training
- Network security
- Contingency planning
- Stock counts/cycle counting

9

Validating Laboratory Information Management Systems

Christopher Evans
Eutech
Billingham, United Kingdom

Laboratory Information Management Systems (LIMSs) are becoming widely used in the laboratories of pharmaceuticals and related industries. LIMS are typically based on client-server technology with a relational database (RDB) as a storage repository. They can be used to manage large amounts of electronic data locally, within a laboratory, or company-wide between multiple sites.

The introduction of a LIMS into the laboratory has been encouraged by Quality Assurance (QA) and Quality Control (QC) laboratory managers in order to provide an automated and efficient means of dealing with the large amount of electronic data produced in the laboratory, from analytical equipment and manual data calculations (see Figure 9.1).The LIMS will also reduce the possibility of errors due to the problems introduced by personnel performing repetitive, scheduled sampling tasks. There have been increasing demands from the regulatory authorities for data integrity, data security and validation of the LIMS in line with the current regulations [1–4].

The success of LIMS integration relies in part on a robust reliable interface between the core LIMS and a wide range of analytical interfaces (e.g., high performance liquid chromatography [HPLC] instruments, gas

Figure 9.1. LIMS Software and Data Flow

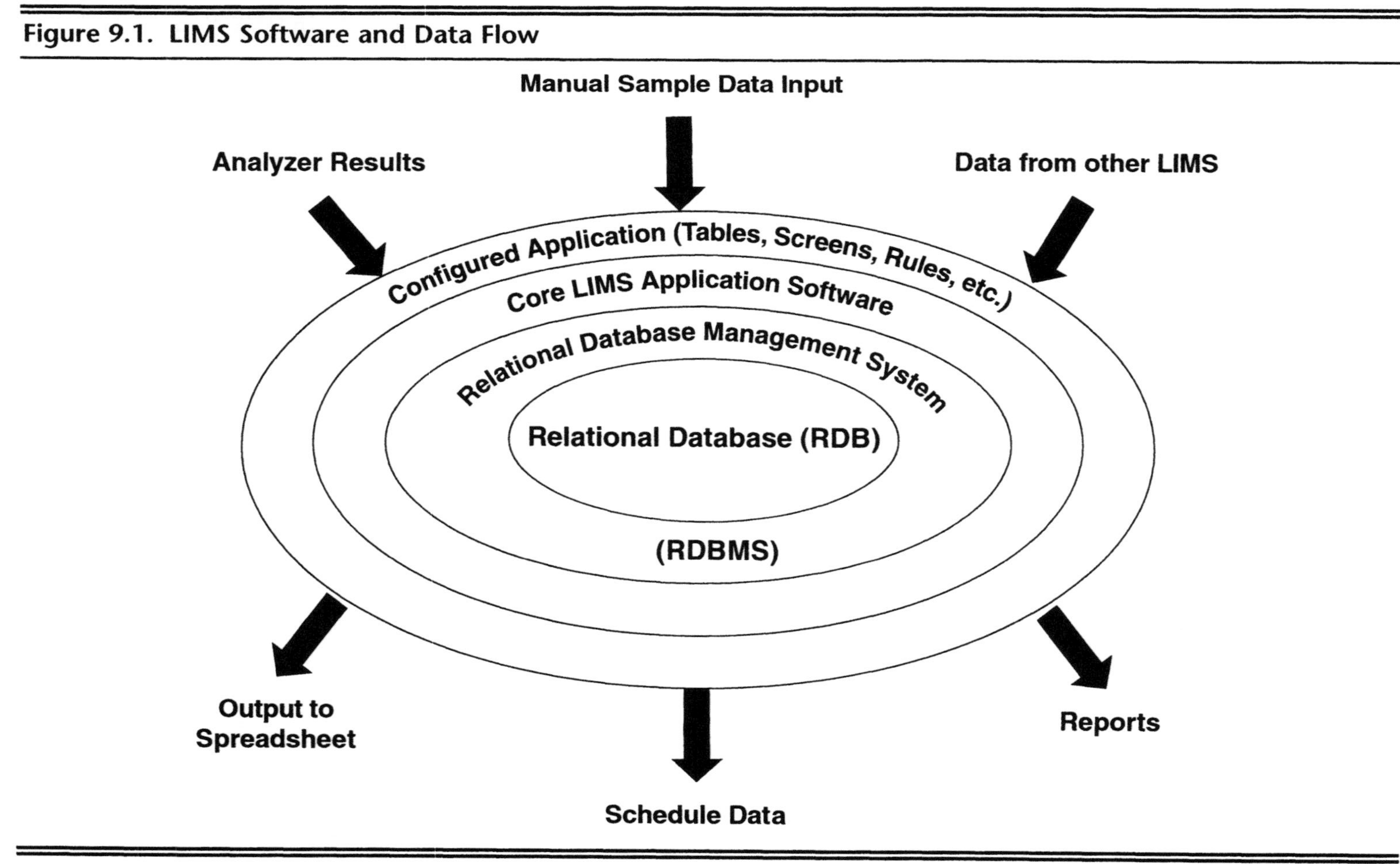

spectrometers, mass balances, etc.). This involves the interfacing of different computer platforms and the inherent difficulties in providing a validated solution as required by the pharmaceutical industry.

A major concern to the pharmaceutical manufacturer is the lack of appreciation by LIMS suppliers for the need for validation. This problem is in part caused by the fact that the suppliers of LIMSs are not dedicated to the pharmaceutical industry and, as a result, are influenced by the standards and approaches taken by other industries. It is therefore essential for the pharmaceutical manufacturer to ensure that the approach to design and testing taken by the LIMS supplier is of an appropriate standard to support the needs of validation.

The typical features of a LIMS are as follows:

- Standard interfaces with analytical equipment
- Monitoring of out-of-specification results using user-programmed limits on a per sample basis
- Import/export of data to/from spreadsheets
- Statistical process control
- Status of materials
- Tracking of samples
- Linkage of multiple samples in one test run
- Production of analyst worksheets and daily/weekly schedules
- Graphical representation of results
- Real-time data input
- Audit trail of events linked to results and laboratory analyst
- Data entry via barcode readers
- Use of modern windows-based screen interfaces
- Configurable reporting

The validation of a LIMS is a potential minefield. Without confidence in the LIMS's ability to provide data that can be used to support regulatory submissions, the system could become a liability rather than a useful tool. This chapter reviews the areas that are critical to the successful validation of the LIMS and suggests an approach which, in the opinion of the author, will lead to a regulatory compliant LIMS installation.

A BUSINESS CASE FOR A LIMS

In order to gain the necessary funding for financing the installation of a LIMS, a business case must be built around the advantages of the proposed installation. This case will normally be based around the ability of the LIMS to reduce costs and possibly manpower and to provide accurate and reliable data to support development activities submissions to the regulatory authorities. The cost of the installed LIMS will be insignificant if it aids the pharmaceutical manufacturer to speed up the route to market for a new, potentially high-earning drug product. Due to the enormous financial rewards of getting a drug to market (a top-selling drug may achieve a sales revenue of up to U.S.$2,000,000 per day), delay is not an option, especially if another manufacturer is attempting to develop a similar drug [5].

It must be clearly understood as part of this business case that it is not simply a matter of installing a LIMS and then "all our problems will be solved". Installation is just the beginning of an ongoing commitment to the maintenance of the validated status of the LIMS and the ongoing cost of that maintenance.

The Regulatory View

The business case is only viable if the principles of the LIMS are accepted by the regulatory authorities. In recent years, there have been an increasing number of citations in the area of laboratories, and the consequent scrutiny of computer systems is making the LIMS an area for regulatory concern. Typical areas of concern have include but are not limited to the following:

- Qualification and testing
- Change control
- Validation approach, quality assurance and auditing
- Electronic records and signatures
- Specification and design
- Reporting
- Standard Operating Procedures (SOPs), including security practice and backup and restoration
- Use of spreadsheets

There are a number of issues that are raised in the approach taken for installing a computer-based LIMS. Some of the issues are associated with the general approach to computing, and others are specific to the LIMS. As part of the specification and setup of the project, these issues must be addressed by using a standard project life cycle. Application of the current regulations and how the authorities interpret them is not always consistent. It is therefore important to provide a measured and well-thought-out approach to ensure that all aspects of the regulations are covered. At present, LIMSs are a target for extra scrutiny by the regulatory authorities; therefore, extra care must be taken. There is general guidance within the regulations but it is not specifically aimed at a LIMS but more toward the general automation system.

It must be clearly understood that however general the guidance, the regulators will have specific interpretations of them. It is therefore sensible to address the LIMS installation to meet the full requirements of this guidance. Table 9.1 indicates the coverage of the applicable parts of the regulations.

THE GOOD PRACTICE APPROACH TO LIMS VALIDATION

Good practice for the use of computer systems in laboratory applications is now well established and based on many years of experience. The LIMS is simply another modern use of the power and flexibility of computer systems and, therefore, may be approached in a similar manner to other systems utilized within the pharmaceutical industry. Specifically, the LIMS is deemed to be a

Table 9.1. Regulation Applicable to a LIMS

U.S. Code of Federal Regulations, Title 21, 211.68 and Part 11	U.S. Code of Federal Regulations, Title 21, Part 58	European Union Directive 91/356/EEC, Annex 11
Personnel	Personnel	Personnel
Building/facility	Facility management	Equipment siting
Equipment design	Equipment design	Validation life cycle
Equipment maintenance	Equipment maintenance	Contracted-out services
SOPs	SOPs	
		Detailed description
Record retention	Record retention	Record retention
Calibration		Software quality
Change control		Change control
Testing	Testing	Testing
Backup files		Error recording
Data storage		Data storage
Hard copy records		
Ongoing evaluation		Contingency Plan
Security		Security
Electronic records		
Electronic signatures		

category of automated systems, and it is therefore recognized that the production of a LIMS should be conducted in line with a defined life cycle such as the quality methodologies identified as Good Automated Manufacturing Practice (GAMP) [1].

Prior to the outset of the validation project, time and energy (and therefore money) must be expended in ensuring that the right project is being undertaken. This means that the real requirements for the validation of the LIMS should be assessed from a regulatory, business and safety point of view. This assessment must be executed as early as possible, preferably prior to purchasing the LIMS; many projects have had problems due to this not being done.

One of the attractive and also risky features of the implementation of a LIMS is that it can be done rapidly. This will, therefore, reduce the amount of time available for ensuring that the appropriate "quality" is built into the system, which is especially the case when implementing a replacement LIMS. The simple answer is that the pharmaceutical manufacturer must ensure that a "quality supplier" is being used for the LIMS. This is easy to state but difficult to do; there are many suppliers on the market who will advise that their systems are the best and that they are also! The pharmaceutical manufacturer must take appropriate steps such as performing a Supplier Audit, following up references and so on before any decision is made.

When there is a nonstandard (but very useful) requirement from the pharmaceutical manufacturer, there will be a desire by the supplier to include this in the LIMS offering, as it will provide future commercial advantage. This, in terms of validation, does not fit in with the old adage, "Never be the first and never be the last . . .". It is a fact that in the world of LIMSs, development is continuing apace, and a structured and controlled approach to validation is clearly required.

Validation Life-Cycle Approach

The validation life cycle is basically split into six main qualification phases as shown in Table 9.2 and all under the control of the Validation Master Plan (VMP). It must be clearly demonstrated that the QA function of the pharmaceutical manufacturer is endorsing the implemented system. In its simplest form, this will require the QA representative to authorize the documentation produced to support validation (e.g., VMP, Source Code Review, qualification documentation, reports, etc.). Table 9.2 shows the linkage between project activities and the qualification process, which is under the control of the VMP.

LIMS Validation Planning

The VMP forms a firm basis for the scope of LIMS validation. The VMP will work in tandem with the Project Quality Plan produced by the supplier of the LIMS.

Table 9.2. Qualification and Project Activities

Project Activities	Qualification Phases
• User Requirements Specification (URS) production	Validation Plan
• Supplier audit and vendor selection	
• Functional Design Specification (FDS) production	Design Qualification (DQ)
• GLP (Good Laboratory Practice) assessment	
• Program specifications	
• Source code review	
• Install hardware (computer system hardware, network, communication links)	Installation Qualification (IQ)
• Install software (system, configured, bespoke)	
• Supporting documentation	
• Structural testing	Operational Qualification (OQ)
• Functional testing	
• Static data load	
• Data verification	
• Dynamic data input	
• Operating procedures and Contingency Plan production	
• Ongoing verification of functionality	Performance Qualification (PQ)
• Validation reporting	Validation Report—authorization for use
• Ongoing support and integrity checks	Validation maintenance—operational compliance

The VMP will define a number of specific areas on which the approach to validation will be based:

- *Responsibilities of personnel/organizations:* The success of any project of this type will rely on the personnel having the right experience in terms of the LIMS and GLP. Key personnel within the pharmaceutical manufacturer's organization who will be associated with the project must be identified. These personnel will have the responsibility for authorizing validation documentation, and executing the validation activities should be within the scope of this plan.

- *Scope of validation:* The boundaries of the validation project must be defined to ensure that there is full coverage. For example, will the analytical equipment interface be validated as part of the project? Will Supplier Audits be required?
- *Standards:* In order to ensure consistency, standards must be used, for example, GAMP [1] and in-house standards and procedures. These procedures will typically cover production of validation documentation, change control, contingency planning, and so on.
- *Validation approach:* The methodology to be used for LIMS validation in the project must be clearly defined in the VMP to ensure that the most efficient and cost-effective approach is taken. This record should provide the project team and the LIMS supplier with sufficient detail to allow the appropriate documentary evidence to be produced so that the supplier can hand over a compliant system. It is important that this document is issued to the LIMS supplier to ensure that there are no surprises regarding the validation requirements at a later stage in the project.

Where the LIMS project is split over a number of different locations (e.g., different laboratories or different sites) but is designed to provide an integrated solution, it may be appropriate to produce individual Validation Plans and associated documentation for each of the LIMS installations. The VMP will then act as an umbrella document for the whole project, with each Validation Plan being identified. In a similar fashion, the documentation for the local installation may be split into the core LIMS and analyzers (see Figure 9.2). Using this method to break down the documentation will allow a much more focused approach to be taken in the production of documentation and will also simplify the problems normally found in obtaining document authorization. The advantage of using individual Validation Plans is that the individual LIMS may be put into productive local operation when the installation is complete, without having to wait for the final Validation Report covering the whole LIMS. Also, when periodic reviews are performed on the LIMS, this may be performed on the local LIMS without needing to review the whole system.

Due to the fact that the VMP will be produced at such an early stage, it will inevitably encompass only the current understanding of availability of personnel and project scope, so this document will need to be further developed during the project. The project team and LIMS supplier must appreciate that this is a live document that will be developed during the project and will continue to be alive until the LIMS is decommissioned.

The LIMS supplier's response to the Validation Plan will be the Project Quality Plan. This document is the LIMS supplier's response to the pharmaceutical manufacturer's Validation Master Plan (Figure 9.3) and will provide the following details:

Figure 9.2. LIMS Physical Layout

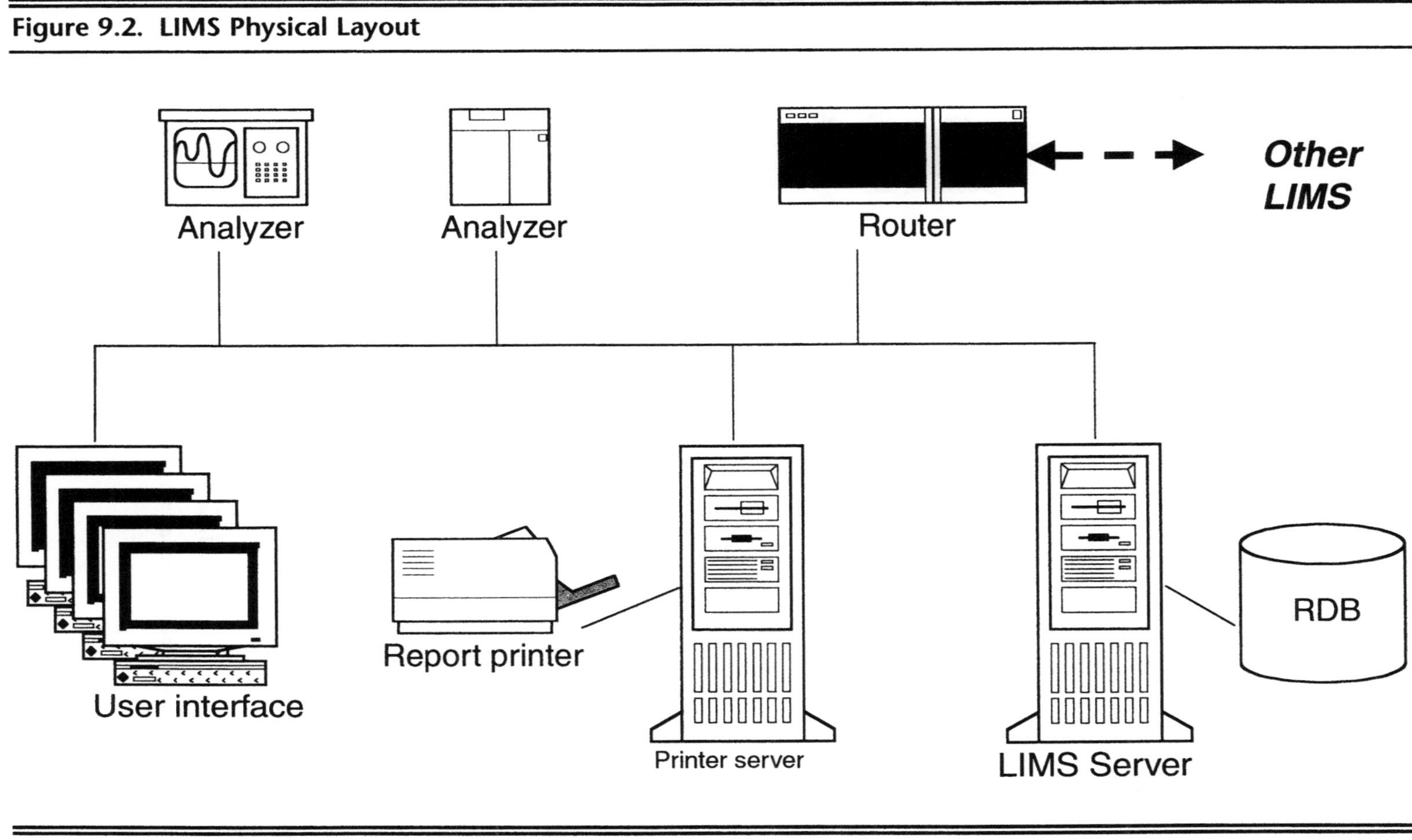

Figure 9.3. LIMS Validation Documentation

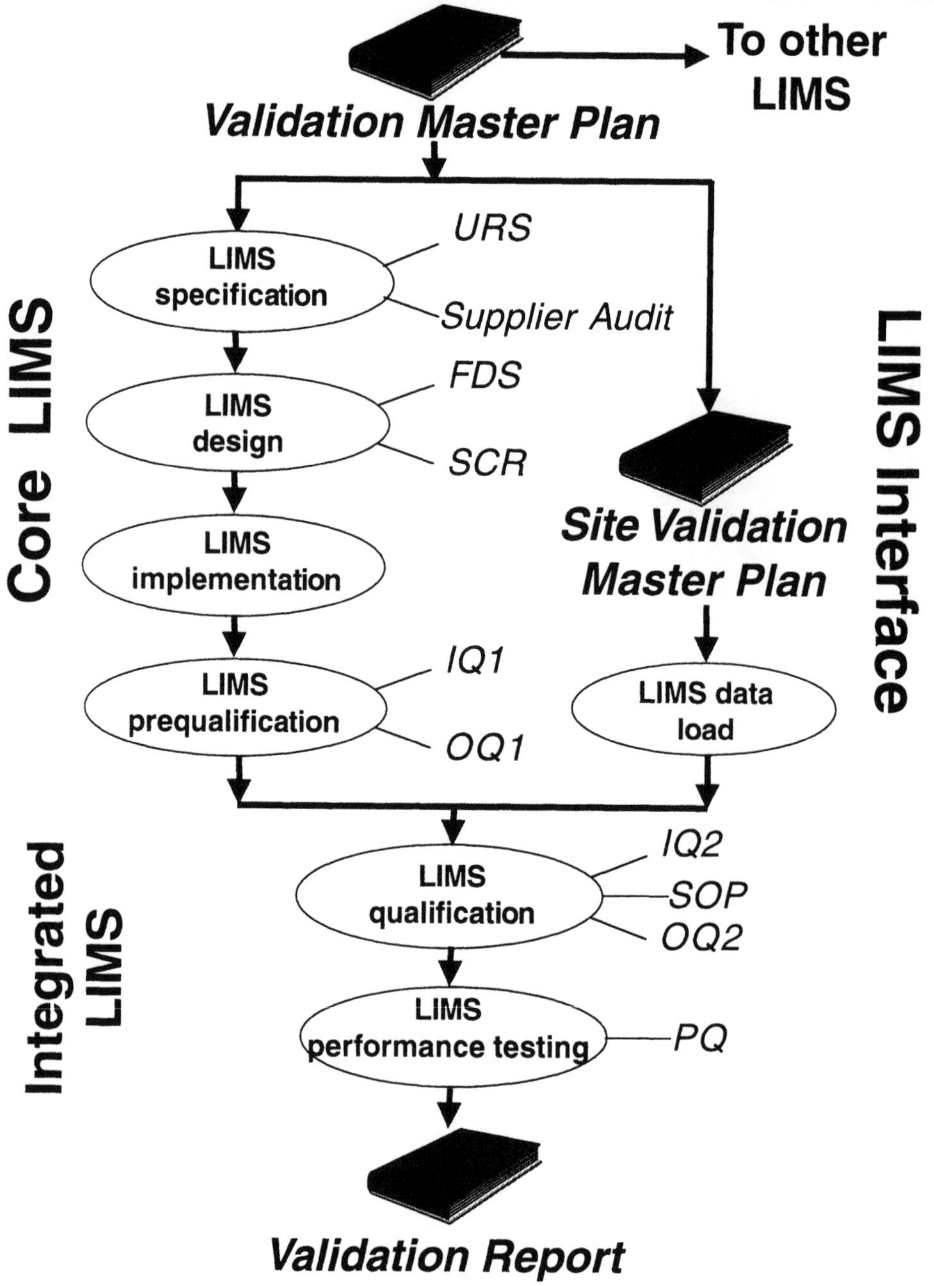

- Organization, roles and responsibilities (internal and external)
- Approach (life-cycle activities and methodologies, e.g., use of GAMP)
- Quality standards and procedures (internal and external)
- Review points
- Communication channels
- Key deliverables (including documentation)
- Project milestones
- Personnel training

Good Laboratory Practice Assessment

It is part of the early decision-making process to assess which parts of the LIMS are subject to full validation and which are covered by normal engineering testing. The location of the LIMS within the company computing architecture may mean that it is utilized for GLP and non-GLP activities. This means that there is a need to assess the criticality of the data in the LIMS database as well as the sources of the data to ensure that the validation effort is targeted at the appropriate part of the system. In the ideal world, the LIMS would be dedicated to GLP activities, however, the economics of the system may mean that it is not practical to separate GLP and non-GLP activities into separate systems.

In order to ensure that the software utilized on the LIMS is controlled, it is good practice to avoid installing software other than that required for the LIMS, e.g., word processors, drawing packages, E-mail on the LIMS access terminals. There may, however, be a need for some standard software packages, for example, spreadsheets (e.g., Microsoft Excel®). Such packages may be regularly updated, sometimes annually (unlike the core LIMS), in which case this must be taken into account when assessing the needs of ongoing validation maintenance.

Due to the important nature of the data stored within the LIMS, thought must also be given to the methodology to be applied to ensure database integrity. This may be achieved by copying any data, produced by the analytical equipment interface or entered by the laboratory technician, to two hard disks within one server or to separate servers. Current technology also allows for disk shadowing between systems, thus ensuring that within a very short time period the data are copied to a second remote location. The concern that is raised by this issue is that we now have two versions of the same data. Which one is the master, and how can we verify that the two databases are indeed identical and are accurately tracking each other? This may be difficult to check with "live data" and would need specific testing to demonstrate recovery following shutdowns and so on as part of the functional testing process.

The assessment of the LIMS ensures that the validation effort is focused toward those system elements with direct GLP impact, such as critical analytical equipment interfaces and the components crucial for the accuracy of data. The GLP assessment will also identify the key elements of the calibration regime.

Typical GLP critical data elements include the following:

- Batch number
- Item number
- Shelf life
- Tester identification
- Material specification
- Sample number
- Sample date
- Sample time
- Sample status
- Sample methodology
- Study reference
- Retest days

User Requirements Specification

The purpose of the URS is to collate the user requirements for how the LIMS application is to operate in both the laboratory and company environment. There may, of course, be areas in the URS that are wishes rather than essentials and also some requirements that are GLP critical. Following review, the first version of the URS will be used for obtaining quotations and proposals from LIMS suppliers, and, as a result, this version will act as the foundation for the project. Time spent in ensuring that the URS is complete, structured and clearly understandable will reap large savings of time and effort (and therefore costs) at later stages of the project and will also reduce the risk of project failure. It is unfortunately human nature to wish to spend as short a time as possible on this phase of the project, as it does not seem to produce anything but documentation, when the users want to see something more tangible.

Production of a meaningful URS is not an easy matter. The pharmaceutical manufacturer will need advice from the LIMS supplier regarding what the current technology is capable of as well as determining that the specific requirements of the user can be implemented in the LIMS. It is unfortunately far too often the case that the LIMS salesman will promise the pharmaceutical manufacturer whatever is asked for, when a feature is not currently available but could be in a future release of the software. It is therefore essential at this early stage in the project to build a relationship with the LIMS supplier to use the knowledge of technical personnel in the development of the URS. For example, the development of operational features such as data entry, sample results verification, tracability of data, reporting of results and so on has all been done before, so why reinvent those particular wheels? Using existing designs will not only reduce the innovative aspects of the system to a degree but will also reduce the risk of problems later in the project.

The approach of bringing the supplier on board at this stage will assist in the structuring of the URS; the supplier will know what is needed in the URS in order to provide a meaningful proposal for a technical solution and how the design documentation is to be structured. It is not common for a close alliance to be established in these types of projects; the potential for a successful project makes putting some effort in this direction worthwhile.

The structuring of the URS to uniquely identify the numerous user requirements will subsequently allow the LIMS supplier to provide similarly structured design and test documentation. This structure will allow the pharmaceutical manufacturer to simply demonstrate tracability of each of the requirements (GLP and non-GLP) through the design and testing process.

The agreed URS should clearly identify the following:

- The number and types of pieces of analytical equipment to be integrated into the LIMS (e.g., gas chromatographs, sampling systems, mass spectrometers, etc.). There will also be a requirement to define the means of communication with these pieces of equipment; this communication link may be typically via RS232 or a local area network (LAN).
- System size and capacity (e.g., numbers of client-server desktop user interfaces, printers, I/O [input/output] count, data storage capacity, etc.).
- Integration of the LIMS with other systems (e.g., MRP II [Manufacturing Resource Planning] systems, typically via a LAN or wide area networks [WAN]).
- Requirements for data presentation, records and reports.
- System performance and availability targets.
- Maintenance and calibration requirements (may require redundancy).
- Documentation requirements (e.g., engineering drawings, user manuals, software listings, database structure, etc.).
- Details of required data manipulation (e.g., calculations).
- Requirements for the use of spreadsheets for the collation or manipulation of data.

The URS is the responsibility of the pharmaceutical manufacturer as the end user; however, it is highly unlikely that the manufacturer will have LIMS experts who can be dedicated to the project. It is strongly recommended that the future LIMS users provide their input to ensure that there is buy-in for the design and functionality of the final solution. This applies in particular to the client-server desktop user interfaces (e.g., display graphics, user entry screens) and reporting mechanisms, which are very important to the user. Client-server desktops are discussed in Chapter 11.

Where expertise on design is required in the form of LIMS suppliers or consultants, these personnel should be included on the project team and work closely with the users. Using the right people at this time will allow the wish list of future LIMS users to be converted into a functional design document that will form the ideal basis for a successful project. When the URS has been agreed, it will act as the cornerstone for the DQ process. Any change to the URS will have ramifications throughout the rest of the project; the later any changes are introduced into the URS, the more significant the effect will be on project cost and timescales.

LIMS Supplier Audit

Due to the expanding nature of the LIMS market, there are a growing number of suppliers moving into the field. If a pharmaceutical manufacturer proposes to use a new supplier or a new LIMS product, it makes a great deal of sense to conduct a Supplier Audit of the proposed LIMS supplier. If it has been some time since a supplier has been audited (e.g., one year or more), it is recommended that a follow-up audit be conducted to confirm that the supplier's Quality Management System (QMS) continues to be followed. It is important that this audit be conducted prior to placing an order for the LIMS, as there may be some concerns regarding the ability of the supplier to produce a quality product.

The purpose of the Supplier Audit is to allow the pharmaceutical manufacturer to review documented evidence of the supplier's QMS. It is also essential to verify that this QMS has been used for projects involving the LIMS that is being considered for the current project, preferably including references to the types of analytical equipment interfaces to be implemented as part of the integrated LIMS. The Supplier Audit will also confirm that the supplier is capable of delivering the correct standard of software engineering and documentation for the LIMS.

It has been shown many times that just because the LIMS supplier is ISO 9000 accredited does not mean the supplier is a quality supplier of software products. This is because the LIMS supplier is often accredited for the implementation and integration of LIMS, rather than the development of software. In these cases, the only way of gaining confidence in the approach to be taken in the software development, testing and change control processes is to visit the LIMS supplier's premises. Where a supplier is accredited to ISO 9000–3 (TickIT), this will give the manufacturer greater confidence that a defined life cycle is followed because it is specifically aligned with the production of software.

The Supplier Audit Report is the documentary evidence approved by the pharmaceutical manufacturer, which identifies any issues raised during the audit and provides recommendations for any corrective actions. It is expected that these corrective actions will be implemented as part of the approval of the LIMS supplier for the project.

The pharmaceutical manufacturer is accountable to the regulatory authorities for any deficiencies in the chosen supplier's capability. The pharmaceutical manufacturer is expected to bridge any gap in standards by assigning its own staff and/or employing consultants to assist the supplier in the validation of the LIMS. As this could increase the validation costs of the project, it is worth taking the supplier's validation credentials into account as part of the selection process. In some cases, it may be appropriate to consider using another company to provide the validation expertise; this has the advantage of demonstrating independence from the LIMS supplier.

Chapters 14 and 15 discuss Supplier Auditing for software, hardware and system integrators in more detail.

Design Qualification

The approach to DQ may be in the form of a DQ Protocol which is formally completed to record the assessment of the various design activities, or as a process which will produce a set of review reports that are then collated as a record. Which of these two approaches is taken is not important; what is essential is that a formally documented approach is followed. A DQ Protocol has been successfully used many times by the author and provides a condensed formal record to support the qualification process, but it does not remove the need for producing reports for software code reviews (SCRs), GLP assessments and so on. It is important to note that the VMP or Validation Plan must clearly document the method to be used and, therefore, the documentation to be produced to support validation.

During design evaluation, a key requirement is evaluation of the system from a maintenance point of view. For example, if the LIMS utilizes analytical equipment interfaces which must be isolated for maintenance, the LIMS must be designed to allow the maintenance activities without there being any effect on the operation of the rest of the system.

The DQ process begins as soon as the URS has been produced and then effectively continues until the implementation of the LIMS in terms of hardware, software and database configuration. The objective of the DQ is to ensure that a high-quality system is ultimately installed, thus giving the pharmaceutical manufacturer a system in which there is a high level of confidence. Quality can only be built into the design and implementation of the LIMS; it cannot be tested in afterward. DQ should therefore be considered as the development of the project, following which the project will move into the testing phase where testing is performed. If the project does not take the opportunity to apply a DQ, there will be an expensive and time-consuming exercise to redesign the LIMS if, as often happens, things go wrong.

Rigorous testing has the potential for finding fundamental faults with the design of the LIMS, which may mean there is a requirement for redesign at a late stage in the project. The resulting redesign may negate design reviews that have already be performed and also have a knock-on effect on any testing that has already been performed. It should be pointed out at this stage that if the pharmaceutical manufacturer feels a need for rigorous testing to provide confidence in the system, then perhaps there was a lack of confidence in the design process.

The whole point of applying a life-cycle approach (see Figure 9.4) as defined in GAMP [1] is to provide a structured, controlled and confidence-building approach to the development and validation of the LIMS and associated analytical equipment interfaces. The purpose of all these activities prior to the implementation of the LIMS is to rigorously assess and verify that the design will meet with the requirements of the users and also the needs of the regulatory authorities. As a result of this process, in-depth testing can be focused where it is needed, thus reducing the overall amount of testing required.

There is a direct relationship between the quality of the documentation produced during this qualification process and the quality of the finally installed system. Clear, concise and accurate documentation will allow the project team to perform the appropriate reviews and allow the generation of suitable test documentation.

Design Approach for Data Integrity and Data Management

The correctly designed, computer-based LIMS will offer a more robust and accurate means of identifying out-of-specification results than can be achieved by a human laboratory technician. With the additional ability to trend, collate and report results, the LIMS has become an important tool within the laboratory environment. The integrity of the data is frequently scrutinized by the regulatory authorities and is often found to be an area of weakness.

Performing statistical analysis on the data maintained within the LIMS is a normal user requirement; the speed with which sophisticated calculations may be performed and the reliability of the LIMS to produce date-time-and-batch-marked results will speed up the production of Study Reports.

Data management in terms of archiving and retrieving of records is an essential feature in the modern laboratory; there are many issues with paper systems, which are alleviated if not removed altogether through the use of computerized data. The facility to search and select data is far more rapid and reliable than a paper-based system. With the increase in regulatory scrutiny of electronic records, it may ultimately become a requirement that data be available in electronic form to support submissions. Other advantages of a

Figure 9.4. LIMS Validation Life Cycle

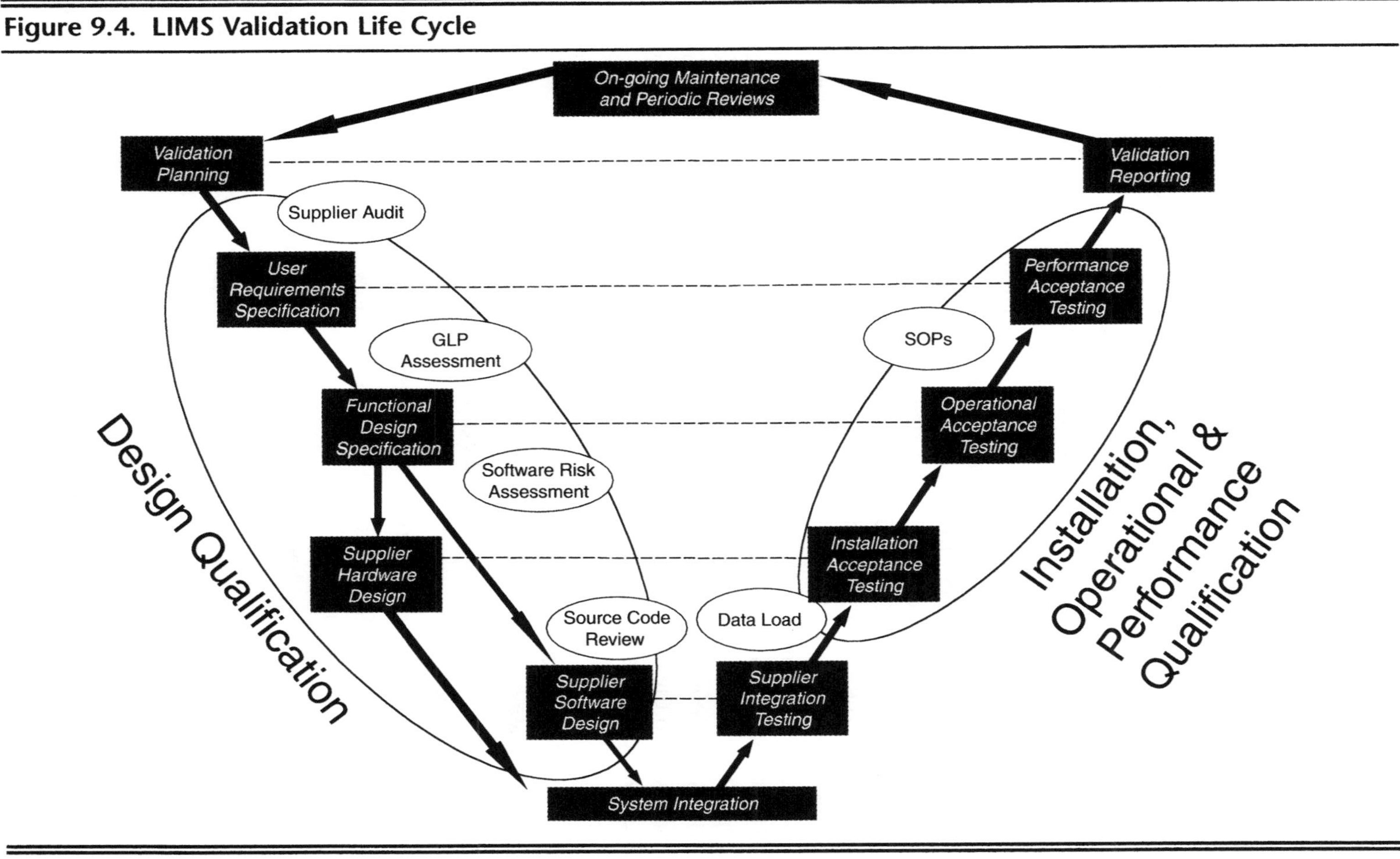

LIMS include integration benefits, allowing various systems to be integrated, and using data interchange to give rapid and consistent reporting of laboratory results. This integration may be between laboratories on the same site or across sites.

Design Specifications

Although the pharmaceutical manufacturer has determined its requirements and recorded them in the URS, by definition these requirements will be generic statements rather than being technically specific to any particular LIMS. The purpose of the design specifications, therefore, is to translate this URS into a proposed technical solution that will fulfill the requirements of the project. The structure of the design documentation, detailing the physical, functional and performance criteria for the LIMS, should be directly traceable back to the URS, to allow the manufacturer to confirm that all requirements have been addressed. This may be achieved by using a cross-reference matrix that identifies each user requirement and references the location in the design documentation where the requirement has been addressed and also holds justification for any requirements excluded or not to be fully implemented. The design documentation will usually consist of an FDS to provide a high-level overview of the proposed LIMS. A more detailed specification for hardware and software, the Hardware Design Specification (HDS) and the Software Design Specification (SDS), are prepared if it is necessary due to the complexity of design. The design documentation should typically specify the following:

- System overview description with diagrams (a concise description covering all interconnected systems)
- System architecture with details of hardware with diagrams
- System interfaces with diagrams (including networks and remote devices)
- List of installed software packages with details of versions and so on
- System performance (e.g., timing, memory storage, availability and spare capacity)
- Software flow diagrams (e.g., Grafset, Yourdon, state transition diagrams, etc.)
- Communications and network protocols (e.g., RS232, TCP/IP [Transmission Control Protocol/Internet Protocol] Ethernet)
- Methodology for data storage and retrieval
- High-level design of the database structure
- Analytical equipment interface, computer and network systems design

- Site environmental conditions, including power supplies
- Client-server desktop user interface software configuration details and terminal emulation requirements
- Peripheral devices (e.g., client-server desktop user interface computers, printers and backup devices)
- Number of users and response time required from the LIMS
- Functional descriptions of LIMS operations
- Contingency Plans, maintenance and operating procedures
- Application software and reports
- Security and access control methods
- Standard software products (e.g., Excel®)

Each of the functions within the FDS should have a unique identifier for tracability in the testing phase of the project. This may be achieved by using the outline function of the word processor used to create the document. The functions identified should be in sufficient detail to allow the LIMS supplier to produce meaningful tests, which will be recorded in the factory and site test documentation.

Where appropriate, the FDS will be supported by more detailed design documentation, such as the SDS and the HDS, and in the case where bespoke software is produced, the Software Module Design Specification (SMDS). These support documents or the FDS itself must provide sufficient detail to allow the core LIMS and analyzer supplier to fully define its respective systems and hence provide suitable test documentation.

Risk Assessment

The risk assessment is a Hazard and Operability Study that considers the requirements for compliance testing for the computer system software and hardware. This assessment is a crucial part of the assessment of the LIMS and will provide guidance for what testing is required and the amount and depth of testing that is appropriate to meet the current regulatory requirements. The results of a risk assessment will also assist in assessing the design and for later contingency planning. For the risk assessment to be valuable to the project, it must be performed in a structured manner that is formally documented. The pharmaceutical manufacturer currently executes risk assessment with assistance from the LIMS supplier.

The general approach to a risk assessment is to identify all of the functions that are performed by the LIMS hardware and software (e.g., manual data entry, automated data entry, report generation) and to assess each of the functions against the effect that they could have on the data that will support regulatory submissions. This approach must be methodical to ensure that functionality is not overlooked. At this stage, it is not appropriate to look at the fine details of how the function is implemented in software, this should be left until the review of the source code.

The general definition of a risk in this context is "a measure of the probability and severity of undesired effects", and more specifically for LIMS software, the critical functions and how they may affect LIMS data. The criticality of the data associated with electronic records and signatures is also a key subject for the regulators at present; this subject is discussed later in this chapter. The intention of the risk assessment is to consider the following areas:

- Analysis of LIMS system criticality:
  - Identification of GLP–critical operations
  - Identification of critical operating limits and parameters
- Review of design to determine potential risk limitation:
  - Consider suitability of LIMS design
  - Consider proposed system hardware in terms of reliability (e.g., failure rates of interface)
  - Consider proposed system services in terms of reliability (e.g., power supplies)
  - Determine requirements for procedural and software/hardware controls
- Future maintenance of the LIMS:
  - Procedural controls
  - Maintenance activities

Each of the risks should be assessed from two angles: (1) the likelihood of the function failing and (2) assessment of the GLP criticality of the function. A scoring system is used to classify the risk; the higher the rating of the risk, the greater will be the need for stress testing the function. It must be accepted that the assessment process can be subjective; therefore, it relies heavily on the experience of the risk assessors.

Source Code Review

Following the development of the software by the LIMS supplier, the software may be subject to a source core review. As part of the process of building QA into the development process, the developed software should be reviewed against the agreed design and Good Programming Practice (GPP). The software applicable to this process according to GAMP [1] is software that is configured or bespoke, with particular emphasis on the software deemed GLP critical by the GLP assessment process. If the Supplier Audit resulted in a low degree of confidence in the supplier's software development process for the core LIMS, then it may be necessary to recommend that the supplier conduct an Source Code Review of the core product.

For the core LIMS, one must be careful not to be lulled into a false sense of security by the supplier's claim that the software is a standard product. What this usually means is that a set of software modules has been used in at least one or perhaps a few other LIMS installations. In this case, we must pose the question, "Is it really standard or in fact bespoke?" The approach that should be taken to these "standard modules" is to determine if they are indeed standard (i.e., they are not revised to fit your LIMS application and have multiple applications for which references are available). If they are not standard, then confirm that the supplier has performed a Source Code Review or a suitable design review. If no review has been performed, the pharmaceutical manufacturer will need to make a value judgment on the risk this imposes (taking into account the results of the GLP assessment) and, if necessary, request that the supplier perform a Source Code Review as part of the project.

Modules are sometimes written specifically for the LIMS because of the need to implement a new feature or to enhance an existing feature. In these cases, there will be a requirement for a Source Code Review to be performed on the resulting software, once again placing emphasis on GLP–critical modules.

In order to implement the LIMS, there will be a need to configure the interfaces between the individual software modules and their data sources/sinks (e.g., client-server desktop user interfaces, database locations, printers, etc.) and also the configuration required for the hardware implementation and interfacing with other equipment. It makes a great deal of sense to inspect configuration data and the database structures together, as it is easier to spot typographical errors and genuine mistakes than to find them via the testing process.

The Source Code Review process will consist of assessing the agreed-on design against printouts of the produced software and data for any bespoke or configured part of the application in order to assess the application's GPP. The aims of the Source Code Review are to

- identify logic errors within the code, with particular emphasis on alarm and I/O handling;
- identify any redundant code (and where appropriate recommend removal);
- identify any dead code (and where appropriate recommend removal);
- verify that the software has been written in accordance with the project-agreed standard;
- verify that the code contains sufficient, meaningful comments to ensure that it can be maintained by a competent software engineer;
- verify that critical algorithms are correct;

- verify the use of version control and change control within the modules, thus verifying that there is tracability of changes; and
- confirm that software listings utilizing this process are complete and accurate.

The above list is applicable for both the core LIMS software and the software driving the individual analytical equipment interfaces.

Following the completion of the Source Code Review, a review report or a number of review reports will be produced to summarize the review findings. These reports may identify actions to be taken by the LIMS supplier, the analytical manufacturer, or the personnel responsible for performing the validation of the LIMS. These actions may range from identifying specific testing that will need to be included in the testing process to revisions of the source code due to deviations from GPP or to correct errors.

TESTING AND QUALIFICATION OF THE LIMS

Supplier Testing Process

Testing normally begins with software module and software/hardware integration testing. It is not normal for the pharmaceutical manufacturer to become involved in this low-level testing unless there have been particular issues raised in the Supplier Audit. It is important, however, that the LIMS and analytical system suppliers formally document this testing process and any inspections as part of Good Engineering Practice (GEP).

Following the satisfactory completion of module and integration testing, the LIMS supplier will be ready to demonstrate system functionality to the pharmaceutical manufacturer. This demonstration will be a form of unit/module/system integration testing of the software on the installed LIMS hardware. The LIMS at this time is still likely to be a development system rather than the final version. This will be the first time that the pharmaceutical manufacturer has received a demonstration of the complete core LIMS or at least a version that will demonstrate the required functionality. The LIMS supplier demonstrating the physical makeup of the core LIMS as well as its functionality will use integration testing (as far as can be achieved within the constraints of the development environment). Due to the complexity of demonstrating a large core LIMS and the logistics of arranging for all of the analytical equipment interfaces to be present at this time, it may well be the case that some simulation of data inputs or connections to other systems may be used. As the LIMS will be made up of the core system and a number of analytical equipment interfaces, it is likely that the analytical equipment will be subject to its own qualification testing process.

The results of these tests will provide the first information for specific reference in the IQ Protocol, for example, version/identification details of the software, identification details of the hardware and a list of any outstanding issues that were raised during the testing. This link will provide tracability through the testing process and will ensure that any failures and outstanding issues are addressed before the end of qualification. A more radical approach would be to use the integration testing as a form of qualification documentation. This means that the testing documentation will become referenced in the IQ/OQ Protocol, which will remove the need for doing general testing when the LIMS is fully commissioned. Consequently, this will reduce the number of tests present in the IQ/OQ Protocols. If such an approach is to be taken, the integration testing must be formally reviewed and approved by competent personnel, and the pharmaceutical manufacturer must also ensure that the completed integration testing documentation is of a quality suitable to support validation. The integration testing should be witnessed; at a minium, the results should be reviewed by the pharmaceutical manufacturer, and a summary report produced.

The summary report will identify any issues raised during integration testing, and the pharmaceutical manufacturer will need to review this information to determine if the LIMS is fit for purpose and therefore suitable for IQ. Confirmation of acceptance of the LIMS will normally be given only if all but minor issues have been resolved.

Installation Qualification

Installation Qualification, as its name suggests, is a process designed to confirm the integrity of the LIMS installation against the design, agreed as part of DQ. IQ should cover all aspects of the hardware, software, documentation, environment and infrastructure for the installed LIMS.

Clearly, before IQ can commence, the component parts of the LIMS must be installed in their final location and the appropriate integration/interconnections completed. At the point that IQ is executed, it is essentially indicating that there is no further need to modify the physical system.

Attempts are often made to "rush into" starting IQ, with the consequential problem of test failure (e.g., the incorrect version of software is installed, parts of the LIMS are still missing, etc.). This not only has the effect of introducing retests but also does not give the regulator who inspects the system confidence that the LIMS has been designed and installed by a quality organization. Because the intention is that IQ will verify that the installation is whole, no failures should be experienced. As there is ample opportunity for a preinspection to be performed by the LIMS installers, it is in fact the case that validation personnel are simply rechecking and recording.

It is essential that the integrated project team be aware of the contents of the IQ Protocol to be used, which will ensure that there are no surprises when this execution takes place.

The LIMS supplier is responsible for all construction and installation activities as defined in the Project Quality Plan. The IQ Protocol will provide coverage of the LIMS hardware and software and all analytical equipment interface hardware and software integrated into the LIMS. The environment into which the LIMS and analytical equipment interface is installed may need to be controlled in terms of temperature/humidity, electrical interference and so on. In addition, the provision of services will need to be assessed, for example, electrical supplies and earthing. The IQ Protocol will typically cover the items listed in Table 9.3.

IQ will also provide the vehicle for confirming that the DQ process has been successfully completed. It is essential that all installation and infrastructure issues be resolved prior to the completion of IQ. It is therefore useful to provide a test within the IQ Protocol which looks back at the results of the DQ process and integration testing and assesses the effects of these issues on the compliance of the installed system.

In order to ensure that the qualification process of the entire LIMS is made as efficient as possible, it is sensible to provide separate IQ documentation for the analytical equipment interface. This will mean that if there is a problem with one analyzer, there will be no delay in the validation associated with the core LIMS. This approach will also be an advantage when a periodic review of the LIMS is performed, as it will allow the review to be broken down into the system components, which will then be individually assessed.

Installation Qualification Report

Following the execution of the IQ tests, there may be issues noted that the LIMS has not been installed in full accordance with the design as agreed at the end of the DQ. It is true to state that there are often issues due to documentation/drawings, environment, interfaces, installation or even missing parts of the LIMS. These will result in failures in the IQ tests, as the LIMS does not comply with the requirements specifically noted in the IQ Protocol acceptance criteria. In these cases, each failure will be recorded in the IQ Report, and a justification made for accepting that the project may be deferred until a later phase or accept that the issue is not to be resolved if this is appropriate. For example, in some cases, the failure may have been caused by typographical errors in the test script or some genuine failure of the test due to a hardware fault, which has subsequently been repaired. Whatever the reason, the project team will need to assess the situation and determine an appropriate course of action. If the failure does not affect operational testing (e.g., a

Table 9.3. The IQ Protocol

Subject	Core LIMS	Analytical Equipment Interface
Software versions for operating systems, utility software, application software, compilers (bespoke and configuration)	✓	✓
Licenses for core software and layered software products	✓	✓
Hardware platform details with unique identification (e.g., serial numbers)	✓	✓
Labeling of hardware platform equipment (including interface cabinets)	✓	✓
Diagnostic self-test for analytical equipment interface	✓	✓
Network compatibility of peripherals (printers, PCs, etc.)	✓	✓
Power-up/power-down tests	✓	✓
Installation of services, e.g., power supplies and earthing	✓	✓
Installation of internal wiring and marking of cabling for maintenance; confirmation that no disconnected wiring is present	✓	✓
Computer room environment testing (temperature, humidity, radio-frequency interference, electromagnetic interference	✓	
Documentation to support the installation (e.g., cable schedules, layouts of cabinets, etc.)	✓	
Network connections to LAN/WAN	✓	
Security access testing	✓	
Calibration certificates for connected instruments		✓

drawing is incorrect), this could be corrected in parallel with operational testing. On the other hand, if the wrong version of the software has been installed, it takes little common sense to work out that this must be corrected prior to moving on to the next phase.

The IQ Report is therefore a milestone in the project that records the project's acceptance that any outstanding issues are of a nature that will not affect the integrity of operational testing, and, therefore, IQ is complete.

LIMS Operational Qualification

When the pharmaceutical manufacturer is confident that the installation of the LIMS has been satisfactorily completed, the project will move on to the stage where the LIMS supplier will provide a demonstration of the functionality of the "final system". OQ is the vehicle for providing documentary evidence of the demonstrated LIMS functionality for the independent parts of the LIMS prior to full integration with analytical equipment. There is also a need at the beginning of this phase to review any issues raised at the end of DQ and IQ. Any issues that will have an effect on operational testing must be resolved prior to the commencement of OQ.

OQ consists of a series of tests that will be based on the FDS of the LIMS and should include reference to the functional tests performed by the LIMS supplier as part of integration testing wherever possible. This may be achieved by utilizing the integration testing documents, which are then used as raw data for OQ, with the OQ including any other specific tests to demonstrate compliance that are not within the scope of the LIMS supplier (e.g., environment tests, internal procedures, maintenance plans, etc.). This will mean that there is no need for duplication of testing and makes the whole process more efficient in terms of time and cost.

Analytical Equipment Interface Testing

As with IQ, there will be a need for the validators of the analytical equipment interface to perform functional testing of the analytical equipment independently of the core LIMS supplier. Only when the analysis equipment has been demonstrated to function correctly should the equipment be connected as a live data source for the core LIMS.

Following connection of the analytical equipment to the core LIMS, formal verification of data values within the core LIMS database, screen displays and reports may be performed. The vehicle for this testing will be the second stage OQ Protocol.

The OQ documents must cover all GLP–relevant functions in sufficient detail, to provide pharmaceutical manufacturer users with a high level of confidence that the LIMS operates in accordance with the agreed design. There should also be challenge testing applied to the LIMS to attempt to stress the system and, therefore, assuming the tests pass, build further user confidence in the installation. It must be accepted that if the pharmaceutical manufacturer is not fully confident of the LIMS at this stage, there is a serious risk that the LIMS will never be satisfactorily implemented and used.

It is recommended that the structure of the OQ should match that of the FDS to provide a simple mechanism for demonstrating that all of the functions of the FDS have been tested. Each OQ test should also contain references to the functions being demonstrated. If the structures of the documents cannot be linked, it will be necessary to provide some form of cross-reference document that will provide this information. The OQ may have one or more functional test scripts for testing each of the functions that are uniquely identified in the FDS. The OQ typically includes the items given in Table 9.4.

It is essential that any checks on inputs to the LIMS from the analytical equipment interfaces be verified. Where test equipment is used, calibration certificates (referencing national standards) must be referred to in the test documentation.

At the OQ stage, there will be a need for the pharmaceutical manufacturer to ensure that the appropriate management systems in terms of procedures and Contingency Plans have been assessed and confirmed as suitable. It may be that as part of the OQ process draft versions of the procedures and Contingency Plans are developed, with the revised documentation being issued following the completion of OQ.

Operational Qualification Report

Following the execution of the OQ tests, there may be issues noted (in the OQ Report) that the LIMS does not function in accordance with the design as agreed at the end of the DQ. As for the IQ Report, the project team will need to review these failures and determine a plan of action or a justification for moving on to the next phase.

Standard Operating Procedures

Qualifying analytical methodologies is a key step in the qualification process. This process will be designed to ensure that the SOPs written to support daily activities (e.g., performing analytical operations on the equipment, calibrations of the equipment, cleaning of the equipment, etc.) are complete and accurate. The operation of the LIMS and, in particular, the analytical equipment interface will need to be specifically documented in the SOPs to act as an aide memoir to experienced users and a training aid to the less experienced. For this reason, the SOPs should be written by personnel knowledgeable in the low-level detail of the LIMS and should be detailed enough for the user to work without reference to other personnel. By using the OQ as a means of trying out test SOPs, any issues of detail should be identified. The first version of the SOPs must be authorized and issued prior to the start of PQ.

Performance Qualification

PQ documents that the integrated LIMS core system and analytical equipment interfaces perform effectively and reproducibly across all of the design operating ranges, and the data management functions meet the specification

Table 9.4. The OQ Protocol

Subject	Core LIMS	Analytical Equipment Interface
Special configuration functions	✓	✓
Testing of bespoke software	✓	✓
Signal diagnostics from linked analytical equipment interface	✓	✓
Special calculations and algorithms	✓	
Verify operating manuals, including challenge testing	✓	✓
Verify sample data and system parameters against source records to ensure the accuracy of data within the RDB (involves checking of data loaded manually or via automated upload from legacy systems)	✓	✓
Software backups and restoration of data	✓	✓
Training records of users	✓	✓
Routine maintenance/calibration routines	✓	✓
Provision of SLAs	✓	✓
Data upload and migration checks	✓	✓
Data integrity checks, e.g., range checks, validation of inputs	✓	✓
Communication driver tests	✓	
Database structure and population	✓	
Disk shadowing demonstration (where fitted)	✓	
Archive and retrieval of documents and records	✓	
User results input and displays	✓	
Analytical report generation	✓	
Audit trail verification	✓	
Demonstrate features supporting the use of electronic records and signatures	✓	

and quality attributes specified in the URS. As with the previous phases, the first part of the PQ will be to determine that there are no outstanding issues from the IQ/OQ which need to be addressed prior to the start of PQ.

The length of time over which the LIMS will be subjected to this PQ testing is typically 4–6 weeks, but each system will obviously need to be assessed to determine if this is an appropriate period. There is an issue regarding PQ that it is only undertaken by the pharmaceutical manufacturer following the handover of the LIMS from the supplier. As this is perhaps the most important phase (i.e., the last one before the LIMS goes live), it is essential that the project team, including the LIMS supplier, are still available in case of problems.

Performance Qualification Report

At the end of PQ, the PQ Report is produced which summarizes the continued operation of the LIMS for the initial period following going live. During this period, the LIMS is expected to have stabilized in terms of user support requests, system performance, changes implemented as a result of the qualification process and network security/integrity. Networks are discussed in detail in Chapter 12.

LIMS Validation Report

The Validation Report for the installed and tested LIMS reviews the results of each of the preceding validation phases. This report will act as a summary of the overall validation status of the entire LIMS. There may have been Validation Reports associated with individual items of analytical equipment and perhaps the core LIMS itself. The overall Validation Report should cover the following:

- Summary of the results of each of the validation activities
- A summary of outstanding issues associated with the analytical equipment interfaces with their references in the change control system
- A summary of outstanding issues associated with the core LIMS with their references in the change control system
- Timescales for future periodic reviews
- Details of the justifications from the pharmaceutical manufacturer for any deviations from the original VMP

The Validation Report is the document that will be utilized in the first periodic review to confirm that any actions recorded have been successfully addressed.

ONGOING OPERATIONAL COMPLIANCE

Maintenance of the LIMS validation status requires a suitable infrastructure to be in place. This infrastructure will consist of a LIMS manager and appropriate SOPs. The LIMS manager will be responsible for controlling any changes to the LIMS interface, analytical equipment interfaces, LAN/WAN architecture, LIMS functionality and the data held within the database.

Responsibilities

The LIMS manager is typically responsible for the daily administration of the entire LIMS (core database, LIMS servers, peripheral devices [e.g., printers, user PCs, etc.], networks). The manager must respond to user requests and problems in agreed-on timescales as he or she is providing a service to the laboratory. The duties include

- addressing user problems;
- adding and deleting users;
- controlling user privileges;
- managing upgrades to the core LIMS, standard software packages and operating systems; and
- managing SLAs.

As a result of the manager controlling the LIMS, the validated status will be monitored and retained during ongoing operation. Where there is a need for change as a result of component failure, upgrades or LIMS development, change control must be in place. Implementing a series of procedures and maintaining access control (both physical and electronic) to the core LIMS and analytical equipment interfaces will assist the maintenance of the validation status over the lifetime of the system.

Change Control

Change control of the LIMS hardware, software and associated documentation (SOPs and operating manuals) is necessary to prevent the system from becoming unmaintainable. A good change control system will allow the LIMS manager to determine what changes have been made to the LIMS, when they were made and what effect they had. It is not acceptable for changes to be made to LIMS functionality without the effect of the change being assessed against the current validated status and current GLP. If the change is necessary and impacts on the validated status, then appropriate revalidation must be performed. This may result in rerunning one or more tests from the IQ, OQ or PQ, or may, in the worst case, result in a rerun of part or all of the DQ.

The LIMS change control system must record

- details of the change,
- authorization for the change,
- assessment of the effects of the change on GLP,
- details of the DQ (if performed),
- date when the change was requested and the date when it was implemented and
- testing performed to verify the operation and reconfirm the validated status and details of the test results.

Where a change is required to the LIMS hardware due to the failure of a component, there are two possible scenarios: (1) The failed component is no longer available and a new design of the part must be installed. In this case, a DQ exercise will be required to assess the effect on the rest of the LIMS, followed by the normal testing approach. (2) The component is a standard offering from the supplier and is therefore a "like for like" replacement. In this case, simple testing of the functionality of the replaced component is all that is required.

The change control system will be utilized in the maintenance phase of the LIMS lifecycle; however, the level of details of the review and the rigor of the testing should be the same as was used in the original validation process. The testing must therefore be carried out by competent qualified personnel, and the records of the testing retained as part of the LIMS validation support documentation.

Upgrades to the LIMS

The rate at which hardware and software are being developed for the LIMS indicates the flexibility and therefore the ease with which changes to the existing installation may be made. The installed LIMS has a fairly short life time (i.e., 2–5 years), after which the hardware will become obsolete and the software will be redeveloped into a more advanced form. This means that the LIMS manager will be constantly inundated with the need to change to the new version of hardware or software. In some cases, the LIMS supplier SLA may be linked to the installation of hardware/software upgrades. As a consequence of these upgrades, the LIMS manager will be responsible for reviewing the validation status of the LIMS. Any change to the validation status is likely to involve documentation as well as hardware/software, each must be viewed, and the effects addressed accordingly. The approach to change should be as follows.

Software Changes

The new software is assessed for compatibility with the existing software, with particular emphasis on any changes made. In terms of system or standard software products, there is normally a "bugs fix" list and details of new, modified and removed features. These documents should be assessed to determine the effects of the change and any appropriate testing performed.

Hardware Changes

The new hardware is assessed for compatibility with the existing hardware. If any differences are identified, these differences should be assessed against the existing LIMS design intent.

Documentation Changes

Document changes will normally result from modifications to the LIMS hardware and software. The documentation supporting the maintenance of the LIMS in terms of hardware and software should be reflected in the existing documentation. Any documentation that is out of date should not be utilized for maintenance purposes.

Software Bug Fixes to the Core LIMS

Upgrades due to developments, including the installation of bug fixes (patches), must be tightly controlled by the personnel responsible for the LIMS. Following the production of the Validation Report, the LIMS is deemed to be in a validated state (subject to any issues raised in the report). It is therefore a risky business to upgrade the operating system on the LIMS or even to upgrade any layered product, as there is great potential for including "new bugs" as a consequence. The difficulty is that the LIMS supplier may insist that the upgrades/patches are necessary for licensing purposes and to allow future maintenance support of the LIMS to be provided.

Examples exist of systems implemented only a few years ago for which replacement parts and software enhancements are no longer available, let alone the relevant software skills to make your own bug fixes or developments. Choices must be made whether to upgrade in an evolutionary way, taking into account the cost of buying upgrade versions and any revalidation requirements, or to upgrade by replacing the present system. Management of the LIMS upgrade path is a key skill in protecting the integrity of the existing LIMS validated status and avoiding maintenance problems in the future. It is also important that in all of the changes described above that the QA function, within the pharmaceutical manufacturer's organization, sign off that the change has been performed in a manner that maintains the validated status of the LIMS.

LIMS Software Backup and Restoration

The core LIMS must be backed up on a regular basis to maintain the integrity of the database. Regulatory inspectors will not accept that data within the database has been lost due to the failure of the server or other incidents (e.g., fire or flood). In order to prevent losses, the pharmaceutical manufacturer is responsible for implementing a reliable and robust documented backup regime, which is normally recorded as an SOP. The frequency of backups must be assessed as part of the GLP assessment process, as this will determine the frequency at which data will be entered into the database and should take into account the risks of data loss. Automation of the backup regime is acceptable, providing that it is adequately tested during the operational testing phase of the project.

In order to demonstrate that the data backed up are available for restoration in the event of a breakdown, it will be necessary to implement a periodic test of the restoration procedure. This procedure will demonstrate that restoration of the data is indeed possible, however, care must be taken to ensure that this activity does not cause live data to be overwritten. This may mean that the data are restored to an off-line version of the LIMS database or to a different LIMS.

Data Archiving

A data archiving system is required as part of the infrastructure that will ensure the future integrity of the backed up data. This system should be documented as part of the backup and restoration SOP. It is important that it can be demonstrated to regulators that appropriate personnel (and their deputies) are managing this system. The archiving process, taking into the account the requirements of the U.S. Food and Drug Administrations (FDA) Electronic Records and Signatures regulation (21 CFR 11), must address the subject of control of GLP–critical data stored on the LIMS.

There are doubts regarding the long-term integrity of storage media (magnetic tapes, floppy disks and even compact disks) that contain records associated with pharmaceutical product development and analysis. The possibility that these media could lose data is of great concern to the regulatory authorities. It is therefore an issue that pharmaceutical manufacturers must address as part of the LIMS infrastructure implementation. Periodic restoration of data verifies that the data are intact and confirms the restore procedure is operable.

There is a further issue regarding the stored data—the obsolescence of devices that use the media and software packages used to format and store the data. Many types of storage devices, such as disk drives and tape drives, are no longer supported by modern computer equipment, which means that the data may not be accessible for presentation to the regulatory authorities in case of issues with a product. Once again, this must be taken into account when establishing the backup and restoration regime.

LIMS Calibration and Maintenance

Where analytical equipment interfaces and instrumentation are utilized as part of a LIMS, the approach to calibration and maintenance will require formal documentation in calibration and maintenance SOPs. Regular calibration is required (the frequency of which must be in accordance with the manufacturer's recommendations) and must cover all the measurements within the operating range of the process. The calibration records must be traceable back to national standards and retained as ongoing evidence of the control of these pieces of equipment.

All analytical equipment interfaces and instrumentation must be installed in a manner which facilitates easy maintenance, adjustment and cleaning (where necessary). In addition, it is necessary to ensure that the replacement of component parts of the equipment is controlled as part of the change control process for the overall LIMS. The identification of spare parts (including consumables) should also be identified as part of the project and included in the ongoing LIMS maintenance strategy.

LIMS Security

The implementation of an effective security regime is very important to ensure that unauthorized personnel do not access the LIMS data. Threats can come from external sources, for example, "hackers" via the Internet/networks, and from viruses.

Control of Security

The pharmaceutical manufacturer will be responsible for providing maintenance of the security aspects of the LIMS. This may take the form of protection due to physical restrictions (locked up or restricted areas) or software protection, for example, passwords to log-on accounts. The management of this function should be in accordance with a formal SOP. The use of passwords and high-level accounts must be strictly controlled to prevent security breaches. Typical examples of control are as follows:

- No more than four high-level users on any major LIMS.
- No shared user accounts.
- No shared user identifiers.
- QA should authorize all users.

Computer Viruses

Thousands of viruses already exist, with more being created every day. Viruses are no longer just transmitted by users using unauthorized floppy disks; they have been introduced by major software suppliers on their installation disks, macros that are part of Microsoft Word® files and via E-mail messages. Continued vigilance is required to prevent the spread of these viruses. Network and local virus protection combats the threat of viruses but requires constant updates to detect new strains. There are in excess of 42,000 viruses, trojans and variants that are known, but the real number may be well in excess of this.

Security Breaches

In order to protect the computer system from the "hacker", the pharmaceutical manufacturer cannot simply rely on passwords to prevent access to the networks; a more sophisticated approach is required. Installing a firewall is one of the most secure means of protecting networks from infiltration via the Internet.

Electronic Records and Electronic Signatures

The FDA is presently training its inspectors in the approach for dealing with the subject of electronic records and signatures. The regulation 21 CFR 11 [3] was introduced in August 1997; the FDA is now considering this topic during computer system inspections. This regulation was introduced to

- permit the use of technology that was not available when the original regulations were written,
- Use technology to improve efficiency and speed up the regulatory process,
- provide equivalence between electronic and paper records,
- reduce costs of storage space by reducing the need for paper records,
- reduce costs for data transmission to the FDA and
- improve efficiency of FDA reviews and approvals of regulated products.

Electronic Records

An electronic record is any combination of text, pictures, data, audio or information represented in digital form that is created, modified, maintained, archived, retrieved or distributed on computer systems. These records were formerly printed, and the printed data became the master copy. The purpose of this regulation is to turn the emphasis toward the electronic version being the master. It must be possible to identify when, where and by whom changes are made to a record. This means that it must not be possible to make a correction that obscures the original entry nor should it be possible to amend the audit trail.

Electronic Signatures

The key requirement for electronic signatures is to ensure that they are as secure and reliable as written signatures. Electronic signatures are the legal equivalent of a written signature. Any electronic signatures must be unique to a single individual and associated with a time stamp which takes into account different time zones. Protection must be provided from unauthorized copying or amendment to signatures. Electronic signatures must appear in full when documents stored in electronic format containing the signatures are printed.

It should be noted that the regulators expect this rule to be applied to both new and existing LIMSs. The 21 CFR 11 regulation is discussed in more detail in Chapter 16.

Contingency Planning

Contingency Plans define the controls that minimize the impact of temporary and long-term loss of all or part of the LIMS. The extent of contingency planning will be determined by the criticality of the LIMS with respect to the GLP operations for which it controls or monitors. Contingency, in the form of stand-by systems and components, should be considered during the development phase for highly critical systems. However, it is also vital that Contingency Plans are established that assume the inevitable—what can go wrong will go wrong. Contingency Plans will define the requirements for system archive, periodic backup, restoration procedures and SLAs. Additionally, Contingency Plans must address the method of system and data recovery, defining the manual operations that may need to be applied in the interim until the LIMS is reinstated.

Training

Training of LIMS users is a key issue that is likely to be the subject of a regulatory inspection. It is therefore essential that a training program be organized as part of the project and maintained as part of the ongoing maintenance of the LIMS. The training records of the LIMS users must be retained along with those of the LIMS support and in-house development personnel. The assessment of training not only applies to the pharmaceutical manufacturer's personnel but also to the LIMS supplier's personnel and validation personnel if they are independent of the LIMS supplier. As part of the Supplier Audit, the pharmaceutical manufacturer must assure that competent, trained personnel are utilized on the project. It is recommended that evidence of this training be provided for reference in the project validation documentation, perhaps in the form of staff resumes or copies of training records.

Periodic Reviews

The validation integrity of the LIMS must be periodically reviewed to ensure that ongoing support systems are effectively in control of the system. The review process is designed to identify trends that may indicate noncompliance with support procedures or weaknesses in the original validation exercise. The review should further examine the original test data sets to determine their applicability to the current computer system configuration and duty. The review should determine if there is a need for further validation of the current LIMS installation.

Periodic reviews typically include assessment of the following:

- System performance
- Maintenance records
- Equipment calibration
- Change control records
- Access privileges
- Network integrity
- SOPs
- Fault reports
- Supplier Audit follow-up

CONCLUSION

This chapter has reviewed one approach that may be taken to the validation of a typical LIMS that will meet the expectations of the pharmaceutical manufacturer and the regulatory authorities. If the LIMS is to be utilized to support research and development or manufacturing laboratories, then the full system must be subjected to validation. It is recommended that a life-cycle approach be adopted for validation based on GAMP [1].

One of the first activities in the validation process is the GLP assessment, which will identify the GLP–critical aspects of the LIMS. These aspects must be fully assessed, designed, tested and documented using a process with a clear audit trail. The GLP assessment will allow the pharmaceutical manufacturer to ensure that effort is concentrated where it is most needed. It is essential that the personnel involved in the validation process are knowledgeable and experienced in the validation of LIMS. This may mean that there is a need to ensure that the LIMS supplier can provide suitable validation personnel or that independent specialists are used. The cost of the system is not the only concern when choosing a new LIMS. If the LIMS supplier is not able to validate the system properly, it will be useless to the pharmaceutical manufacturer. Once a validation project is complete, the system must be under the control of the LIMS manager who will be responsible for ongoing operational compliance. Ongoing operational compliance not only covers the issues of maintaining the validation documentation up to date and managing all aspects of change control but also refresher training for existing personnel and ensuring that any new personnel are fully trained.

REFERENCES

1. UK GAMP Forum. 1998. *Supplier's Guide for Validation of Automated Systems in Pharmaceutical Manufacture,* Version 3. Available from International Society for Pharmaceutical Engineering.

2. European Union. 1992. Good Manufacturing Practice for Medicinal Products in the European Community, Annex 11: Computerized Systems.

3. U.S. Code of Federal Regulations, Title 21, Part 11: Electronic Signatures; Electronic Records (revised 1 April 1999).

4. U.S. Code of Federal Regulations, Title 21, Part 58: Good Laboratory Practice for Nonclinical Laboratory Studies (revised 1 April 1999).

5. Wingate, G. A. S. 1997. *Validating Automated Manufacturing and Laboratory Applications: Putting Principles into Practice.* Buffalo Grove, Ill., USA: Interpharm Press, Inc.

10

Validating Electronic Document Management Systems

Roger Dean
Richard Mitchell
Pfizer Limited
Sandwich, Kent, United Kingdom

The Shorter Oxford English Dictionary [1] defines the word *document* as "that which serves to show or prove something". In the pharmaceutical industry, documentation is an essential component of the way business is conducted. Examples include submissions to regulatory authorities to gain approval to market a drug; in manufacturing operations, documentation provides instructions on how to make a drug and subsequently how it was made, the resulting quality and how it was distributed.

Other documentation which is not directly applicable to Good Manufacturing Practice (GMP) is also very important. Regulations governing safety at work and impact on the surrounding environment also require documentation to demonstrate assessment of risk, working practices and results.

Documentation is a valuable asset to a company, as it can contain information that may not be recoverable if lost; even if it is possible to recover, it is likely to cost a considerable sum to restore the information to a usable condition.

Traditionally, documentation has meant paper. GMP has, when been incorrectly applied, quite rightly become a euphemism for "great mounds of paper". Even when correctly applied, the result is, inevitably, a large amount of documentation (ergo paper) to serve as evidence of properly conducted work commensurate with the requirements of GMP. For GMP documentation, systems have been developed to manage paper and ensure that the right pieces get to the right place at the right time. These have been generally based on the multicopy approach, with controlled copies being distributed to known locations and being withdrawn as required. This is expensive and time-consuming, and paper can be easily lost or damaged. The problems associated with the management of paper can be overcome by implementing an Electronic Document Management System (EDMS).

WHAT IS AN EDMS?

EDMSs control and retain documents from creation to archiving and all stages in between. Thus, a word processing package used to prepare a document for use in its paper form would not be part of an EDMS. However, if the same package is integral to a system in which a document is created, reviewed, approved, viewed, superseded and archived, then it is part of an EDMS. It would be wrong to restrict the term *document* to the output of a word processing package, as documents can contain a variety of formats, including diagrams, pictures, spreadsheets and the like.

A quick search of the Internet will identify a number of EDMS providers. Some require a lot of customization to present a package that will suit the requirements of your business. Others are particularly geared toward the pharmaceutical industry and have built in much more of the functionality required to meet the GMP regulations. However, it is almost certain that some customization will be required, but the quantity is rapidly decreasing with time as the EDMS providers understand and implement GMP requirements.

The needs of the organization will determine the type of EDMS to be implemented. Systems can be local or, more commonly, distributed throughout the company to maximize the sharing of information and, hence, benefits. This distribution can be site wide or even intersite using local area networks (LANs) and/or wide area networks (WANs). The validation of such networks is covered in Chapter 12 of this book. Many systems are client-server based and may use a Web browser to provide read-only access into the system. The latter is changing with the introduction of Web-based interactive packages. Figure 10.1 provides an example of EDMS architecture for a multiuser distributed system.

EDMSs can be configured in many different ways to support the way documents are managed. Figure 10.2 shows an example life cycle for a document, such as a Standard Operating Procedure (SOP).

Figure 10.1. Example of System Architecture

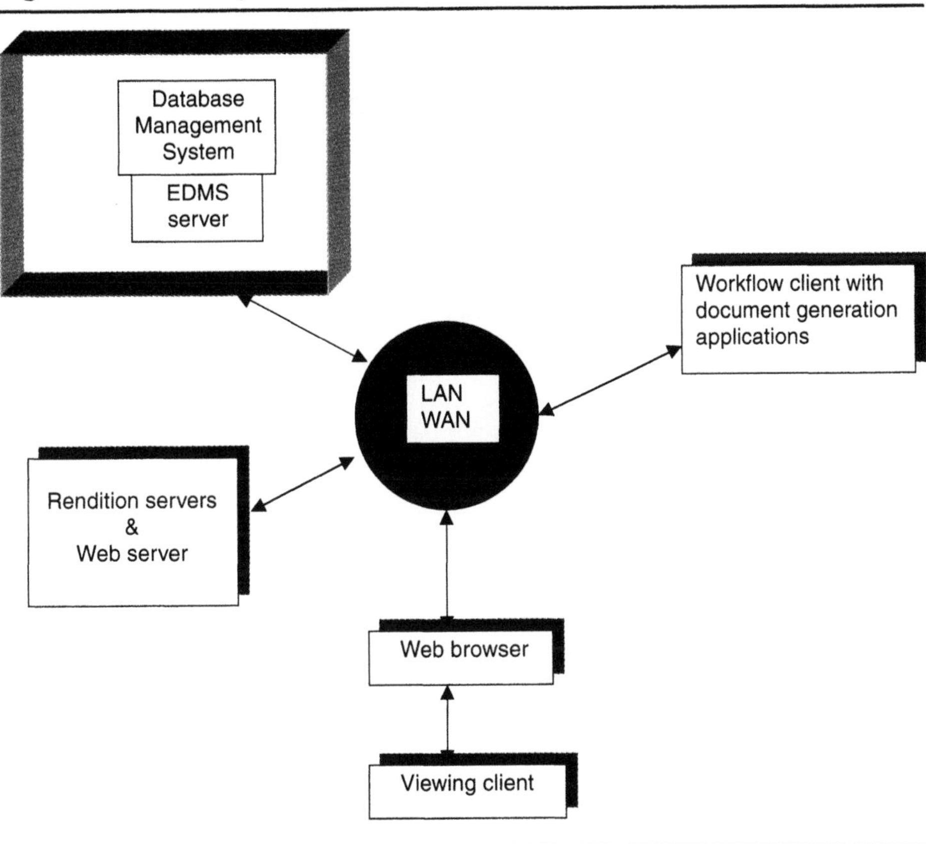

THE REGULATORY ENVIRONMENT

The regulations governing the manufacture of pharmaceuticals demand that documentation on the manufacturing and associated processes be in place. Regulatory inspections use this documentation as the primary source of evidence of compliance or otherwise. A summary of the GMP regulations involving documentation and documentation management as well as some example citations in this area can be found in *Documentation Systems: Clear and Simple* [2]. These clearly illustrate that the management of documentation is a key GMP function.

The regulatory environment applicable to electronic systems, such as an EDMS, became clearer with the publication of the U.S. Food and Drug

Figure 10.2. Example Life Cycle for a Controlled Environment

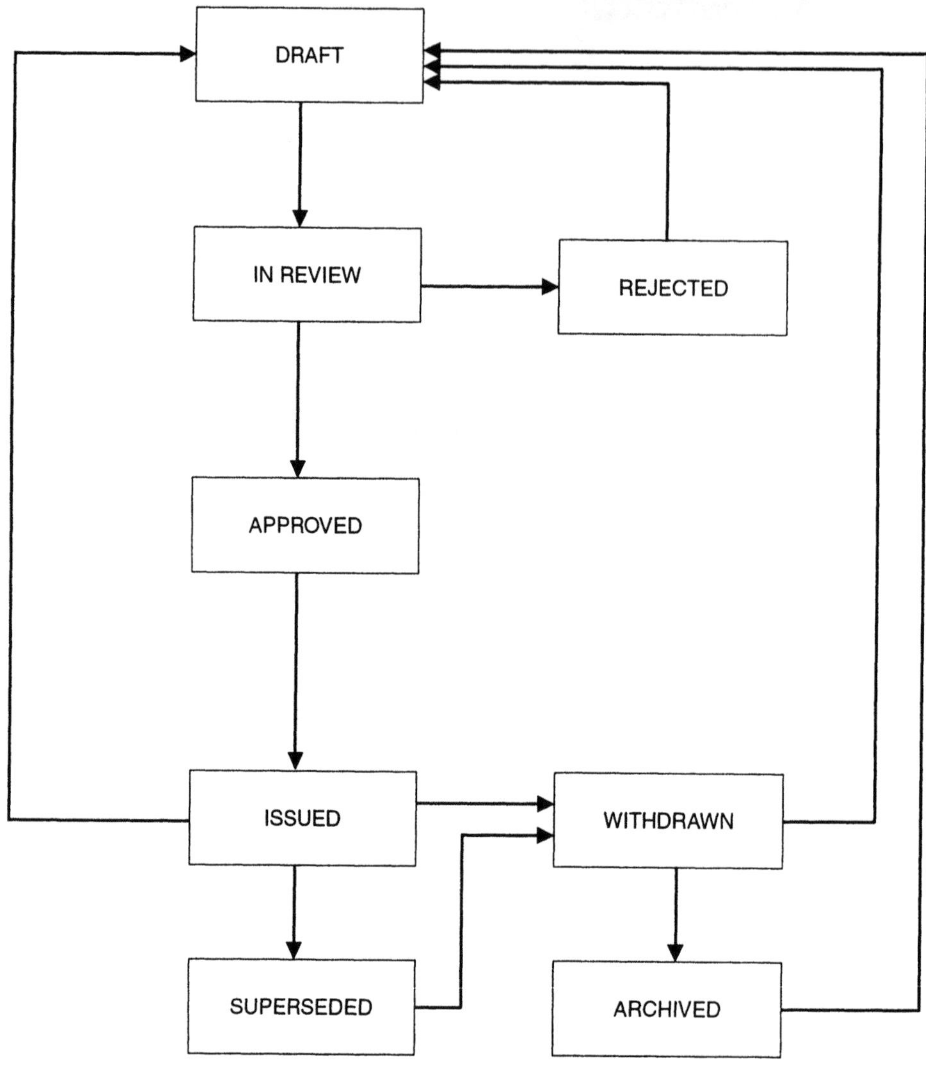

Administration (FDA) electronic records and electronic signatures rule, 21 CFR 11 [3]. Although only applicable to systems subject to inspection by the FDA, it gives guidance on what should be considered when developing and implementing an EDMS. The details of this rule are covered in Chapter 16 of this book.

IMPLEMENTATION AND VALIDATION OF AN EDMS

Implementing a site-wide or intersite EDMS is a major undertaking that requires a significant investment of time, resources and money. The opportunity should be taken to review current document management practices to determine if they are still appropriate for the business and for use with an EDMS. An EDMS offers the capability to streamline document management practices, and some lateral thinking may help to drive a positive change into the organization. Consideration should be given to the part the EDMS will play in the overall information systems strategy of the organization. Failure to do so could require costly modifications as other elements of the strategy are implemented.

One of the major factors on resources and costs when implementing a GMP–compliant EDMS is validation. It permeates all stages of the implementation process, as described below.

Project Team

The project team should consist of a core team of representatives from the key project areas. This should include Information Technology (IT) to provide expertise on the EDMS and IT infrastructure, user groups to ensure the system meets their needs, Quality Assurance (QA) and Validation to assure quality. The project team should meet regularly to review quality practices and set up regular communication sessions with the wider user base to keep them aware of all facets of the project and so that critical decisions can be made in a timely manner.

Life-Cycle Validation

The basic validation approach is no different from that applied to other information management systems, such as a Manufacturing Resource Planning (MRPII) system or a Laboratory Information Management System (LIMS). An approach based on the validation life cycle in GAMP [4] is appropriate. Figure 10.3 shows how validation documentation relates to the project activities.

User Requirements Specification

A User Requirements Specification (URS) is a very important first step because it forms the basis of the System Definition and the selection of the supplier and influences the approach to validation. It should be the mechanism by which users have the opportunity to express their needs. Dividing the needs into musts and wants provides the project team with an indication of how to weigh the requirements during supplier selection.

Figure 10.3. Project Phases with Validation Deliverables

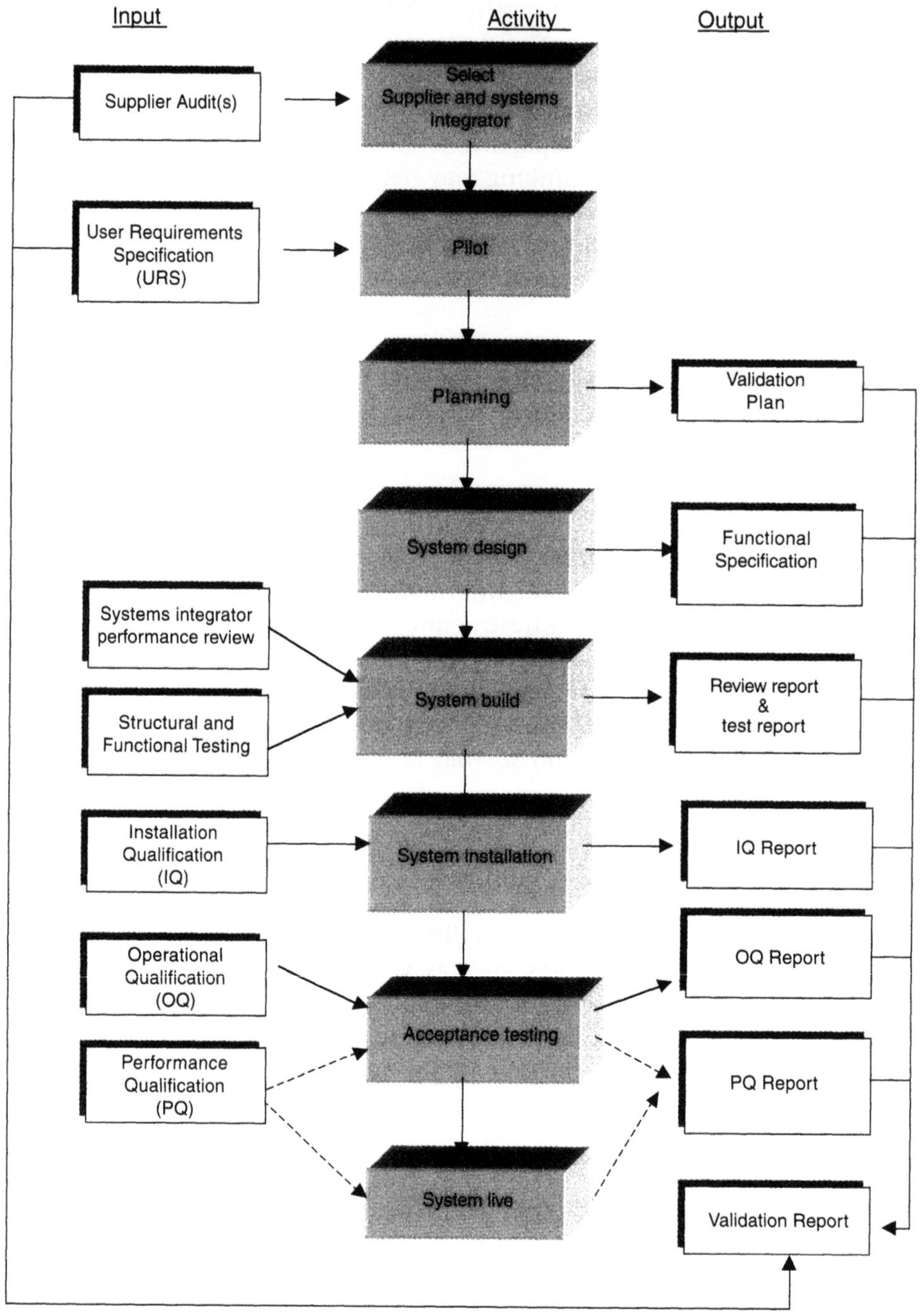

Input

Activity

Output

Supplier Audit(s)

Select Supplier and systems integrator

User Requirements Specification (URS)

Pilot

Planning

Validation Plan

System design

Functional Specification

Systems integrator performance review

Structural and Functional Testing

System build

Review report & test report

Installation Qualification (IQ)

System installation

IQ Report

Operational Qualification (OQ)

Performance Qualification (PQ)

Acceptance testing

OQ Report

PQ Report

System live

Validation Report

Examples of items covered in the URS are as follows:

- Potential number of users
- Easy-to-use viewing tools
- Easy-to-create documents in the company's standard packages, e.g., Microsoft Word®, Lotus AmiPro®
- Easy to update, withdraw or archive documents
- Easy to print or prevent the printing of controlled documents
- Speed of access to the system and document retrieval times
- Good search facilities
- Presence or absence of hypertext linking
- Human machine interfaces (HMIs) required and their design
- Interfaces to other computer systems
- Good audit trail facilities
- System availability requirements
- Impact of the existing IT Infrastructure

The URS should then form the basis of the evaluation criteria for suppliers of the EDMS.

Selection of a Supplier and Systems Integrators

In large multisite organizations, the system to be implemented may be governed by a corporate standard. This gives advantages both at the implementation stages and throughout the system's life in terms of knowledge of the product in the organization, availability of skilled implementation teams and, of course, cost.

Where a new supplier is being selected, a number of factors need to be considered, and a detailed supplier selection process may be undertaken. Evaluation may consist of gathering information from various sources, such as the supplier, companies who already have a system installed and trial demonstrations. Where possible, a fixed duration trial (pilot) should be set up on-site for the project team and user representatives to run through some scenarios of how the system may be used.

In order to validate the EDMS, a Supplier Audit is essential to determine if the system has been developed and will continue to be managed within the framework of an adequate Quality Management System (QMS) and to good software engineering standards. Auditing the suppliers of electronic systems is becoming a specialized field in its own right; however, there are a number of references that give guidance in this area [4,5]. A positive Supplier Audit gives the user confidence that the system can be validated and that once implemented, the supplier's activities will not adversely affect maintenance of the validated state. Inadequacies found in the QMS may lead the user to include special provisions in their own systems, such as incident management and change control procedures.

The supplier selection process may result in the selection of a system that requires a lot of customization in order for it to meet stated requirements and GMP. If this is the case, then a systems integration partner may be required for the implementation. The QMS of the systems integrators should also be audited to ensure that their methodology will result in a validated system.

Pilot Trial

The pilot is a short trial of the most probable supplier's system set up approximating the requirements outlined in the URS and involving a cross representation of the user community. The pilot system is, by definition, temporary and, hence, will not be built to the same standards as the actual EDMS. It is meant to convey a feeling of the system. The pilot is important as it will

- provide evidence that the system will meet the users' needs;
- allow the team to form a better understanding of what the EDMS in question will provide;
- assess possible configurations;
- identify potential pitfalls of the technology; and
- expose the users to the screen interfaces, thus allowing any problems in this area to be highlighted at an early stage.

The concept of a pilot can also be used to compare possible suppliers. However, this decision should not be taken lightly, as conducting a useful pilot requires considerable resources and time on the part of the team, plus the setting up of a pilot system by the supplier may involve a large cost.

The completion of the pilot will result in a decision to continue or abandon the EDMS project. If the decision is to continue, then it is essential that the URS be refined in light of the experience gained, both adding and deleting functionality and expanding detail as necessary. It must be remembered that these changes to the URS will affect the supplier's quotation, which must be resubmitted based on the revised URS.

Planning

Completion of the pilot is an appropriate time to prepare a Validation Plan for the project. There is a case for preparing a version of the Validation Plan at the start of the pilot, but the effort may be abortive if the project is subsequently abandoned.

Completion of the Validation Plan will usually accompany the preparation of a detailed Project Quality Plan. The quality representatives on the project team should ensure that key validation activities are included and have adequate resources assigned. For systems that need to be customized and require systems integrators, planning is very important to get the best value for money out of an expensive resource.

System Design

The initial part of the system design should be to examine the document management processes and identify possible improvements, rather than just mimic the current manual system. This can be achieved by creating user discussion groups.

The next stage is to expand the basic requirements in the URS from what is required to how the users want that functionality to look and behave. The users must be involved in this stage, both to give them ownership of the product and, more importantly, to obtain the benefit of their experience and knowledge of documentation management. This process, however, must also utilize experts on the EDMS, such as the systems integrators, to facilitate the user group discussions. The involvement of systems integrators helps the team achieve a realistic and practical set of requirements by identifying key deliverables from the system whilst advising against functionality that is unworkable or that will require significant amounts of customization or even bespoke code. Discussion groups can also highlight inconsistencies between requirements of different user groups, especially when the system is designed for used by all departments rather than a small select group, such as Documentation Management. An example of this is print protection, where one department may want free access to printing, whereas another group, normally QA, insists that the printing of certain documents must be limited. Hence, a compromise must be reached which provides the disputed functionality but in a manner that complies with GMP. In some cases, however, GMP requirements will dictate the system design with no room to provide the requested functionality. From these discussions, a detailed Functional Design Specification (FDS) can be prepared that can be compared against GMP requirements; if deemed to be acceptable, the FDS can be approved by the project team including the quality representative. The FDS must be cross-checked against the URS to ensure that it encapsulates all of the users' requirements. The creation of the FDS is very important in terms of validation, as the Operational Qualification (OQ) test scripts will be written against this document, ensuring that all implemented functionality is tested.

Example elements of an FDS include the following:

- Hardware configuration
- Software configuration
- Performance criteria and system availability
- Document database structure

- Document types supported
- Document workflows
- Viewing capability
- User group configuration
- Access control, e.g., passwords
- Audit trail
- Search features
- On-line help
- Interfaces to other systems
- Training requirements
- Maintenance functionality

System Build

The EDMS should be configured from the FDS. Wherever possible, the project team should review the system as it is being configured to ensure that it meets the business requirements. Redesign and reconfiguration (performed under change control) at this stage will significantly reduce user dissatisfaction, delays and cost compared to similar activities after the system has been implemented.

To enable the system to be developed and user tested simultaneously, it is useful to create separate EDMSs for the systems integrators to configure (a development system) and for users to try out the functionality during its development (test system). Once the development of the system reaches an implementation stage, then a validated EDMS is required. This may be the test EDMS put under strict control to prevent unauthorized changes, or, more usefully, a separate validated EDMS to ensure that the validation is carried out under controlled conditions and not affected by users "testing" the system. Upon completion of the OQ, the live EDMS can be created and implemented after conducting user acceptance testing to ensure it behaves as per the validation system. The provision of several EDMSs requires a lot of space on the platform. As such, it may not be possible in all cases. However, it is essential to have an EDMS in addition to the live one in order to correct and validate any faults found without endangering the integrity of the live database and to remove the need to take the live system out of use. The above process requires strict software/configuration version control to ensure that the various systems are using the appropriate version.

It is also important to audit the integrators during the configuration stage to ensure that they are complying with their own QMS. This gives the added assurance of good practices being adhered to and that the system is validatable.

Comprehensive structural testing should be carried out by the systems integrators. If this is carried out properly, the number of faults found during OQ should be significantly reduced. This testing should be documented.

Installation Qualification

Installation Qualification (IQ) is a phased process for these systems. Prior to OQ, the hardware platform must be qualified against the specified design. This is no different from standard IQ, and checks on the following may be included:

- The installed hardware
- The installed operating system software
- Environmental conditions
- Support procedures
- Maintenance agreements

The second phase is the installation of the configured software. This is also little different from a standard IQ for information management software and may include checks on the following:

- Installed core software
- Installed configuration
- Installed bespoke code (if appropriate)
- Software licenses
- Support procedures
- Maintenance agreements

This should put the system in a suitable state for OQ to commence.

At the completion of testing and when the system is ready to go live, a final IQ is required. This consists of moving the tested EDMS to the production system (if differing EDMSs are used) and rolling it out to the users. Rolling the EDMS out usually requires some activities, such as setting up desktop icons for connecting to the system, setting up user passwords and so on. A user IQ should be performed for each client by starting the system and accessing a known test document from each user station to ensure that the installation has been set up correctly. This should all be recorded.

If the development and validation have been carried out on a different server from that to be used for the live EDMS, then the installation of the software on the live server must be validated by migration testing. If the servers are not identical, then the full OQ should be repeated on the "live" server. However, the two servers are usually identical. In this case, it is only necessary to perform a subset of the OQ to ensure that the system works as intended. This obviously also applies when upgrading servers once the system is live.

Acceptance Testing

As part of any project implementation, the systems integrators will perform unit and structural testing. The more effort that can be put into ensuring that the system works as designed, the easier user testing becomes.

The user's first formal phase of acceptance testing is the OQ. All functionality as defined in the FDS should be tested to demonstrate that the system is fit for purpose. It is important that the OQ tests the functionality as a whole rather than just checking that isolated modules behave correctly. This will involve taking a document through all workflows from start to finish; if the workflow is a continuous cycle, then at least two cycles should be tested. It is equally important to identify and test for functionality that should *not* happen as well as checking that the system works as expected (e.g., in a workflow that involves parallel review followed by approval, the test should check that the document is not forwarded to the approver until all the parallel reviews have been completed). Unless the users have been provided with a test EDMS, this will be the first time the full system is available to the project team, and so there may be a tendency to try and make improvements to the system as it is tested. The project team must be clear whether such enhancements, or in the systems integrators language "functionality creep", will be implemented with the associated cost and delays or whether only major concerns, such as noncompliant GMP functionality, will be corrected. If faults are found during OQ, it is better to complete the full protocol, if possible, in order that all required changes are identified and resolved prior to running the OQ again.

There are differing schools of thought on whether Performance Qualification (PQ) is performed before the system goes live or afterward. In the former, PQ may consist of testing the system in a "live" environment with a restricted user base but using the system as envisaged when rolled out to all users. Alternatively, PQ may be used to assess the system after it goes live, checking system attributes that cannot easily be tested as part of the OQ. Testing here may include

- system availability, including ability to log on and access documents;
- system access times and document retrieval times with the full user base, network traffic and expected number of concurrent users;
- performance of the server;
- review of training;
- ability of the users to use the system;
- number of incidents and change requests; and
- password management.

This type of qualification may be termed *ongoing assessment.*

A summary report should be prepared after each stage of acceptance testing, highlighting outstanding issues, their criticality to the project and assigning responsibilities with a timescale for completion.

User Procedures

A validated system must have written procedures that have been formally reviewed, approved and issued. These procedures should be reviewed by someone from the intended user base who has an appropriate level of expertise in document management and has been trained on the EDMS.

Database Population

Early in the project, it must be decided whether existing documents will be imported into the EDMS. If the decision is to bring documents into the system, there are a number of ways of doing it. For example, either the electronic files can be imported into the EDMS, or hard copies of the document can be scanned and the resulting file imported. Generally, it will be necessary to employ a mixture of methods, particularly where old documents on obsolete word processing packages are involved or where not all of the electronic files are available. For GMP–critical documents, a validation program should be established to ensure that the version of the document in the EDMS is a true representation of the regulated document. For electronic files, it is possible that they have not been managed to the same standard as that for the the paper system. Care must be taken to ensure that the correct document, i.e., the *current* approved and issued version, has been imported into the EDMS and that the file has not been corrupted or changed. For scanned images, the validation of the document in the EDMS should check that

- it is the *current* approved and issued document,
- all the pages of the document are present and in the right order and orientation,
- there are no erroneous pages and
- the image is legible.

The importance of this exercise cannot be over emphasized. If the system contains incorrect information, there could be GMP compliance issues. From a practical perspective, users quickly become disillusioned with systems if they cannot rely on the information they contain. Hence, it is important that the information is also maintained during the implementation phase to ensure that any documentation updated in the hard-copy system is also updated in the EDMS.

In addition to validating each individual imported document, a check should be made to ensure that all required documents have been imported. Failure to do this could result in critical documents missing from the EDMS. As part of this final check, the documents should also be checked to see that the EDMS contains the current version of all the documents in question in case documents have been updated since importation. On completion of this validation step in a full life-cycle system, the management of the documents in question should then be transferred to the EDMS.

Training

Acceptance and the continued use of a system is reliant on the perception of the user base of its usefulness. Training is key to helping users have a positive impression and ensure that they know how to use the functionality that they require in their job function. All training must be documented.

The timing of user training will depend on whether all users will use the system immediately it goes live or whether there will be a phased roll out of users. The former obviously demands that all user training be completed prior to implementation of the live system, whereas the latter means that each individual user must be trained before being allowed access to the live system. Training should use the procedures that will be available for the system. This not only checks the procedures to determine if they are correct and that they are easy to follow but also familiarizes the users with them. If the user base is large, it may be useful to train a group of people who can provide on-the-job support to their colleagues.

Training is also required for the administrators of the system so that the EDMS can be maintained. This is usually provided by the system vendors.

Validation Summary Report

At the completion of implementation, a Validation Report is required to summarize all of the validation activities. It should summarize the outcome of each step identified in the Validation Plan and review the progress of any outstanding actions from the IQ, OQ and PQ task reports. There may also be issues from the Supplier Audit or the review of compliance of the systems integrators against their own QMS that may need to be assessed. The report should then assign a validation status. It should also set a time for when a quality review of the system should be conducted. This is usually one year; if the project is being implemented in phases, the report may defer assigning a review date until all of the subsequent phases have been completed.

MAINTAINING THE VALIDATED STATE

Getting to a validated state requires significant expenditure of time and money. As well as being required for regulatory compliance, it makes good business sense to retain the system under control. A formal set of procedures and systems is required, including the following:

- *Change control:* A system that manages change—changes to hardware, version changes of the core software or local configuration changes—is critical to maintaining control of the system. The identification of categories where changes can impact the system can help to decide the degree of revalidation required.
- *Access security:* Control of access to the hardware, software and to the system via the user interface is very important. Access to system administration functionality should be controlled, particularly where a user performs significant events, such as the creation or modification of user accounts.
- *Incident reporting mechanism:* An easy-to-use system for users to report unexpected events with the system is an important monitoring tool.
- *Version control:* In modern systems, the interface software, including customizations, is often on the client or user's desk and, hence, is open to the possibility that the wrong version of the software is installed, e.g., not updated during an upgrade operation, or there is interference by the user. Some automatic means should be found to check that the correct software is installed and preventing use if this is not the case.
- *Contingency Plans:* In the event of system unavailability, Contingency Plans are particularly important if the EDMS manages the instructions on how to make product. Paper copies of the instructions may need to be held with some mechanism to prove that they are official copies and are true representations of those that are held on the system.
- *A Disaster Recovery Plan:* is required in the event of a major failure to the server or other crucial elements of the system. A risk assessment should be performed to determine the criticality of the system to the business. The higher the degree of risk, the more comprehensive the plans should be to quickly restore the system. This adds to the cost. Disk mirroring, platform mirroring, backup strategy and identifying a business partner who will provide a similar platform in an agreed timeframe should all be considered. As well as considering actions in the event of the unavailability of the main platform, recovery actions due to failure of other crucial elements of the computer infrastructure, such as networks, should be included.

- *Backup strategy and media storage:* How backup is done, records to demonstrate that the procedure is being followed, how the backup can be restored and the shelf life of the storage media all need to be considered. It is also prudent to prove that the restore procedure works before it is required.
- *Maintenance agreements* with the hardware supplier and the systems integrators should be considered.
- *Quality reviews* of the system are required at the frequency assigned in the Validation Report. They will include reviews of the change control and incident reporting methodology, training, procedures and any outstanding actions from the Validation Report or previous reviews.

SUMMARY

EDMSs are integral in the drive toward a paperless manufacturing environment. EDMSs also provide a useful tool to share information in a way that minimizes duplication and ensures that it is easily accessible when required.

In the pharmaceutical industry, validation is a prerequisite to their use for GMP purposes. The validation methodology used is similar to that used for other information systems. As with all systems, the more attention that is devoted to the design, validation and ensuring that the users will be happy to use the system, the greater will be the benefit to the business.

REFERENCES

1. Brown, L., ed. 1983. *The Shorter Oxford English Dictionary,* vol. 1, 3rd ed. London: Oxford University Press, p. 589.
2. Vesper, J. L. 1998. *Documentation Systems: Clear and Simple.* Buffalo Grove, Ill., USA: Interpharm Press, Inc., pp. 10–23.
3. U.S. Code of Federal Regulations, Title 21, Part 11. *Electronic Records; Electronic Signatures* (revised 1 April 1998).
4. UK GAMP Forum. 1998. *Supplier's Guide for Validation of Automated Systems in Pharmaceutical Manufacture,* Version 3. Available from the International Society for Pharmaceutical Engineering.
5. Garston-Smith, H. T. 1997. *Software Quality Assurance: A Guide for Developers and Auditors.* Buffalo Grove, Ill., USA: Interpharm Press, Inc.

11

Compliance for the Corporate IT Infrastructure

Peter Wilks
Glaxo Wellcome
Uxbridge, Middlesex, United Kingdom

A modern information technology (IT) department, supporting computer rooms, networks and the desktop environment, is a very exciting place to work. There has been and continues to be a proliferation of new technologies and innovative ideas in both hardware and software. The evolution of IT technologies has led to changes in the structure and culture of support organizations and the new skills needed by individuals to design, maintain and operate these systems.

But how does this relate to the specific regulatory needs of the pharmaceutical industry? A highly regulated industry requires well-developed and validated systems, documented evidence of fitness for purpose and trained people following approved procedures.

This chapter discusses the necessary steps to ensure the corporate IT infrastructure can adequately support the high quality expectations of validated applications. It is based on work of an infrastructure special interest group within Good Automated Manufacturing Practice (GAMP) and on presentations provided to ISPE (International Society for Pharmaceutical Engineering) and GAMP meetings.

IT INFRASTRUCTURE TODAY

Ten years ago, users became heavily dependent on their computers (e.g., word processing systems, applications to run stock control or order processing systems, and electronic mail to communicate more effectively with one another). A system failure caused inconvenience and in some instances departments had to revert to their paper-based systems whilst problems were resolved.

Today, word processing systems are being superseded by electronic documentation management systems. Applications are now integrated to allow data, once stored electronically, to be used for many different purposes locally or remotely. Electronic mail systems are being used to communicate messages and documents globally. This huge progress in modern applications is largely due to a more sophisticated IT infrastructure made up of many hardware components and software layers. The principal infrastructure elements to support applications are as follows:

- Equipment to control the environment
- Hardware platforms and peripherals
- Operating systems and software tools
- Networks (cabling, hardware, communications software)
- Desktop environment

The pharmaceutical industry is an extremely data-intensive and information-rich industry. Computers are used in every aspect of research, development, clinical trials, manufacturing and distribution. The success of a pharmaceutical company will largely depend on how efficiently it is able to manage and use that data to deliver the desired business benefits. The key business requirements include

- reliability,
- integrity,
- security,
- capacity and
- performance.

The consequence of a system failure today or in a system being declared noncompliant to current good practice by a regulatory authority is likely to be serious and may bring an entire site or a geographic region to a standstill.

Pharmaceutical organizations that do not align their IT departments with current good practices will put in jeopardy all the good work undertaken by projects and users to validate individual applications. The need to have a compliant IT infrastructure should not be underestimated.

CULTURAL CHANGES

In most enterprises, infrastructure is the responsibility of the IT or Computer Services groups to whom the concepts of validation are still likely to be quite unnatural. At the same time, Quality Assurance (QA) and Validation groups in the company, who are familiar with validation regulations, are likely to have very high expectations.

This cultural chasm should be addressed through training programs for IT personnel in current good practices, requirements for validated applications and the need to maintain quality procedures. Similarly, Validation and QA groups require an appreciation of the processes within the IT group and an understanding that, within the operating environment, decisions have to be made in a timely manner to provide uninterrupted service.

ORGANIZING FOR VALIDATION

Compliance of the IT infrastructure will only be achieved through careful planning, organization, and communication with business project teams. The work involved in meeting compliance is called validation. Validation activities are usually initiated by new or updated applications, infrastructure or services. In some instances, it may be necessary to validate existing systems.

To validate a computer system, one should ensure compliance of the application software, the computer hardware environment (hardware, peripherals and operating software), data, procedures and people.

A common approach to validation is to set up a project team focused on the application. The configuration of the infrastructure directly interfacing with the application is documented and qualified. Such an approach, if realistic systems are not in place to capture infrastructure configuration changes and maintain qualification, will rapidly lead to the validation status of the application becoming meaningless. Figure 11.1 shows a validation project as seen by one application team.

A better approach is for the IT department to maintain quality management processes for configuration and change control and to own the qualification processes for the infrastructure that they manage. This approach enables the compliance status of the infrastructure to be maintained in the future. To support this work, IT departments should consider having their own quality staff to help in the review and approval of procedures and documents.

An important consideration for the ongoing maintenance of the compliance status is the mechanism for change management. The impact of infrastructure changes on the application and vice versa must be evaluated and managed. Figure 11.2 shows multiple validation projects relying on a qualified infrastructure. It also shows change management needing to encompass both application and infrastructure changes.

Figure 11.1. Application View of Validation

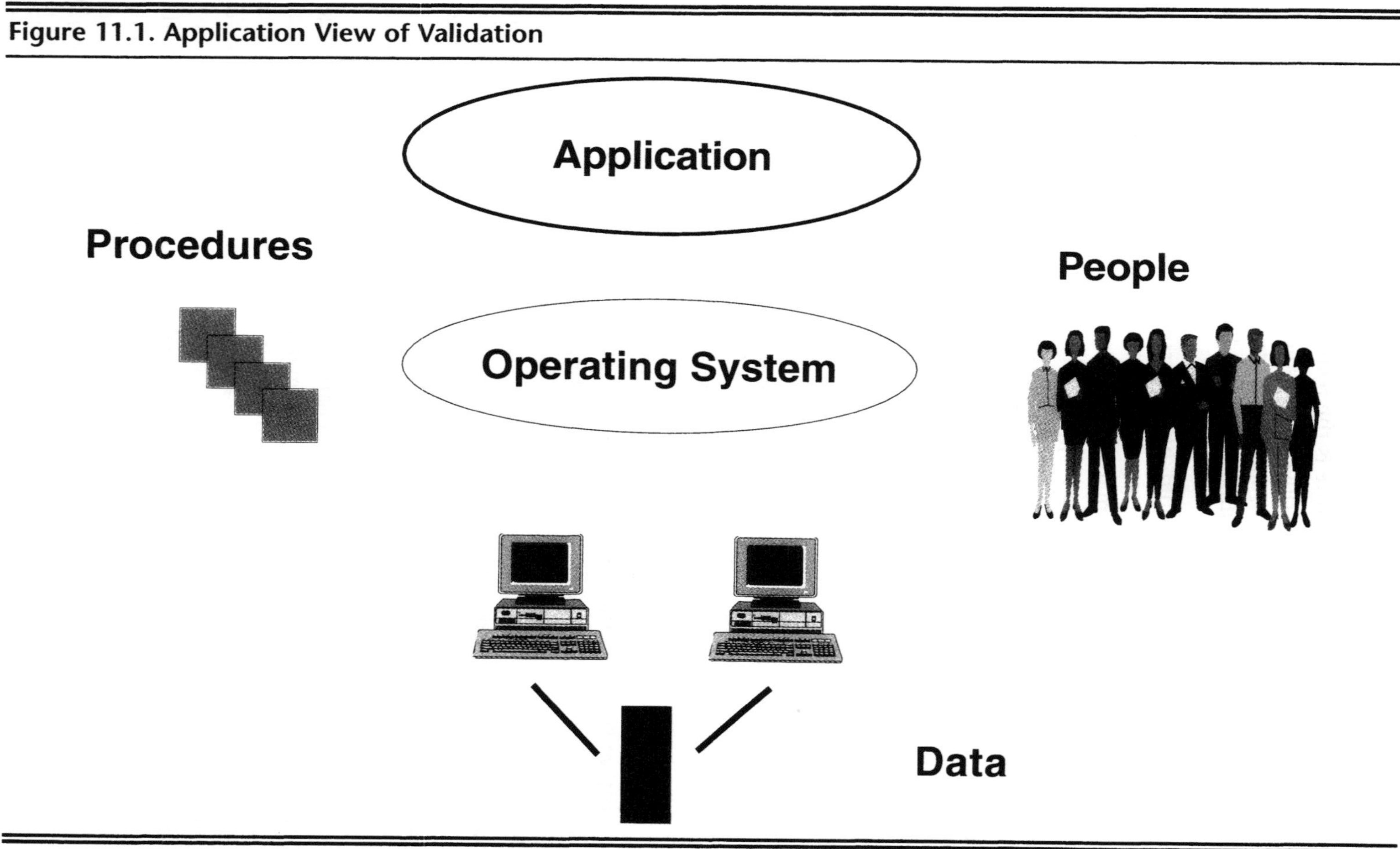

MANAGEMENT PLANNING

Formal plans and specifications are required to describe the extent of the current infrastructure and the plans for future improvements.

A high-level overview of the infrastructure topology and an infrastructure master list should exist. The purpose of this information is to describe the infrastructure to someone, such as an auditor, not familiar with the company's organization. This may be achieved by high-level diagrams showing the division of infrastructure into geographical areas, platforms or other logical parts.

Quality plans to manage corporate IT infrastructure for regulatory compliance should include or reference the following subjects:

- Corporate IT strategies, policies and standards
- High-level overviews and qualified status of computer rooms, servers and networks
- Organization of the IT group
- Development life cycle for infrastructure
- Detailed inventories and configurations
- Computer operations procedures
- Service delivery procedures
- Service support procedures

Documents describing the above should be approved, controlled and available during an inspection.

CLASSIFYING INFRASTRUCTURE

The objective of infrastructure compliance is to produce documented evidence providing a high degree of assurance that the infrastructure will consistently work correctly when used.

Classifying infrastructure software is useful when planning compliance activities. The objective is to make compliance both meaningful and cost-effective. There are two questions that need to be answered: (1) "Do we need to audit the supplier?" and (2) "Is white box testing required?"

Examples of recognized classifications are the International Organization for Standardization (ISO), the Open System Interconnection (OSI) reference model, and GAMP software categories. The GAMP software categories provide a guideline for the activities required for different types of software. These are described in GAMP Guide Volume 1, Part 1, Section 9 [1].

Figure 11.2. Infrastructure View of Validation

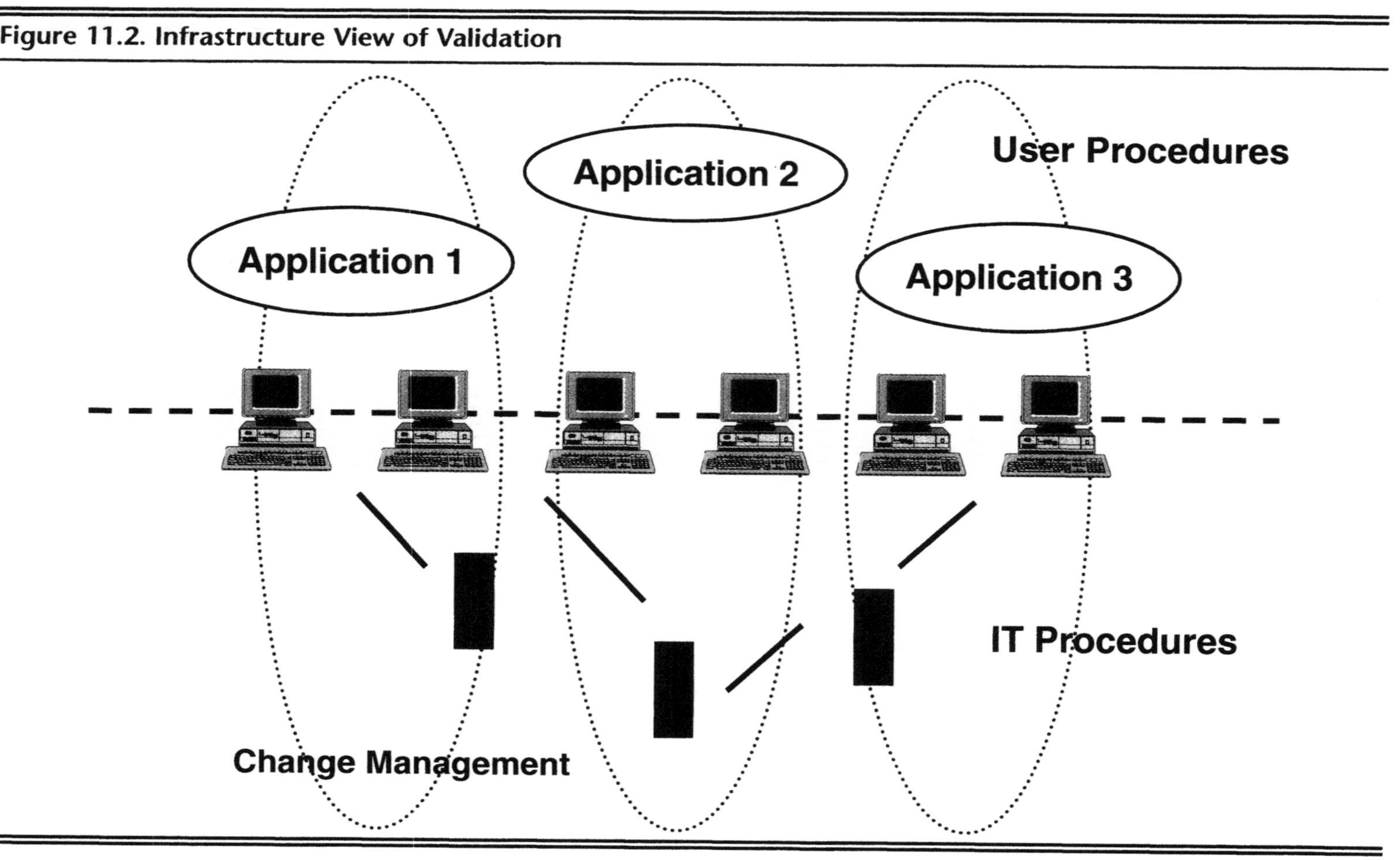

The OSI Reference Model

The OSI reference model defines an architecture of seven layers, with each layer responsible for communicating with the same protocol layer running in the opposite computer and also providing services to the layer above it. The layers are described in Table 11.1.

Software Categories

The GAMP software categories can be correlated with the OSI reference model as follows:

- Firmware controlled devices (e.g., intelligent bridges and routers) belong in layers 1, 2, 3 and 4 of the OSI model. These devices relate to GAMP category 2 software. Configuration settings for baud rates should be documented.
- Network operating systems deal with layers 5 and 6 in the OSI Model. Operating systems relate to GAMP category 1 software.
- Layer 7 network standard software packages relate to GAMP category 3 software.

Table 11.1. OSI Layers

Layer	Layer Title	Layer Description
7	**Application**	Interfaces directly with the application programs running on the network. This layer provides services such as file access and transfers, peer-to-peer communications and resource sharing.
6	**Presentation**	Translation of data formats to enable computers using contrasting languages to communicate. Data encryption is handled in this layer.
5	**Session**	Establishes bidirectional communication between applications using conversational techniques or dialogues.
4	**Transport**	Ensures reliable message delivery and the control of data between systems in a flow of packets.
3	**Network**	Standardization of the addressing mode between multiple, linked networks and services to ensure packets of information arrive at the correct destination.
2	**Data Link**	Defines the control of communication between two devices directly linked together and the packet and framing methods.
1	**Physical**	Defines the mechanical components, type of medium, transmission method and rates available.

- Layer 7 network configurable software packages relate to GAMP category 4 software.
- For software packages which require bespoke scripts to be written, then the software package itself should be considered as either GAMP category 3 or 4. The scripts should be treated as GAMP category 5 software requiring specifications, programming standards, Source Code Reviews and testing.

GAMP software categories 1 to 3 are most common within the infrastructure. However, attention should be given to configurable support software or software requiring bespoke scripts.

DEVELOPMENT LIFE CYCLE FOR INFRASTRUCTURE

A life-cycle approach should be used when managing the IT infrastructure. Documented evidence is needed during planning, specification, selection, design and testing, installation, qualification and operation. An overview of these stages is provided in GAMP Volume 1, Part 1, Section 7 [1].

PROJECT PLANNING

There are many ways to segment planning. As described in previous sections, compliance activities are usually initiated by new or updated applications, infrastructure or services, and one or more project teams manage this work. All compliance activities to be undertaken should be detailed in a plan.

Where new infrastructure is being installed to support a new or updated application, the infrastructure is considered to be qualified and part of a larger project. Where a new infrastructure service is being provided, it may have its own plan. Infrastructure itself does not require validation. It is important that compliance steps and terminology are made clear.

The roles that each group plays in the validation of the total system should be clearly defined. This includes a description of who is responsible for qualifying each element of hardware and software. Similarly, it must be clear which groups are responsible for maintaining, reviewing and approving the procedures.

Supplier Selection

Corporate IT places considerable reliance on many suppliers for infrastructure products and services. Where possible, the infrastructure should be constructed from proven standard components from approved suppliers. Caution is required when introducing new technologies, which may involve more in-depth trials and testing.

It is the responsibility of the pharmaceutical company to ensure that the products and services that are purchased are fit for their intended purpose. This should be achieved through auditing of the life-cycle development for bespoke products and adherence to quality procedures for services.

The supplier selection process should be defined. This may include corporate strategies on architecture, platforms, software and services. It should also cover quality requirements and auditing requirements for potential suppliers.

Technical Specification

For each element of the infrastructure, technical specifications should detail what is to be installed, the purpose of the element being installed, the key operating parameters, the interfaces to other infrastructure and the key constraints of the system. These documents, which may include drawings, diagrams and some supplier documentation, must be reviewed, approved, formally controlled and kept up to date.

Design and Testing

In the majority of cases, the manufacturer will complete detailed design and "white box" testing activities for the individual infrastructure elements. The results of these activities are not normally available for detailed inspection, and the pharmaceutical manufacturer's IT department would not be expected to have the necessary expertise to perform a meaningful review.

Installation Plan

Installation plans should be created for the infrastructure hardware and software. They should identify the configuration settings and any dependencies or constraints.

Installation Qualification

Installation Qualification (IQ) will confirm that the correct infrastructure and associated operating system software are installed. It will also confirm that the configuration of the infrastructure is documented and that systems are in place to maintain the configuration data. As appropriate, confirmatory tests should be documented to confirm the correct installation and operation of the infrastructure element. If commissioning documentation is referenced, then this must be reviewed and approved to be complete.

Appropriate approaches to different elements of the infrastructure are discussed in the following section.

QUALIFYING INFRASTRUCTURE ELEMENTS

Computer Rooms and Environmental Control Equipment

Environmental requirements for computer systems and media storage areas must be documented. In particular, computer rooms should be designed such that computers and data are secure and maintained within the manufacturer's recommended environmental limits for temperature and humidity. IQ must demonstrate adequate security, environmental controls and maintenance programs.

Hardware Platforms and Peripherals

All hardware platforms with their associated peripherals and operating system software must be defined in a technical specification. Detail regarding resilience against system failure, such as mirroring, should be included.

Installation plans and IQ must be executed to demonstrate the correct installation. Operating procedures must be approved, and operators following these procedures must be trained.

Networks

Both the physical and logical attributes of the network should be defined in technical specifications. Textual descriptions supported by diagrams, addressing OSI layers 1 and 2, should be produced, defining all networks at both local and wide area levels. Documentation should specify all physical components. References should be made to the industry-wide standards that were followed.

Full descriptions should be provided of the logical methods of data transmission, addressed by layers 3, 4 and 5, throughout the network. The protocols described should clarify how devices exchange information, any routing used, in what form the information is transmitted and how multiple devices communicate simultaneously but successfully.

Boundaries of open and closed systems should be defined. Methods for protecting data in open and closed networks, such as access controls, firewalls and cryptographic techniques of data, should be specified.

Installation plans and IQ should demonstrate that the network has been correctly installed and that communications between desired nodes are effective. Where documentation refers to commissioning tests, these should be reviewed and approved.

The Desktop

Standard desktop configurations or desktop builds should be established and imposed wherever possible. Departures from the standard should only be permitted if it can be clearly demonstrated that the additional costs of maintaining that desktop in a validated state can be justified by its role in the business.

Each standard desktop build should be documented and tested with the combination of validated applications for which it is intended. Procedures must be in place for authorizing user access, protecting against and managing viruses and monitoring or auditing the ongoing desktop environment.

OPERATING AND SUPPORT PROCESSES

Validation of the infrastructure includes ensuring that processes are in place to control the operation and maintenance of infrastructure. These processes are implemented by formally approved and controlled procedures. The following sections describe the key areas that should be specified and tested.

Computer Operations

Computer operations include but are not limited to the following:

- Data center management
  - Defines the security of data centers
  - Defines the environment management of the data center
  - Defines fire protection and safety management
- Systems management
  - Defines how to manage platforms, networks and the desktop
  - Defines who is responsible for the services
  - Defines any necessary preventative maintenance
  - Defines how to manage fixes and patches to system software
  - Defines power-up, starting, closure and power-down of services
- Job scheduling
  - Defines the assigning of priorities for running jobs
- Event management
  - Defines the monitoring of system events and the actions to be performed when events arise
- Data management
  - Defines the procedures for data backup, recovery and archiving
  - Defines the conventions for labeling and storing media
- Security

- Defines the physical security of machine rooms or other sensitive areas
- Defines the processes for logical access into networks and applications, including the authorization, management, review and removal of user identities and passwords
- Defines the methods and tools to protect against and detect software viruses
- Defines the process to investigate security breaches

Service Support

Service support activities include the following elements:

- Configuration management
  - Defines how the inventory of the hardware, software and its configuration is maintained, including model numbers, versions, patch levels and system parameters
- Help desk and call handling
  - Defines how calls are logged and progressed
- Problem/escalation management
  - Defines the process for analyzing problems
  - Defines how decisions and resolutions to problems are agreed on
- Change management
  - Defines the process for logging change requests, reviewing and approving changes and progressing these changes through to implementation

Service Delivery

Service delivery activities include the following:

- Contingency planning
  - Defines recovery action required for a major disaster
- Contract management
  - Defines the Service Level Agreements (SLAs) between the IT group and the business, which defines the level of service, in terms of availability, support, quality and performance
  - Defines the managing of suppliers and third-party relationships
- Capacity and performance management
  - Defines the monitoring of capacity and performance

Good Practice

It is good practice for an IT department to adopt standard ways of working and to employ generic procedures wherever possible. However, the greater the diversity of hardware platforms, communications and desktop builds, the greater the number of procedures required by the IT department.

Testing Procedures

The necessary organization for supporting processes, such as the help desk and change control coordinators, must be in place before the procedures can be tested. These tests should be documented in an operational acceptance test protocol.

A useful technique in testing is to walk specified events through the procedures with the actual operations staff who will be performing the defined roles. This testing is particularly useful for service support and delivery processes. Physical testing should also take place for the computer operation processes defined above.

Training

Technical and managerial staff must be trained and able to demonstrate that they are competent in the procedures that are applicable to them. They must also have an appreciation of current good practices and how these impact their daily work.

Training records must be maintained for all staff involved in the implementation, operation or support of the infrastructure. This will include contract staff and third-party arrangements.

SUMMARY

Pharmaceutical organizations today are totally dependent on accurate data stored and manipulated by validated business applications. The IT infrastructure is the platform for business applications and must be qualified and managed following approved procedures.

IT departments have traditionally not been directly involved in validation projects, hence, there is a lack of understanding in current good practices. Expectations from user and QA groups will not be achieved unless there is a training program and culture change within the IT department.

To meet compliance requirements, fitness for purpose must be demonstrated through specification, installation, qualification, procedures and trained personnel. The consequences of not meeting current regulatory requirements are immense. However, compliance of the corporate IT infrastructure can be achieved through careful planning, organization and communication.

REFERENCES

1. UK GAMP Forum. 1998. *Supplier's Guide for Validation of Automated Systems in Pharmaceutical* Manufacture, Version 3.0. Available from the International Society for Pharmaceutical Engineering.

12

Validating Local and Wide Area Networks

Nicola Signorile
Hoechst Marion Roussel (Gruppo Lepetit)
Localita Valcanello, Italy

Modern IT (information technology) applications would not be possible if it were not for innovations in communication network technology. Indeed, major IT system vendors and consultancy firms, including Digital Equipment, were saying at the end of the 1980s: "The network is the system". Over the past 20 years, IT systems have evolved from the development of centralized systems, through main frame systems, to distributed computing environments. Supporting network developments include client-server technology and Intranet/Internet technology. These developments have offered users even more flexibility and functionality but at a price—the networks systems supporting IT applications have become more and more complex.

Within the pharmaceutical industry, IT applications are increasingly being used to support the manufacture of drug products. These applications must be validated to fulfill GxP regulations, including any support networks. Chapters 6 to 10 discussed a range of IT applications and how they might be validated. This chapter considers the validation of communication networks. The integrity of data being carried by a network must not be compromised. An IT application may be perfectly functional, but as the IT fraternity say, "garbage in, garbage out". Networks usually have a potential GxP impact on IT applications. Validation of IT applications should take a "systems approach" and not ignore supporting networks. This chapter introduces network terminology before presenting a validation strategy, with practical advice on documentation and common issues.

LOCAL AREA NETWORK AND WIDE AREA NETWORK

Networks link collections of independent computers and devices (such as printers), providing a shared communication medium over which the computers can transfer information to one another. Prior to the development of networking technology, individual machines were isolated, and their range of applications was limited.

Local area networks (LANs) are those networks usually confined to a small geographic area, such as a single building, a group of localized buildings, or a site. LANs are not necessarily simple in design; some may link many thousands of systems and service hundreds of users. The development of various standards for networking protocols and media has made possible the proliferation of LANs worldwide for business and manufacturing applications.

Wide area networks (WANs) are those networks installed on a wide geographical area, typically linking multiple sites. They can cross national boundaries and join continents. LANs are often connected to WANs to create a communication web. The combined LAN/WAN topology is akin to the human body's central nervous system.

NETWORK APPLICATIONS

Two examples of IT applications that require their supporting network to be validated will be described here. The first example describes a Manufacturing Execution System (MES) based on a LAN, and the second example describes an Enterprise Resource Planning (ERP) application that uses a WAN.

A Site MES Application

The architecture of the MES application is shown in Figure 12.1. The system provides electronic batch record (EBR) functionality interfaced to a Laboratory Information Management System (LIMS). The system is a typical client-server application that has its own main functions distributed on the LAN. The main functions are as follows:

- *Database Server:* UNIX®-based computer with an Oracle® database that manages "manufacturing" data (all data used at and coming from the shop floor and all specification data, such as bills of materials, specifications, batch records, etc.).

Figure 12.1. Site Manufacturing Execution System

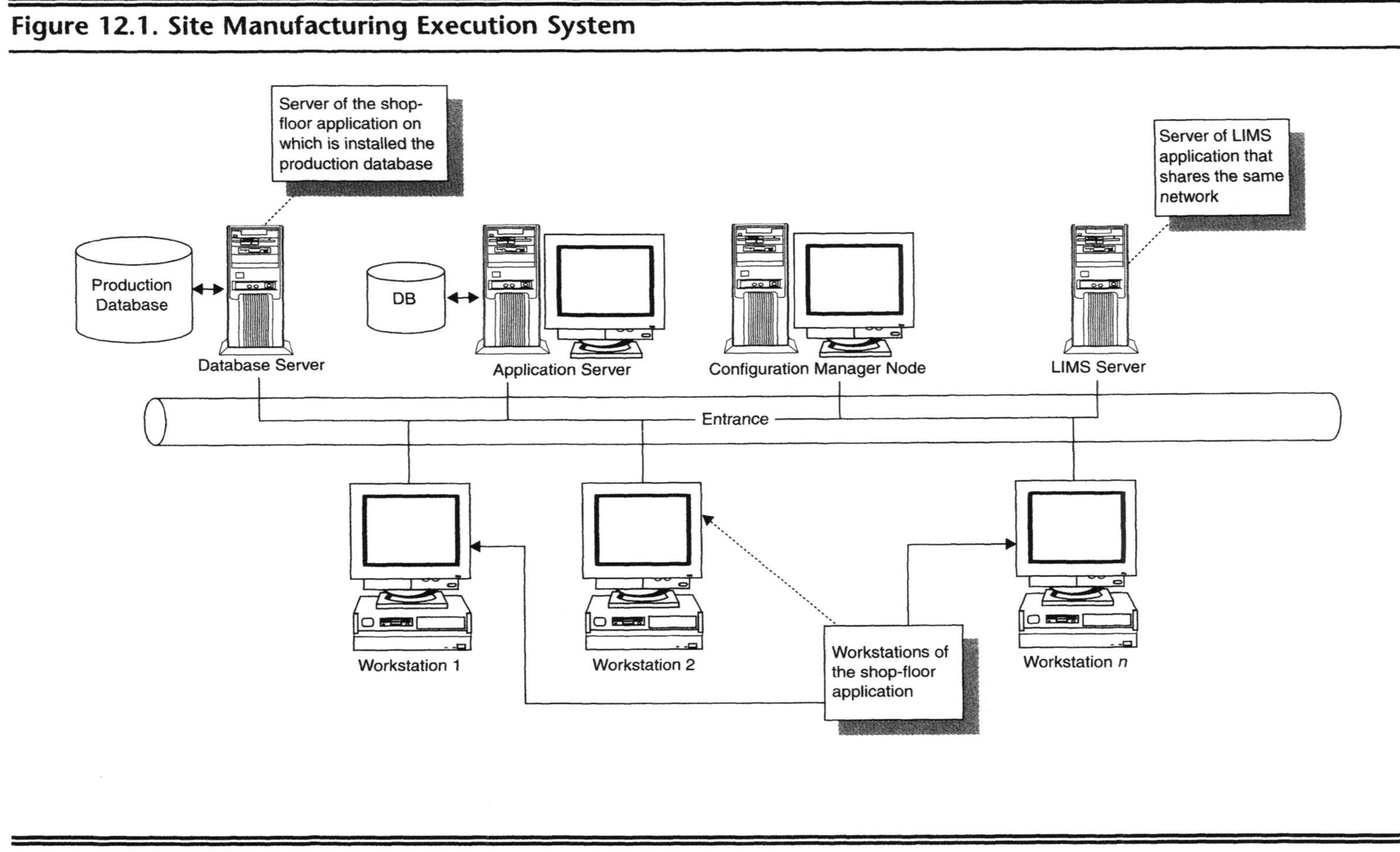

- *Application Server:* An OS/2-based computer that manages the software that is distributed to all application workstations. It takes care of communication between workstations and database server.
- *Configuration Manager Node:* An OS/2-based computer with a DB/2 database that manages "application" data (i.e., all data specifying the setup of each workstation and associated security functions).
- *Workstations:* OS/2-based computers that provide the user interface to the application (e.g., display operating instructions, mimics and alarms/messages to the operators). About 100 workstations are distributed on the shop floor and connect to equipment such as scales and Programmable Logic Controllers (PLCs).

The system is interfaced with a local LIMS, and data are exchanged back and forward between the systems. In this example, the GxP nature of the data managed by the MES and LIMS and data exchange between the systems requires that both systems and the LAN be validated.

In this example, and generally speaking in all cases in which more than one application shares the same LAN, it is convenient to proceed with a separate network validation project. All of the applications that need to be validated can refer to the validation package of the network, avoiding duplicated work during validation of the IT applications.

A Multisite ERP Application

A multisite ERP application is presented in Figure 12.2. The ERP application is based on a single SAP R/3 production instance installed on a computer located at the company's HQ (headquarters) offices. The HQ users, as well as users from four different manufacturing sites geographically distributed, access the SAP R/3 application. All users access the application through the WAN.

There are about 700 application users, with an average of 300 concurrent users, and 4 MESs exchange data with the SAP R/3. Each manufacturing site installed a different MES based on the topology described in "A Site MES Application" example above. The ERP application passes Good Manufacturing Practice (GMP)–relevant data (including production orders, bills of materials, materials allocation in the warehouse, materials consumption data, and materials' status) back and forward with the MES.

Due to the client-server nature of the SAP R/3 product, the client side SAP Graphical User Interface (GUI) must be aligned at the server version. To guarantee that all 700 users of the application, distributed on the five sites, receive client version upgrades at the same time and align with the server version concurrently, an application has been installed to automatically distribute the software through the network (Microsoft SMS® application).

To validate the application SAP R/3, it is necessary to validate the infrastructure on which the application is built on, the WAN.

Figure 12.2. Distributed ERP Application

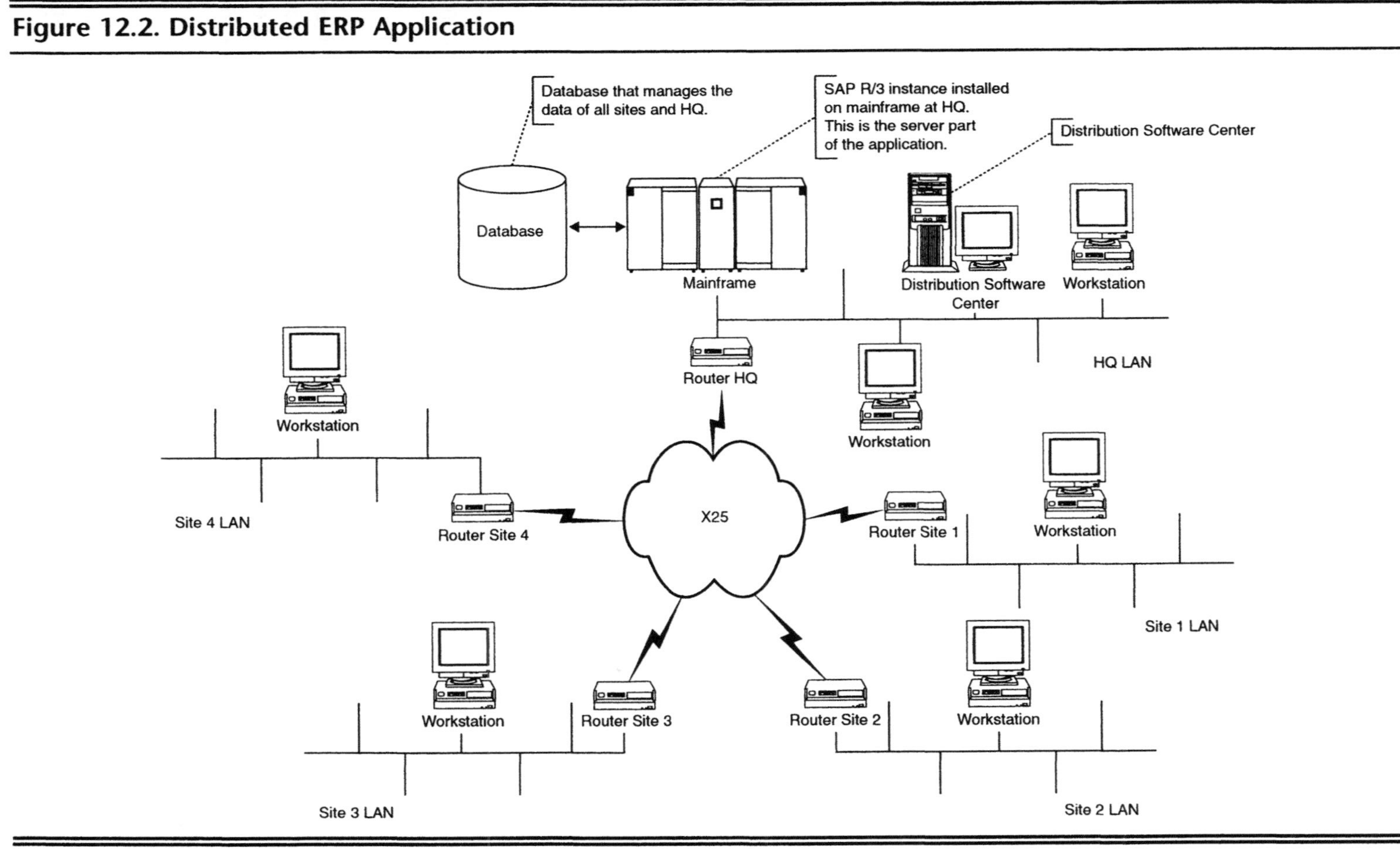

NETWORK COMPONENTS

Common network terminology will now be introduced for those who are unfamiliar with the components that constitute a LAN or a WAN.

Protocols

Network protocols are standards that define how computers communicate. A typical protocol defines how computers should identify one another on a network, the form that the data should take in transit and how this information should be processed once it reaches its final destination. Protocols also define procedures for handling lost or damaged transmissions. Transmissions are sometimes described as "packets" of information. The most common network protocols include TCP/IP (Transmission Control Protocol/Internet Protocol), IPX (Internet Packet Exchange), AppleTalk, and DECnet (Digital Equipment Corporation network).

Network protocols use physical cabling in exactly the same manner, thus allowing protocols to peacefully coexist. This concept is known as "protocol independence", meaning that the physical network does not need to concern itself with the protocols being carried. The network builder can use any of the protocols supported by an item of equipment. The final choice may depend on personal preference, a defined operating philosophy or perhaps more arbitrary criteria.

Ethernet

Ethernet is the most popular LAN technology in use today. Other LAN types include Token Ring, Fiber Distributed Data Interface (FDDI), and LocalTalk. Each has its own advantages and disadvantages. Ethernet strikes a good balance between speed, price and ease of installation. These strong points combined with wide acceptance in the computer marketplace and the ability to support virtually all popular network protocols makes Ethernet the perfect networking technology for most computer users today.

The Ethernet standard is defined by the Institute for Electrical and Electronic Engineers (IEEE). IEEE Standard 802.3 defines rules for configuring an Ethernet and specifies how elements in a network interact with one another. Networks, equipment and network protocols that utilize and adhere to the IEEE standard will operate in the most efficient manner.

Media and Topologies

An important part of designing and installing an Ethernet is selecting the appropriate Ethernet medium for the problems at hand. There are four major types of media in use today: thickwire, thin coax, unshielded twisted pair and fiber optic. Each type has its strong and weak points. Careful selection of the appropriate Ethernet medium can avoid recabling costs as the network grows.

Ethernet media can be divided into two general configurations or topologies: "bus" and "point to point". These two topologies define how "nodes" are connected to one another. A node is an active device connected to the network, such as a computer or a piece of networking equipment, for example, a repeater, a bridge or a router.

A bus topology consists of nodes strung together in series, with each node connected to a long cable or bus. Many nodes can tap into the bus and begin communication with all other nodes on that cable segment. A break anywhere in the cable will usually cause the entire segment to be inoperable until the break is repaired.

Point-to-point media link exactly two nodes together. The primary advantage of this type of network is reliability. If a point-to-point segment has a break, it will only affect the two nodes on that link. Other nodes on the network continue to operate as if that segment were nonexistent. Obviously connecting only two computers together makes for a very limited network. Repeaters may be used to bind groups of point-to-point segments together. See "Repeaters" for more information on how to connect both point-to-point and/or bus segments together to make larger, more useful, networks.

Thickwire

Thickwire, or 10Base5 Ethernet, is generally used to create large "backbones". A network backbone joins many smaller network segments into one large LAN. Thickwire makes an excellent backbone because it can support many nodes in a bus topology, and the segment can be quite long. It can be run from workgroup to workgroup where smaller departmental networks can be attached to the backbone. A thickwire segment can be up to 500 m long and have as many as 100 nodes attached.

Thickwire, as the name suggests, is a thick, hefty, coaxial cable and can be expensive and difficult to work with. A thick coaxial cable is used because of its immunity to common levels of electrical noise, helping to ensure the integrity of network signals. The cable must not be cut to install new nodes; rather nodes must be connected by drilling into the media with a device known appropriately as a "vampire trap". Nodes must be spaced exactly in increments of 2.5 m apart to prevent signals from interfering with one another. Due to this combination of assets and liabilities, thickwire is best suited for but not limited to backbone applications.

Thin Coax

Thin coax, or 10Base2 Ethernet, offers many of the advantages of thickwire's bus topology with lower cost and easier installation. Thin coax coaxial cable is considerably thinner and more flexible than thickwire, but it can only support 30 nodes, each at least 0.5 m apart. Each segment must not be longer than 185 m. Subject to these restrictions, thin coax can be used to create backbones, albeit with fewer nodes.

A thin coax segment is actually composed of many lengths of cables, each with a BNC (Bayonet Neil-Concelman) type connector on both ends. Each cable length is connected to the next with a T connector wherever a node is needed. Nodes can be connected or disconnected at the T connectors as the need arises, with no ill effects on the rest of the network. The low cost of thin coax and its reconfigurability and bus topology make it an attractive medium for small networks, for building departmental networks to connect to backbones and for wiring a number of nodes together in the same room (such as a laboratory).

Twisted Pair

Unshielded twisted pair (UTP) cable offers many advantages over the thickwire and thin coax media. Because thickwire and thin coax are coaxial cables, they are relatively expensive and require some care during installation. UTP is similar to, if not the same as, the telephone cable that may already be installed and available for network use in a building.

UTP cables come in a variety of grades, with each higher grade offering better performance. Level 5 cable is the highest and most expensive grade, offering support for transmission rates of up to 100 Mps (megabits per second). This grade of cable is unnecessary for ordinary 10BaseT applications with 0 Mps. Level 4 and level 3 cables are far more popular for current 10BaseT configurations; Level 4 cable can support speeds of up to 20 Mps, Level 3 up to 16 Mps. Level 2 and level 1 cables are the lowest grades and least expensive wire, designed primarily for voice and low-speed transmissions (less than 5 Mps); these should not be used in the design of 10BaseT networks.

A UTP, or 10BaseT Ethernet, is realized with a point-to-point topology. Generally, a computer is located at one end of the segment, and the other end is terminated in a central location with a repeater or hub. Since UTP is often run in conjunction with telephone cabling, this central location can be a telephone closet or other area where it is convenient to connect the UTP segment to a backbone. UTP segments are limited to 100 m, but UTP's point-to-point nature allows the rest of the network to function correctly if a break occurs in a particular segment.

Fiber Optic

Fiber optic, or 10BaseFL Ethernet, is similar to UTP. Fiber optic cable is more expensive, but it is invaluable for situations where electronic emissions and environmental hazards are a concern. The most common situation where these conditions threaten a network is in LAN connections between buildings. Lightning strikes and current loops due to ground potential differences can wreak havoc and easily destroy networking equipment. Fiber optic cables effectively insulate networking equipment from these conditions, since they cannot conduct electricity. Fiber optic cable can also be useful in areas where large amounts of electromagnetic interference (EMI) are generally present, such as on a factory floor.

The Ethernet standard allows for fiber optic cable segments up to 2 km long. Remote nodes and buildings that otherwise would not be reachable with LANs can be brought into the fold.

An investment in fiber optic cabling can be a wise one. As network technologies evolve and demands on the network increase, FDDI and other technologies faster than Ethernet can be run on the same cable, avoiding major rewiring.

Transceivers

Transceivers are used to connect nodes to the various Ethernet media. Transceivers, also known as media attachment units (MAUs), attach to the Ethernet cable and provide an Application User Interface (AUI) connector for the computer. The AUI connector consists of a 15-pin D shell type connector, female on the computer side, male on the transceiver side. Virtually all Ethernet compatible computers provide such an AUI connector. The transceiver is generally attached directly to the computer's AUI connector, or the transceiver may be attached to the computer with a specially shielded AUI cable, which must be less than 50 m long. In addition to an AUI connector, many computers also contain a built-in transceiver, allowing them to be connected directly to Ethernet without requiring an external transceiver.

Repeaters

Repeaters are used to connect two or more Ethernet segments of any medium type. As segments exceed their maximum number of nodes or length, signal quality begins to deteriorate. Repeaters provide the signal amplification and retiming required to connect segments. Splitting a segment into two or more segments with a repeater allows a network to continue to grow. A repeater connection counts in the total node limit on each segment. For example, a thin coax segment may have 29 computers and 1 repeater, or a thickwire segment can have 20 repeaters and 80 computers.

Ethernet repeaters are invaluable with point-to-point media. As pointed out earlier, a network with only two nodes is of limited use. A twisted pair repeater allows several point-to-point segments to be joined into one network. One end of the point-to-point link is attached to the repeater, and the other is attached to the computer with a transceiver. If the repeater is attached to a backbone, then all computers at the end of the twisted pair segments can communicate with all of the hosts on the backbone.

Repeaters also monitor all connected segments for the basic characteristics necessary for the Ethernet to run correctly. When these conditions are not met on a particular segment, for example, when a break occurs, all segments in an Ethernet may become inoperable.

Repeaters limit the effect of problems to the faulty section of cable by "segmenting" the network, disconnecting the problem segment and allowing unaffected segments to function normally. A segment malfunction in a point-to-point network will generally disable only a single computer, where the same problem in a bus topology would disable all nodes attached to that segment.

Just as the various Ethernet media have segment limitations, larger Ethernets created with repeaters and multiple segments have restrictions. These restrictions generally have to do with timing constraints. Although electrical signals inside the Ethernet media travel near the speed of light, it still takes a finite time for the signal to travel from one end of a large Ethernet to another. The Ethernet standard assumes it will not take more than a certain amount of time for a signal to propagate to the far ends of the Ethernet. If the Ethernet is too large, this assumption will not be met, and the network may not perform correctly. Timing problems must not be taken lightly. When the Ethernet standard is violated, packets will be lost, network performance will suffer and applications will become slow and may even fail.

The IEEE Standard 802.3 specifications describe rules for the maximum number of repeaters that can be used in a configuration. The maximum number of repeaters that can be found in the transmission path between two nodes is four. The maximum number of network segments between two nodes is five, with a further restriction that no more than three of those five segments have other network stations attached to them (the other segments must be interrupter links which simply connect repeaters). These rules are determined by calculations of maximum cable lengths and repeater delays. Networks that violate these rules may still be functional, but they are subject to sporadic failures or frequent problems of an indeterminate nature. Bridges are recommended for networks where many repeaters are required; they can limit the amount of traffic on each segment and improve performance.

Bridges and Routers

An Ethernet may eventually become too large. It may not be possible to add additional nodes without violating Ethernet standards, or traffic on the network may be causing such a high load that performance suffers. Then it may be necessary to split the Ethernet into two or more separate Ethernets with a bridge or a router.

Each of the resulting, smaller, Ethernets can be expanded with more repeaters and segments because the IEEE Standard 802.3 specifications then apply to each of the new Ethernets, not both Ethernets combined. Bridges and routers allow hosts on these two new and distinct Ethernets to talk to one another by using a technique known as "store and forward".

In store-and-forward devices, packets are gathered off one Ethernet and saved in memory. When the bridge or router senses that the other Ethernet is available, it transmits the packet. To each of the two Ethernets, the bridge or router looks just like any other host, since the bridge or router obeys the same rules for accessing the Ethernet. The major difference between a bridge/router and repeater is that repeaters do not store packets; they simply clean up the signal on the network and send the signal out to all other ports.

Bridges and routers can reduce the network load if used intelligently. Bridges listen to all traffic on the network, "learning" where various hosts reside. If a bridge detects a packet on one Ethernet destined to a host on another Ethernet, it will forward the packet to the Ethernet to which the destination host is attached. If a bridge detects a packet on an Ethernet destined to a host on that same Ethernet, it does not forward it. The second Ethernet is thus spared from receiving the packet, which was not of any use to its hosts, and overall load is reduced. Bridges are protocol independent; they can store and forward packets for any network protocol type without regard for the information contained within.

Bridges read an entire packet before they compare it to their address list; this is done so that short or illegal packets, packets with bad cross-redundancy checks (CRCs) or packets with late collisions may be automatically filtered out of the network. Obviously, this means that there will be a small delay factor between the time the bridge is finished reading a packet and the time it takes to forward it on. For the benefit of having any bad packets filtered out, most bridge users are willing to incur the very small delay for full packet examination.

A new class of bridging devices, called ether switches, offers users another option. Ether switches read only enough of a packet to determine the source and destination addresses for filtering purposes and then send on the packet at that point. This process speeds throughput but does not filter out illegal or bad packets, unless the problem is evident in the first few bytes. The speed advantage of these devices must be weighed against the need to filter.

Routers work in a similar fashion to bridges, except routers are protocol dependent. Routers know about the inner workings of the protocols that they support. This intimate knowledge allows routers to do sophisticated packet forwarding and can provide a great reduction in network traffic by filtering extraneous packets. The price paid for this intelligent forwarding capability is usually additional configuration and cost.

Some routers offer bridging services as a supplement to their primary capabilities; these routers are referred to as "B routers". B routers offer such standard bridge features as source/destination address filtering and automatic filtering of bad packets in addition to their protocol-specific routing functions.

Terminal and Print Servers

As their names suggest, terminal servers and print servers support the use of terminals and printers on networks. They support modems and other devices as well. The primary difference between them is that terminal servers are bidirectional devices, while print servers are unidirectional devices, at least as far as data transmissions are concerned. Unlike transceivers, repeaters, or port multipliers, terminal servers and print servers are intelligent devices that have their own network addresses and perform more than just a physical connection or signal forwarding function.

REGULATORY REQUIREMENTS FOR NETWORKS

As pharmaceutical manufacturers increasingly integrate manufacturing operations on a national and international basis, their reliance on networks increases. Pharmaceutical manufacturers are required to validate these networks. Validation in this context has been defined by the U.S. Food and Drug Administration (FDA) as "establishing documented evidence that provides a high degree of assurance that a specific process will consistently produce a product meeting its predetermined specifications and quality attributes" [1].

The validation of computer networks is clearly important, but how do we go about validating them? How much detail is required? The FDA gave the following advice in 1983 on the topic of computer networks supporting manufacturing operations [2]:

If the firm is on a computer network, it is important to know

(1) *what output, such as batch production records, is sent to other parts of the network;*

(2) *what kinds of input (instructions, programs) are received;*

(3) *the identity and location of establishments which interact with the firm;*

(4) *the extent and nature of monitoring and controlling activities exercised by remote on-net establishments; and*

(5) *what security measures are used to prevent unauthorized entry into the network and possible drug process sabotage.*

It is possible under a computer network for manufacturing operations conducted in one part of the country to be documented in batch records on a real-time basis in some other part of the country. Such records must be immediately retrievable from the computer network at the establishment where the activity took place.

The FDA cites clause 180 in 21 CFR 211 in relation to computer networks, which deals with records and reports for manufacturers of finished pharmaceutical products. The concern is that the computer network must maintain the integrity of data passed through the network. This links to the recent issue of 21 CFR 11 in 1997 which deals with electronic records and their security. Further information on this topic can be found in Chapter 16.

The UK GAMP Forum provides some advice in their latest guide [3]. Basically, the same validation methodology should be followed as for other automation and IT systems: Categorize software components and follow a V life-cycle approach. The GAMP Guide identifies five categories of software:

1. *System software:* Record version of software.
2. *Firmware:* Record configuration.
3. *Standard software:* Validate application.
4. *Configurable software:* Audit application; validate application and any bespoke code.
5. *Bespoke software:* Audit supplier and validate complete system.

Networks are largely made up of standard components (system software, firmware and standard software); there is little bespoke programming other than configuration and perhaps some specialist interfaces.

The application of the FDA and GAMP guidance is discussed in the following sections of this chapter and is based on the practical experience of validating networks within an international pharmaceutical manufacturing company. Supplier audits are not discussed here, but further information in regard to this aspect of validation can be found in Chapters 14 and 15.

A VALIDATION STRATEGY FOR NETWORKS

To define a validation strategy, we first must consider the current status of the system. If the system has already been installed, the validation will be *retrospective*; if it is a new network, proceed with *prospective* validation. This chapter considers a prospective approach to validation; however, most of the concepts can also be used for retrospective validation. It should be noted here that retrospective validation is usually a much more expensive and timely task compared to prospective validation. It has been suggested that retrospective validation can be in excess of five times more expensive than prospective validation [4].

Now that the kind of validation to be performed on the system has been defined, it is necessary to clarify the scope of the system. Agreeing on the scope is extremely important for two reasons: (1) We can adopt the appropriate variant of the V model life cycle depending on the use of different categories of software. (2) It establishes what is part of the system and what not is part of the system—what will be validated and what won't be validated.

What is in and out of scope must be clearly visible to the user. If a network server or network control system are, for example, determined to be out of the scope of one project, then they must be covered by another validation exercise. Other network scope issues might include who is responsible for the network interface cards or the firewalls between interconnected networks. It is quite common for networks to fall between projects and not to be validated until a regulatory inspection identifies this as a GxP nonconformance. It is in the interest of pharmaceutical manufacturers to avoid the embarrassment of a regulator identifying absent validation and any consequential official warning or sanction. Determining who is responsible for validating a particular network can be assisted by identifying what data are transferred over the network and who are responsible for that data.

Validation and the System Life Cycle

A life-cycle approach should be adopted when validating networks. In a simple form, this might consist of a "cascade" development methodology, forcing the definition and approval of each document produced in the prior phase before proceeding with the next phase. The cascade approach is certainly applicable to the development of network systems, but an "incremental" approach is usually adopted. The incremental approach facilitates the construction of complex network systems from configurable software and hardware equipment. Configurable packages provide a means of easily modifying a network system by reconfiguring the software and/or equipment to reflect any changing requirements without developing an entirely new replacement network system.

Figure 12.3 shows the phases within a "cascade" approach and an "incremental" approach. Each box shows a different phase of the life cycle. The descending arrows show passage between one phase and the next (enabled only after the approval of "deliverables" related to the prior phase); the ascending arrows show system acceptance step by step (after the execution of the tests related to the specific phase). Note that the same life-cycle phases are still valid inside those different development methodologies.

The V model commonly used in the pharmaceutical industry for computer systems validation is presented in Figure 12.4. The model illustrates the cascade approach and the relationship between specifications and testing.

The V model can be developed to fit the "incremental" approach (Figure 12.5).

The validation activities that should be executed during a network development life cycle are described in Figure 12.6. The right side of the diagram lists the network development activities and for each activity, or group of activities, the related validation activities.

Figure 12.3. "Cascade" and "Incremental" Approach

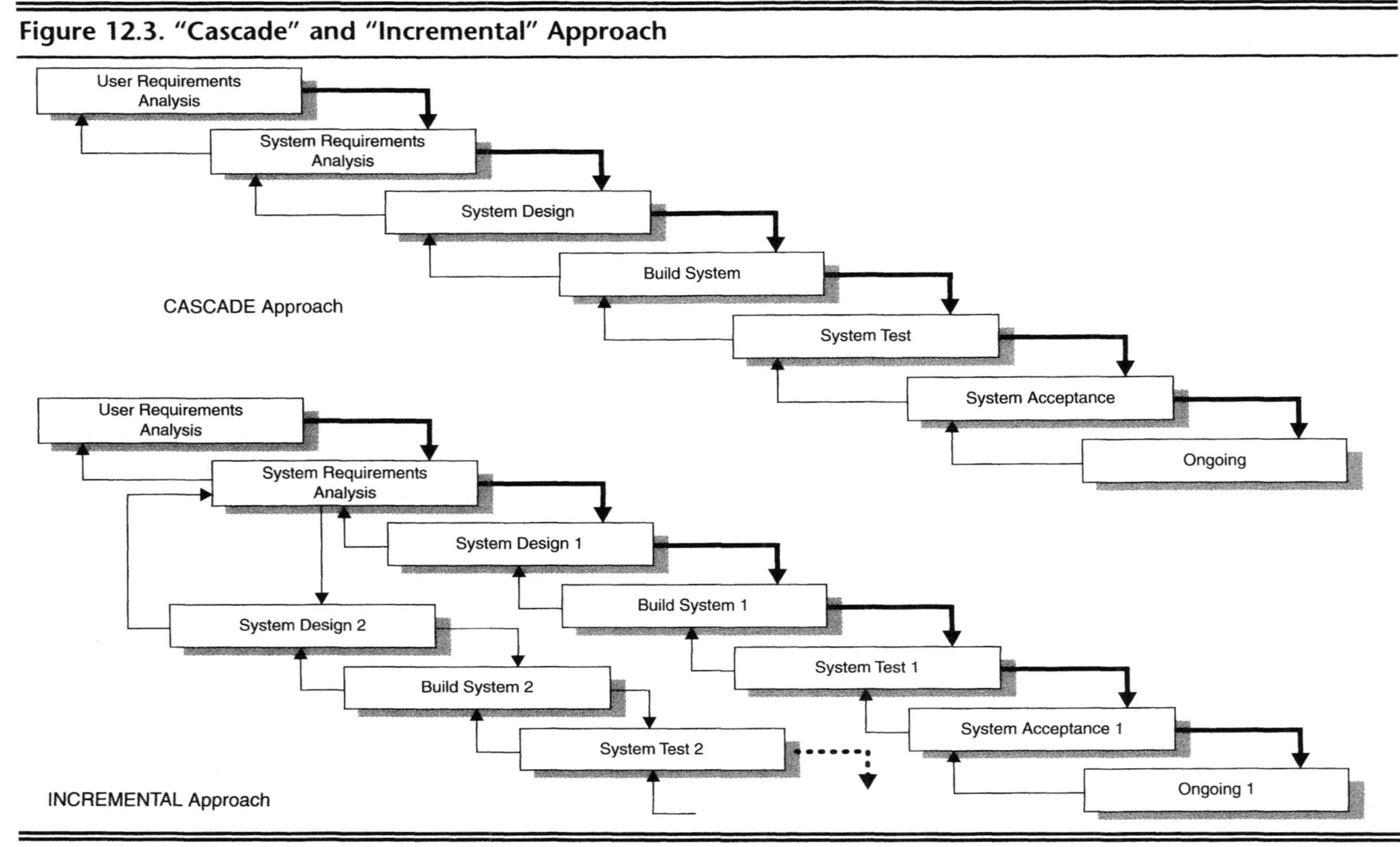

Figure 12.4. Relationship Between Specifications and Testing

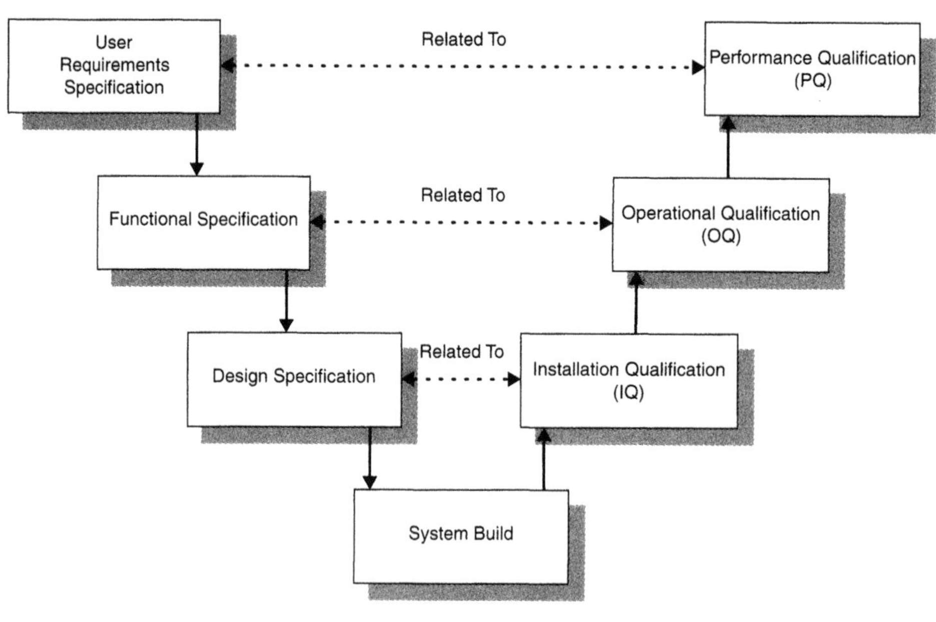

Figure 12.7 identifies the documentation that should be produced during the development of a network system. Of course, the specification and testing stages could be different depending on the complexity of the network design. In particular, the network wiring and electrical design specification (Document A), the network hardware design specification (Document B) and the network software design specification (Document C) chould be merged into one document divided into three sections. The cabling and labeling specification (Document D), in reality, should be a set of drawings that shows how the data and electrical cables are physically located. The cabling and labeling specification will include piping/ducting layouts and the location of the network equipment. The equipment installation details (Document E) and software module installation and configuration details (Document F) could also be merged into one document. The data wires acceptance testing, hardware acceptance testing and software acceptance testing (Documents G, H and I) could again be merged into one document that collates all of the tests that must be executed on the network. Separate documents can nevertheless be appropriate when project logistics require the installation and testing of electrical supplies, network equipment and network software at different times.

Figure 12.5. Relationship Between Specifications and Testing in the Incremental Approach

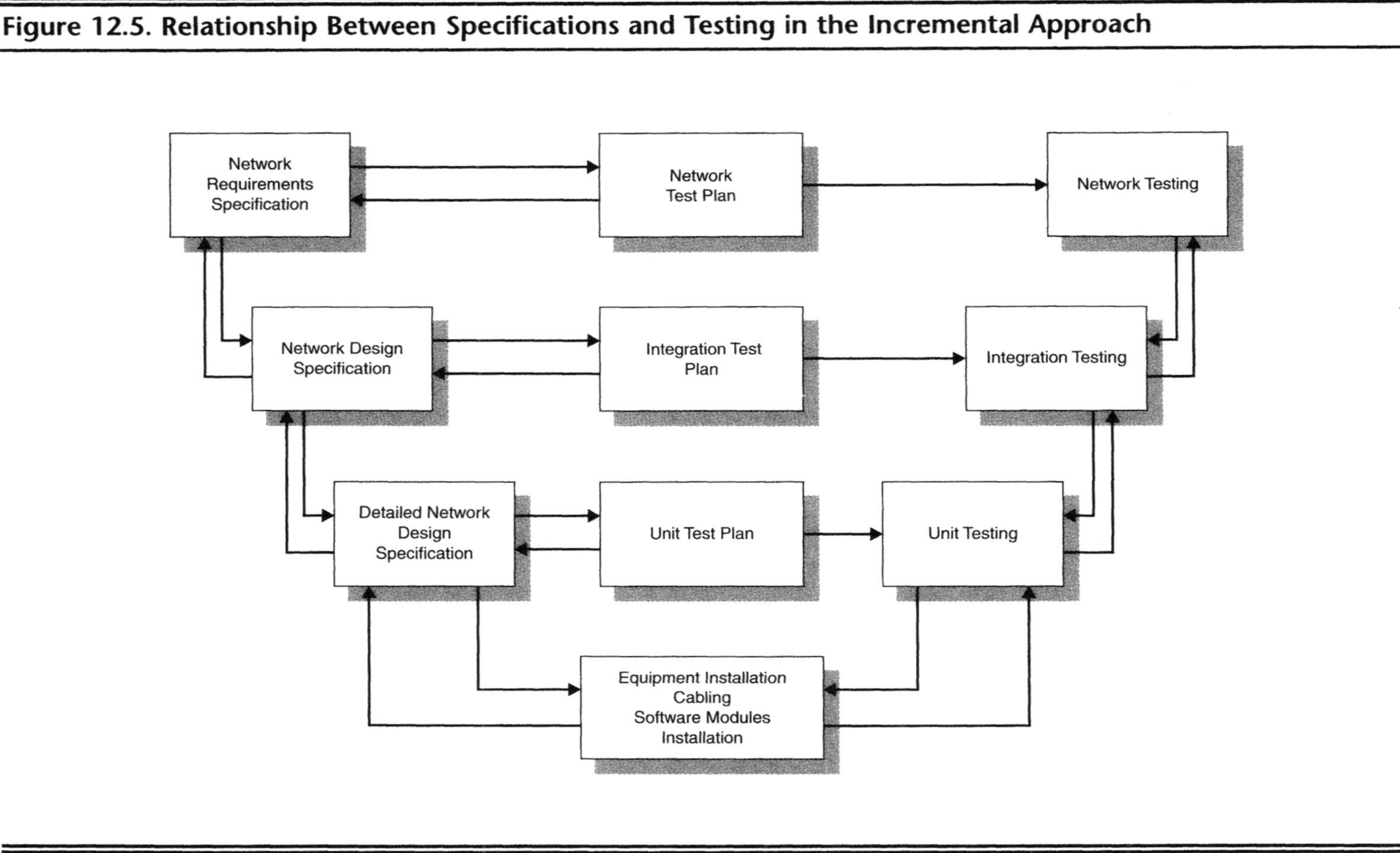

Figure 12.6. Validation Activities for a Network

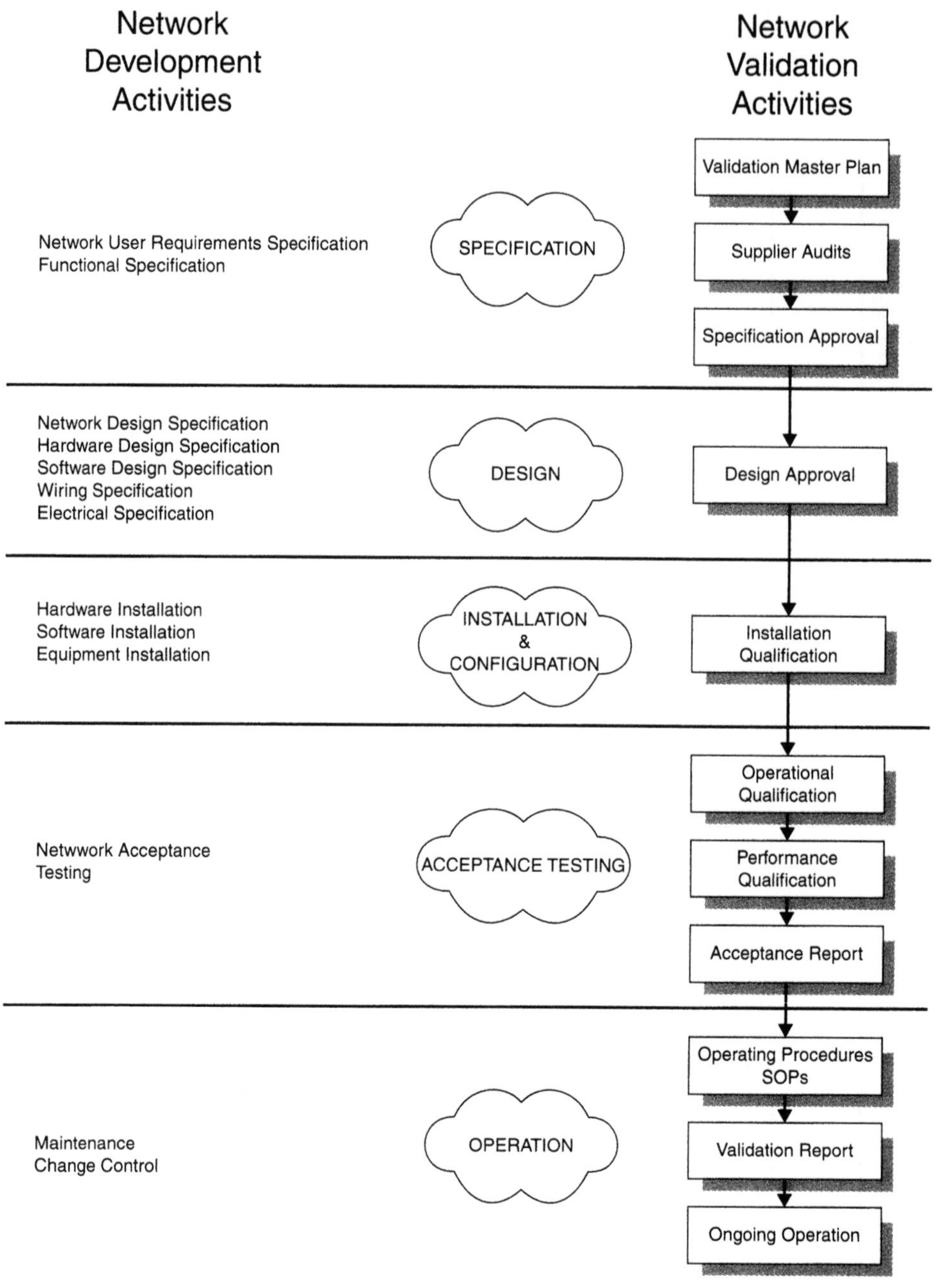

Figure 12.7. Documentation in the Life Cycle

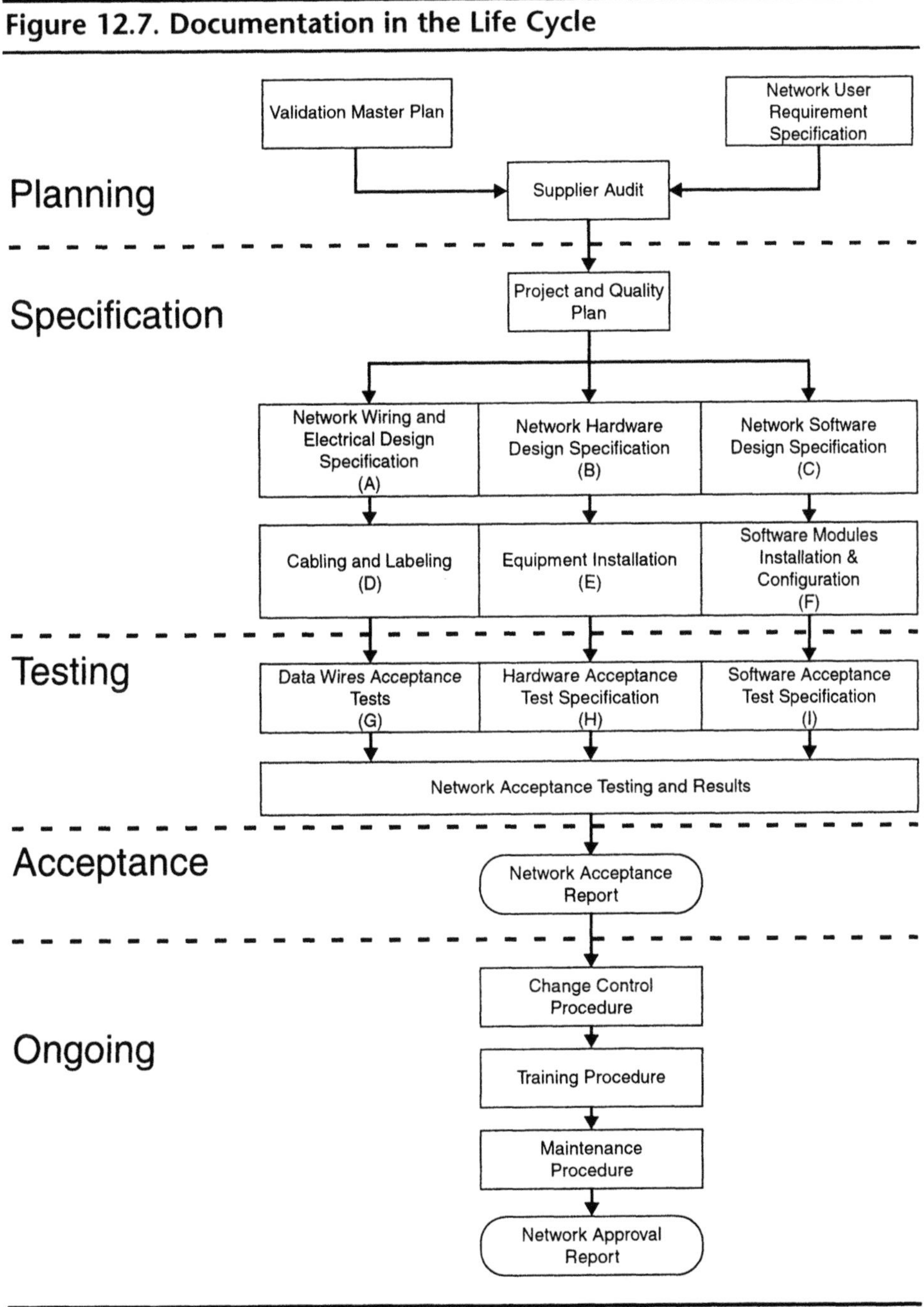

Specifications of Networks

The following specifications are based on the GAMP Guide [3].

The LAN Design Specification

The contents of a LAN design specification are outlined below. It is strongly recommended that the design specification include a map of the network and the systems interconnections.

- **Introduction:** This section contains information on who produced the document, under whose authority and for what purpose. The relationship to other documents should also be reported.
- **Design Overview:** This section should briefly describe the design of the LAN. It should introduce the basic concepts used in the design, and discuss the rationale of the proposed solutions. The following subsections should be included:
    - *Site/Area Description:* It may contain a drawing of a simplified site layout (area on which the LAN will be installed), identifying building and area classifications (e.g., manufacturing units, warehouse, laboratories, etc). For each area, a description of the other network systems should be available (areas or buildings that require wireless sub-LAN, or areas that must sustain a large number of high bandwidth connections simultaneously because they are dedicated to videoconferencing, multimedia devices, etc.).
    - *A summary of peculiarities* of a site LAN environment:
        - Temperature
        - Humidity
        - External interference
        - Physical security
        - Radio-frequency, electromagnetic and ultraviolet Inter ference.
    - *The design/solution:* Demonstrate how design requirements meet, or do not meet, the User Requirements Specification (URS). Include how operational environment needs are fulfilled.
- **LAN Architecture**
    - *LAN General Description:* This subsection briefly describes the design of the LAN. It may include a drawing of logical schema of the LAN with the major network components and how they interact with the environment. A description of these topics should be listed:

  - Connectivity
  - Network redundancy
  - Routing capability
  - Equipment/device naming conventions.
- *LAN Detailed Description:* This subsection contains a list and description of all LAN components. Those components may be classified as follows:
  - Higher Components: Components that fall into the application presentation and session OSI (Open System Interconnection) layer, e.g., NOS (network operating system). It should list all application services.
  - Network Components: Components that fall into the session and transportation OSI layer, e.g., routings, transportation protocols.
  - Physical Components: Components that fall into the data link and physical layers, e.g., terminal servers, hub management cards, and so on.
  - Cabling Infrastructure: List of all wiring standards that must be taken in account, e.g., wiring standard EIA/TIA-568. In this subsection may be specified, for each area, the related wiring concept, e.g., horizontal wiring, and so on. It should list all types of cables that must be used. Screening and shielding, labeling and tools and equipment should also be considered.
- *Electrical Supplies:* All electrical supply requirements for the LAN are addressed. Elements to be considered include:
  - earthing,
  - loading,
  - filtering,
  - uninterruptible power supply (UPS),
  - disconnection by fault and
  - electrical safety.
- *Network Management:* This subsection defines, if part of the design, the network control system and its functions.
- *Security:* This subsection defines the security requirements and physical and logical access.

- **LAN Detailed Design**
  - *Exact configuration of each component of the network:* This subsection defines the exact number of each component of the net-

work. It should consider the following elements for each hub/working group:

- Reference with geographical location on layout
- Number and type of equipment/cards
- Number and type of connection gates available
- Hardware/software parameters to be used
- Address for each network component.

— *A series of drawings showing the exact location of all network equipment and cables:* In this subsection should be attached, as minimum,

- the complete layout of the site showing the backbone cable path and location of main network objects, e.g., hub end working groups;
- a detailed drawing showing, for each area or building, the location of each network component and the cable path from the hub to the faceplate on the wall.

The WAN Design Specification

Below is a definition of which sections should be included in the WAN design specification. Due to the particular objects of the design, diagrams and drawings are strongly recommended to define WAN.

- **Introduction:** This section contains information on who produced the document, under whose authority and for what purpose. The relationship to other documents should also be reported.
- **Network Overview:** This section contains a general description of the WAN. It may describe the different implementation phases of the WAN, the interconnection points and geographic location of the interconnection points. This section should contain information on
  - — connectivity,
  - — services,
  - — access points,
  - — security and network management and
  - — network dimensioning and performance.
- **The WAN Architecture:** This section briefly describes the design of the WAN.
  - — *General Description:* The basic concepts used in the design of the WAN are introduced. It may include a drawing of the logical schema of the WAN with the major network components and

how they interact with the environment. The description of these components should include

- network features and characteristics,
- connectivity,
- network redundancy and
- dynamic routing capability.

- **Detailed Network Design:** This section should contain a description of each component of the network, both hardware and software. It should describe how these components are connected and configured, including
  - hardware and software requirements,
  - protocols supported and
  - configuration.
- **Access Points Architecture:** This section should describe the access points architecture, in terms of users and technology that must be used to connect to the WAN. It may address
  - different remote access categories and
  - cabling configuration.
- **Application Services:** This section should describe the application layer services supported by the WAN, including the following services: electronic messaging systems and Simple Mail Transfer Protocol (SMTP).
- **Network Management:** This section should describe all of the management functionality available on the WAN, including
  - configuration management,
  - performance management,
  - fault management and
  - security management.
- **Security:** This section should describe all security issues involved in the operation of the WAN, including
  - authorization granting rights to access resources,
  - access control mechanisms,
  - access control policies,
  - routing control mechanisms and
  - security protocols.

Qualification of Networks

A network test specification must be written and approved before testing can begin. The test protocol should not introduce any new specification details but instead reference the network design specification. Raw data should be collected during testing to provide evidence of test outcomes. This evidence should be retained with the test specification and a test report summarizing the results of individual tests, listing test discrepancies and failures and identifying corrective actions and any necessary or recommended repeat testing. It is this test report which will conclude whether or not the network is for for purpose. Table 12.1 is a sample test protocol for network qualification.

PRACTICAL ISSUES

Network validation is only necessary where that network is used to convey controlling instructions or GxP data. This said, it is important to recognize that networks will often be installed without a validation requirement and then later in its life be requested to support a GxP application. It can be very difficult to retrospectively validate a network. It is much better from the outset of installing a network to establish good IT practice and keep appropriate records detailing work done. Then if the use of a network changes, there is some documentary evidence in place that can be used to supplement validation. In particular, good IT practice for networks should include

- configuration management,
- Installation Qualification and
- change control.

The content of the network specification and test protocols will depend on the complexity of the system. For small systems, it may be possible to incorporate the design of the network as a special section within the Functional Specification and similarly incorporate the network test cases as a special section within the Operational Qualification.

Supplier audits for network hardware and software components are not usually necessary as these are normally an industry standard.

Once a network is validated, care must be taken to maintain its validation status. Tests should be formally recorded when new systems are connected to a validated network, even if the new systems themselves do not require validation. The addition of a new system to a network will alter traffic loading and, hence, could compromise the responsiveness of the network. Security, too, could be compromised. Networks often employ firewalls to protect sensitive portions of a network from interference or abuse.

Another issue that has practical implications on validation is the use of third parties to maintain and support networks. In these circumstances,

Table 12.1. Sample Test Protocol for Network Qualification

Introduction

The Validation Plan, network design specification and validation procedure used for testing are reference here.

Scope

Define the scope of the qualification program to be undertaken.

- Visual check of components
  - Against design specification
  - Against standards
  - Against statutory requirements
  - In accordance with manufacturer's instructions
- All equipment and materials undamaged, clean, new and correctly installed (refer to installation records).
- Any requirements for hazardous areas are met.
  - Capacity testing
  - Software version checking
  - Electrical supply and interference testing
  - Manufacture diagnostic testing
  - Power on-off testing
  - Operational environment
  - Configuration/system testing (each user port tested for connection to network)

Test Plan

Describe the overall testing philosophy.

- Specific areas not tested and why
- Any logical grouping or ordering of tests
- Personnel required for test groups

Testing Prerequisites

- Hardware requirements (systems setup)
- Test equipment requirements (including simulation tools)
- Test data requirements
- Reference document (e.g., operating manual, vender data sheets)

Test Procedures

Details of all the test cases. Each test case should be on a separate page. The test case should collectively provide 100 percent coverage of the network design specification.

contracts must be established with the suppliers defining what procedures will be used by contract staff and what records will be maintained and/or handed over to the pharmaceutical manufacturer. The pharmaceutical manufacturer, not the supplier, is accountable to the GxP regulatory authorities for validation. The operational terms of the contract are usually defined in a Service Level Agreement (SLA).

The new regulation 21 CFR 11 on electronic records and electronic signatures impacts the use of networks. A detailed assessment of this regulation is beyond the scope of this chapter. A discussion on 21 CFR 11 can be found in Chapter 16.

REFERENCES

1. FDA. 1995. *Glossary of Computerized and Software Development Terminology*. Rockville, Md., USA: Food and Drug Administration.

2. FDA. 1983. *Guide to Inspection of Computerized Systems in Drug Manufacturing: Reference Materials and Training Aids for Investigators*. Rockville, Md., USA: Food and Drug Administration.

3. UK GAMP Forum. 1998. Supplier's *Guide for Validation of Automated Systems in Pharmaceutical Manufacture,* Version 3. Available from International Society for Pharmaceutical Engineering.

4. Wingate, G. A. S. 1997. *Validating Automated Manufacturing and Laboratory Applications: Putting Principles into Practice*. Buffalo Grove, Ill., USA: Interpharm Press.

13

Maintenance and Support of Validated IT Systems

Erna Koelman
Kees de Jong
Kees Piket
Solvay Pharmaceuticals
DA Weesp, The Netherlands

This chapter addresses the various aspects that are important to keep validated information technology (IT) systems in a validated state during—and after—the operational phase of their life cycle. As most systems are now connected to a computer network, communication technology–based components should be included when considering the configuration of a system. Therefore, it is now common to speak of information and communication technology (ICT)–based systems. IT staff are challenged to maintain and support these systems within the continuously changing environment in which these systems are operating.

Many events may jeopardize the validated state of a system. These can be technical or organizational events. Technical events include renewal of PC (personal computer) hardware, migration to a new version of an operating system and all kinds of software changes (e.g., to cope with a new millennium, implement a changed information security policy). Organizational events include renewal of IT staff or outsourcing IT activities. In order to keep systems validated, procedures must be defined and understood by all staff involved. Modern systems are often maintained by more than one organizational unit. Local organizations may include centralized helpdesks, third-party network service providers, competence centers and so on. In addition, there may be decentralized internal staff and/or external partners (e.g., service providers and ICT suppliers) who are playing a role in maintaining and supporting current systems. Wherever the term *support organization* is used in this chapter, it can be any combination of the above-mentioned internal staff and external partners.

MANAGEMENT SUPPORT ORGANIZATION

Systems, System Components and Changes

The following systems or system components can be distinguished: local area network (LAN), wide area network (WAN), network operating system (NOS), operating system (OS), application software, database management software, standard application software, hardware, documentation and Standard Operating Procedures (SOPs). Table 13.1 provides examples of systems and components, and the types of changes that may occur with each system or component.

When a computer-related system is validated, its components must be clearly defined. Changing one or more system components affects the validated state of the system. Change control procedures should assure or reestablish the validated state of the system when a change is applied.

Tasks and Responsibilities in the Support Organization

The structure of the support organization must reflect the tasks and responsibilities it has been assigned. This is usually defined, for example, as something like providing and maintaining appropriate IT facilities to support the information needs of the business processes. In general, its scope can be defined in two areas:

1. The area of system acquisition, development and implementation
2. The operation and maintenance of the ICT components and the ICT infrastructure, whose evolution is generally independent of individual IT projects.

Production process automation is often not within the scope of the IT department.

System Acquisition, Development and Implementation

Although this chapter is about the operational phase of systems, the basis for proper support and maintenance of the systems should be established during the acquisition, development and implementation phase. During these phases, the operational characteristics of the new system need to be revealed

Table 13.1. Examples of Systems, System Components and System Changes

System/Component	Examples	Examples of Changes
LAN	Network within the site for which the access is fully controlled by (or contracted out but still under the responsibility of) the company	Any change in the LAN infrastructure (e.g., OS, NOS, servers, workstations, applications, file servers, network equipment)
WAN	Connections between different LANs and public networks	Addition of a new connection, changes to firewall configuration
NOS	Windows NT, UNIX®, Open VMS and associated tools	Bug fix, patch, new version, new feature enabled, change system parameter in the NOS
(Client) OS	Windows, OS/2, MacOS	Bug fix, patch, new version, new feature enabled, change system parameter in the OS
Application software	LIMS, EDMS, MRP II system, spreadsheet macros	Bug fix, patch, new version, changed macro in the application software
Database management software (central and distributed)	Oracle®, Microsoft Access®	Bug fix, patch, new version of the DBMS
Standard application software	Spreadsheets, statistical packages, word processor	Bug fix, patch, new version in the standard application software
Hardware	Computers (clients/servers), disk systems, interfaced instruments, network equipment (e.g., hub, switch, router, firewall)	New disk system, replacing equipment
Documentation and procedures	Validation documentation, user and technical manuals, operation and maintenance, backup, security procedures, network management system, incident/problem management system	Validation Plan; updates of the user manual, system description or SOPs

VMS = Virtual Memory System; LIMS = Laboratory Information Management System; EDMS = Electronic Document Management System; MRP II = Manufacturing Resource Planning; DBMS = Database Management System and should be assessed in relation to the existing ICT infrastructure. The support organization has to be involved in the early stages of the life cycle. The role of the support organization, before adopting the system in its maintenance and support processes, can be defined as follows:

- Complete the User Requirements Specification (URS) document with ICT–related issues that are important for the company (i.e., security issues, fitness for inviting to tender, standards to be met).
- Translate the URS into a Functional Specification.
- Contribute to a System Integration Test Plan that will test the ICT–related requirements.
- Complete the Acceptance Test Plan with ICT–related tests.
- Prepare the integration of system components in the existing ICT–infrastructure.
- Prepare and agree on the maintenance and support activities for the new system.

Operation and Maintenance

Much attention has always been given to the system development process. But gradually the importance of the daily operation and maintenance activities of validated systems is recognized. Different information systems (IS) are more and more integrated, often sharing the same layered components in the infrastructure, some of which need their own version. It must be emphasized that the support organization, *not* the development group or the system manager, should formally accept the introduction of a new system in order to ensure that the infrastructure is able and equipped to handle this new functionality. Operational management must be aware of the impact the new system will have on the infrastructure. Maintenance and support procedures of the system must be ready before it is released for use.

The key jobs that can be distinguished within a modern support organization are shown in Table 13.2; most of them are usually found in the central IT department. A comprehensive set of required services should be defined, including service support and delivery. Service support includes the helpdesk, problem management, change management, configuration management and software distribution. Service delivery includes service level management, capacity management, contingency planning and cost management. These services may be provided by different IT organizations (internal and/or external), but operational procedures need to be aligned for a seamless overall service.

Interdepartmental Responsibilities

Once a computer-related system has been validated, the validated state has to be maintained. This requires an adequate maintenance system and SOPs that are incorporated in the relevant Quality Management System(s) (QMS), which should lead to

Table 13.2. Key Roles and Responsibilities Within the Support Organization

Role	Area of Responsibility
Operational manager	• Change management • Problem management • Helpdesk
Infrastructure architect	• Configuration/asset management • Business continuity planning • Network capacity
Security manager	• Network access policy • System integrity
Test manager	• Test facility management • Test procedures • Test base administration
Quality manager	• Translating company quality policy into ICT procedures • Quality system support organization • Qualifying third parties • Standardization
Contract manager	• Composing request for proposals • Service/Performance Level Agreements (SLAs/PLAs) • Cost management (internal and external)
Business analyst (can also be part of business unit)	• Information planning for the business processes • Translating GxP requirements into ICT service levels • Project management

- operation in line with policies and procedures;
- proper access control (if electronic signatures are used, the identification system must be validated, and system security should be equivalent to or greater than the security in a comparable manual system);
- appropriate training programs developed, conducted and documented on an ongoing basis;
- approved and valid user manuals (which are kept current);
- well-documented problem handling;

- a proper change control process;
- a proper maintenance process; and
- ongoing evaluation (consider possible retirement, if appropriate).

Senior management has overall responsibility for setting up, implementing and maintaining a company policy and a derived set of procedures on the validation and use of IT systems. Without executive management support, however, business pressures can often lead to validation practice being compromised.

Keeping systems validated is the responsibility of the system owner; however, all staff involved in the operation and maintenance phase and in the retirement of the relevant system have delegated responsibilities.

Key players that can be identified in this process are as follows: all levels of management, the persons responsible for the processes and activities to be performed under GxP, suppliers (external suppliers as well as internal system developers), IT/system managers (including service providers, if applicable), system users and Quality Assurance (QA). It is highly recommended to also involve the Purchasing and/or Legal departments in the process if new modules, updates or extensions are required. Table 13.3 indicates the basic roles, responsibilities and tasks for the key players in order to validate systems and to keep them validated. Most of the responsibilities and tasks are equally relevant for the period prior to initial validation and for keeping the systems validated.

MANAGEMENT PROCEDURES

In the operation and maintenance phase of a computerized system, the main objective is to keep the system operational and in a validated state. In order to meet this objective, procedures need to be in place to ensure that the results of the operation and maintenance phase have a high probability of being of acceptable quality.

The following subjects need to be covered by the procedures:

- Use of the system (procedures and/or user manual)
- Backup and restore
- Periodic evaluation
- Preventive maintenance
- Corrective maintenance (problem report procedure)
- Authorization, safety and security
- Training
- Disaster Recovery and Contingency Plans
- Business continuity

Table 13.3. Basic Roles, Responsibilities and Tasks in Keeping Systems Validated

Function/Role	Responsibility	Main Tasks
Management	Overall responsibility for company policy.	• Set up and maintain a validation policy and procedures in line with GxP regulations and current standards. • Make sure that all staff are qualified and trained. • Ensure implementation of validation strategy/policy. • Make available sufficient resources to guarantee that systems are kept in a validated status. • Monitor compliance/define QA responsibilities. • Make sure that proper measures are taken to guarantee integrity, completeness and retrievability of the data of retired systems.
Persons responsible for the GxP processes/activities/studies (e.g., preclinical study director, process owner, clinical project leader or sponsor/investigator)	Validated data acquisition, data processing and archiving procedures related to the GxP activities under his/her control.	• Identify the systems used in the GxP activities. • Verify the validated state of the systems used for GxP activities. • If applicable, request revalidation. • Make sure that only trained staff are involved.
System owner	Final responsibility for keeping the system validated until after retirement of the system.	• Make sure that formal change control procedures, including impact analyses, exist and are followed for the system. • Rerelease the system after changes have been made and accepted. • Make sure that the required periodic reevaluation takes place. • Authorize continued use of the system after reevaluation. • Keep training available for all relevant staff. • Make sure that proper measures are taken to guarantee integrity, completeness and retrievability of the data of retired system.

Table 13.3 continued on the next page

Table 13.3 continued from the previous page

Function/Role	Responsibility	Main Tasks
Suppliers (including internal system developers)	Design and produce systems in line with current standards and GxP regulations.	• Implement a proper quality system which covers (sponsor and) regulatory requirements. • Make use of recognized standards (e.g., ISO, IEEE, TickIT, GAMP). • Ensure that all staff are adequately qualified and trained. • Supply adequately updated system and user documentation. • Supply qualification and validation support
IT/system managers	Maintain systems and the generated data in line with current standards and GxP regulations.	• Implement procedures for GxP–compliant access control to all levels of system components. • Implement GxP–compliant procedures for the maintenance of the system. • Implement GxP–compliant change control procedures. • Implement proper problem handling procedures. • Implement proper procedures for archiving of electronic data and, if applicable, for retired systems. • If applicable, make sure that the service providers comply with current standards, the GxP guidelines and company policy on validation.
System users	Use the system in line with GxP regulations.	• Obtain/receive GxP training. • Obtain/receive system training. • Use the system in line with system manuals and procedures.

Table 13.3 continued on the next page

Table 13.3 continued from the previous page

Function/Role	Responsibility	Main Tasks
QA	Make sure that the required quality standards are met.	• Know the basic principles of software engineering. • Request to apply system life-cycle principles and proper standards. • Check compliance with standards, regulations, company policies and procedures. • Advise and give guidance on, review and approve the necessary validation activities and documentation. • Monitor the validation process, including adherence to approved plans. • Audit the use of the systems, central computer services facilities and suppliers (if applicable). • Monitor and approve the system retirement processes and documentation.
Purchasing and/or Legal Department	Advise on supplier selection and/or legal matters for new modules, updates and so on.	• Collect information on expected reliability and continuity of suppliers and on the costs of their products (including support). • Negotiate about financial and/or legal/licensing matters • Advise on supplier selection processes on experiences with known suppliers

ISO = International Organization for Standardization; IEEE = Institute for Electrical and Electronic Engineers; GAMP = Good Automated Manufacturing Practice

- Change control
- System retirement

This section describes the minimum requirements for the necessary procedures for most of these subjects; change control and system retirement are, however, described in more detail in separate sections of this chapter.

Use of the System

A set of procedures or a user manual needs to be in place that describes how the system must be used.

Backup and Restore

In order to guarantee the availability of the stored data, procedures on backup and restoration of software programs, data entered and operational data need to be in place. The following items need to be covered in the procedure(s):

- What is the frequency of backup?
- Verify retrieval of the backup before storing the backup (e.g., random checks).
- At least two generations of each backup set should be kept.
- There should be a system in place to manage the availability of the backup within an appropriate period of time.
- Stored backups should be checked for accessibility, durability and accuracy at a frequency appropriate for the storage medium (frequency to be specified in the SOP).
- Backup copies should be kept separate from the computerized system in a fire-protected area.
- Following changes to the system, change control should ensure the availability and integrity of the backup copies by restoring the data on a trial basis.

Periodic Evaluation

Computerized systems should be periodically (frequency to be specified in the SOP) reviewed by the system owner or his or her delegate to confirm that its validated status has been sustained. This review should be documented. Topics for consideration in this procedure should include, but are not limited to, the following:

- *Operating and maintenance procedures (or SLA):* Check if they are still current and that they are followed.
- *Change control:* Check what has changed within the computerized system and whether it is sufficiently documented.

- *Security:* Review physical access and use of passwords, including frequency of change and list of active users and unauthorized access attempts.
- *Training:* Check if all users are trained, and that training on changes has been applied.
- *Performance:* Check on documented performance metrics of the computerized system (e.g., response time, level of utilization of memory, network load). Check on incidents and corrective actions. Check the use of any helpdesk facility.
- *Revalidation:* Examine the need to revalidate, e.g., if many changes have been made or if regulatory requirements have altered. Check whether any unauthorized changes have been made.

Preventive Maintenance

For each system, procedures for preventative maintenance (including periodic calibration or revalidation) should be established. The objective of preventative maintenance is to check, calibrate or replace critical system components to prevent future failure and/or breakdown of the system, as this could have a negative effect on the quality of the product (or data). The following items need to be covered in the preventative maintenance procedure(s):

- Critical system components should be listed.
- The procedure should describe what should be done, when it should be done (frequency) and what the qualifications of the person performing the task should be.
- System components should be labeled with the date of the last and/or next maintenance (it should be clear which date is indicated).
- All maintenance should be documented (e.g., maintenance logbook). The documentation should include the activities performed, the persons who performed the activities, when they were performed and the results confirming that the maintenance was satisfactory.
- In case preventative maintenance leads to changes, the change control procedure must be followed.

Corrective Maintenance (Problem Report Procedure)

When errors or problems in software and/or hardware are detected, their cause must be investigated. It might be necessary to decide to perform corrective maintenance. Corrective maintenance is usually conducted through a change control procedure. Any changes resulting from emergency repairs should be subsequently reviewed, tested, documented and approved in accordance with the change control procedure. The problem report procedure should assure that the problem is documented (in logbooks or on separate forms) as detailed as possible. Errors, problems and decisions whether or not to perform corrective actions and the effects on the product or the integrity of data have to be documented and archived. For large computerized systems, this can be organized with the use of a helpdesk facility (including proper procedures); for smaller systems, a separate procedure should be established.

Authorization, Safety and Security

Procedures should describe how the information on computerized systems is protected. There are two basic ways of protection of the information, which should both be covered:

1. *Authorization:* How is access to the system or to information managed, e.g., the use of passwords?
2. *Security and safety:* This refers to the protection of the information. How is the information stored on the system? How is the system protected against illegitimate access, corruption and loss through unauthorised persons? How is the information protected against corruption and loss from faults in the system?

Training

All personnel using computer-related systems in GxP areas have to be trained adequately. Staff, including contractors, should be trained according to a Training Plan. Training must be documented in the training records. The way training is arranged and documented may be described in a general company, departmental or system-specific procedure, or a combination of some or all of these, as part of the validation process and documentation.

Contingency Plan

The Contingency Plan identifies the actions to be taken in the event that the system fails to function for a period of time, and how the business (process) will be managed during this emergency. The actions taken to bring the business (process) back into use may involve temporary solutions (e.g., using manual records or using another computer) or the decision to stop the process until the system is functioning again. In all circumstances, it is important to consider what actions (like testing) are needed to ensure that the solution is functioning in a controlled manner. The Contingency Plan should detail the procedures for continuing the activities in case of system unavailability (e.g., quitting and restart later, quitting and continue later, full paper and backup procedures). When the computer systems return to service, there may also be a catch-up to input GxP data collected manually when the systems were unavailable.

Disaster Recovery

The most common "disasters" are a power failure and a hard-disk failure; however, other disasters also need to be considered, if appropriate. A Disaster Recovery Plan should be available for each system as well for the central computer facilities. Such a plan (or SOP) should describe all activities required to restore a system to the conditions that prevailed before the disaster. If applicable, actions to be taken in case of a disaster should be described at the departmental level. Disaster recovery procedures should be tested initially and, if applicable, also periodically.

Business Continuity

The system owner should assure as much as possible that the business continues independent of the continuity of the supplier(s). For source code of software, this can be arranged by a so-called software escrow agreement. Where documentation is archived by a third party, an agreement should be made to keep access to this documentation. If old versions of systems need to be available in the future, continued access to system software and documentation must be assured.

Change Control

All changes to validated computerized systems must be reviewed, authorized, documented, tested (if applicable) and approved before implementation. Changes due to emergency repairs, however, should be subsequently reviewed, tested, documented and approved in accordance with the change control procedure.

The procedure for change control can be divided into four phases (see also Figures 13.1 and 13.2):

1. Request for change
2. Change evaluation (impact analysis) and authorization
3. Testing and implementation of the change
4. Change completion and approval

Request for Change

For every system, a system owner should be (have been) appointed according to the system documentation (or Validation Plan). A proposed change should be directed to the system owner who is responsible for ensuring that all changes to the system are reviewed, authorized, documented, tested (if applicable), approved and implemented in a controlled manner. The system owner may delegate this responsibility if it is documented.

Figure 13.1. Change Request, Evaluation and Authorization

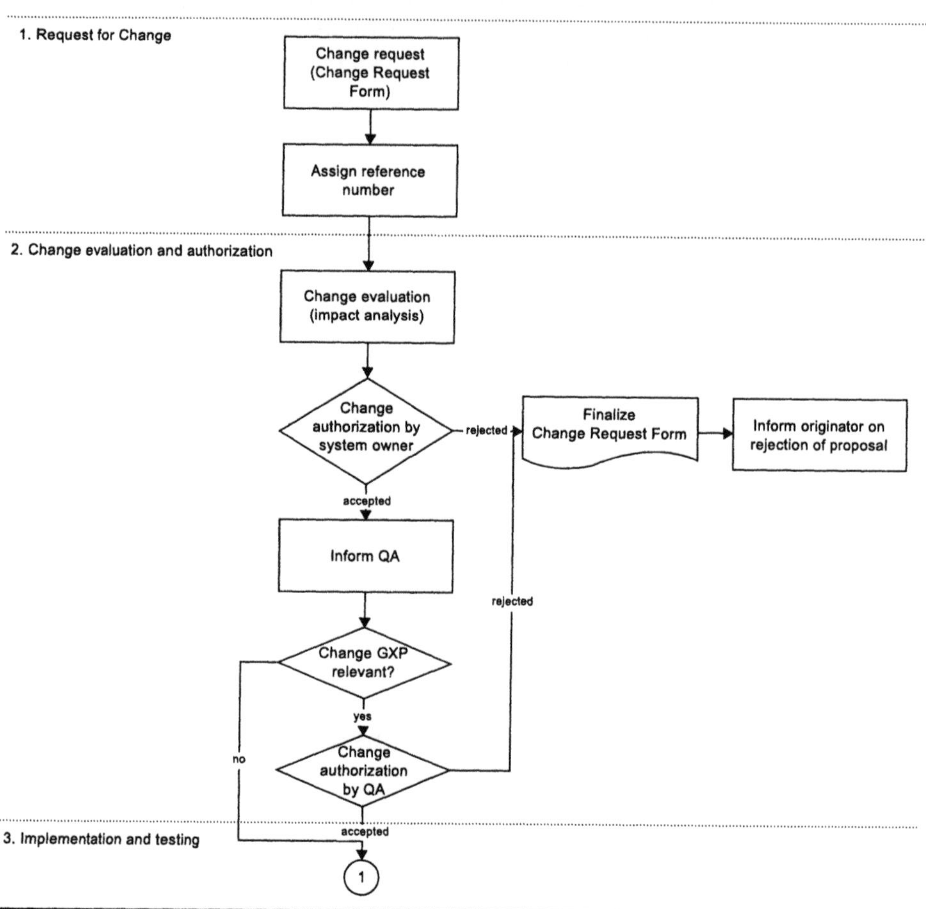

Figure 13.2. Implementation, Testing, Completion and Approval of Changes

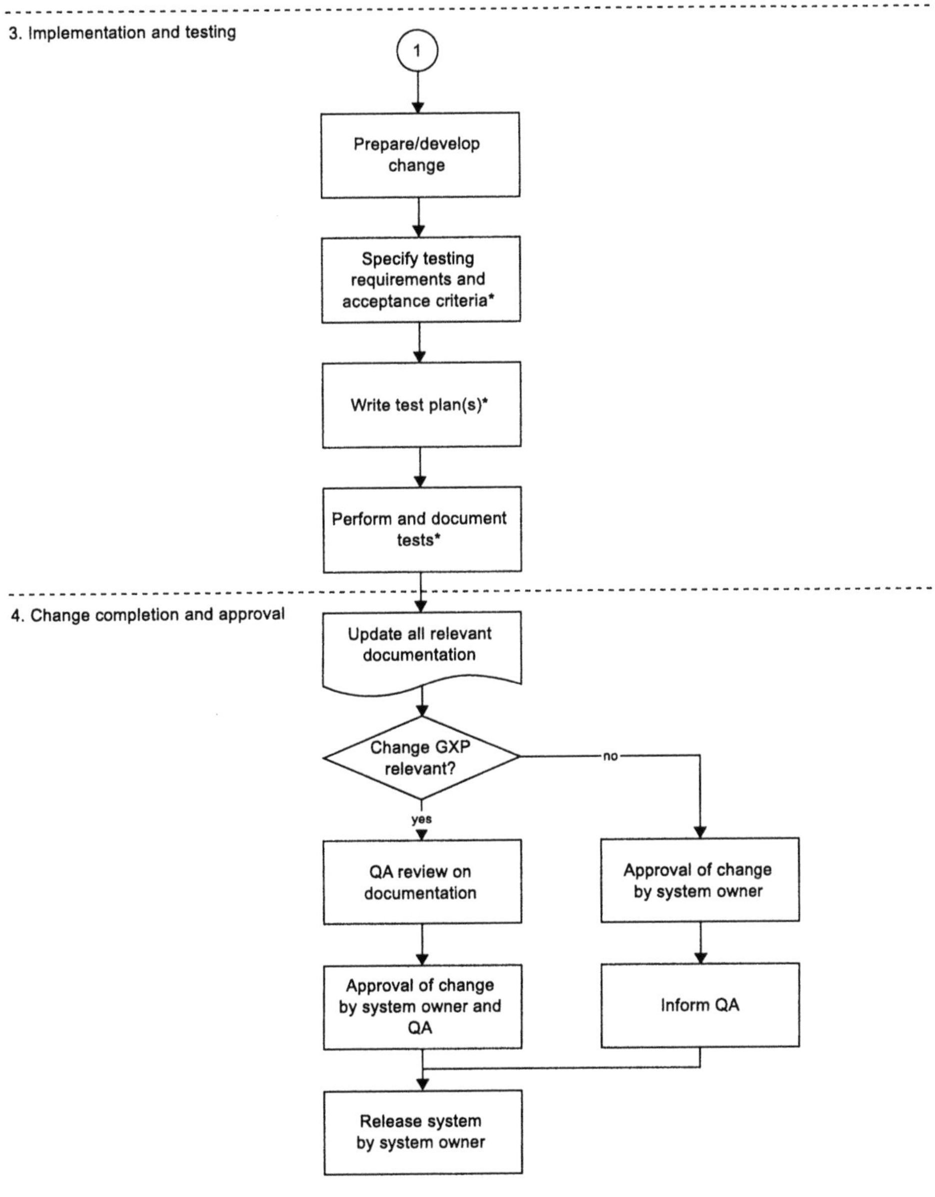

*Approval by QA if change is GP relevant

Any proposed change should be requested by submitting a Change Request Form. An example of such a form is given in Appendix 13.1. The Change Request Form should include at least the following items:

- Requester name
- Origination date
- Identification of component or software module to be changed
- Description of the change
- Reason for the change
- Unique reference number, to be assigned by the system owner or his or her delegate using a logging mechanism

Change Evaluation (Impact Analysis) and Authorization

Each change request raised must be reviewed and evaluated (accepted or rejected). In principle, for systems used for GxP–related activities, QA should be involved in the change control process.

For changes to a small systems like stand-alone analytical systems (e.g., HPLC [high performance liquid chromatography]), it is very straightforward which departments are affected by the change. For changes to large systems like an EDMS, a LIMS or upgrading the hardware (e.g., server) or the software (e.g., operating system) of the network, it is difficult to predict the impact on the (other) applications. For this kind of change, a good impact analysis is very important. This impact analysis includes issues as urgency of the change, risk, schedule, cost (time and manpower), safety and performance.

For some types of changes—e.g., minor changes with no expected effect on the outcome—the process might speed up by specifying (per system) a list of changes that are allowed to be executed without QA involvement for each separate change. The prerequisites are that this list and the simplified procedure to be followed result in equally good documentation. It should be established in an SOP what is approved by QA; adherence to a simplified change process should be subject to QA audits.

For changes to systems with a scope wider than the originating department, the change request is circulated to those departments, identified by the system owner or his or her delegate, that may be affected by the change. The departments that will be affected by the change should review the change request with their appropriate technical, management, QA and user personnel.

An impact analysis should be documented on, or attached to, the Change Request Form. It should list the alternative solutions, potential impact on other systems or applications and the required changes to system documentation. The affected departments should give a recommendation to the system owner for acceptance or rejection of the change. Once the impact of the proposed change is assessed, the system owner or his or her delegate must make a decision whether to accept or reject the proposed change.

After a change is accepted by the system manager, QA should be informed about the change. QA will review the change on the GxP relevance. QA can determine whether an additional authorization (accept/reject) is required. This will depend on the GxP relevance of the system or the impact of the change to the validation status of the system. Future QA involvement in the rest of the change process depends highly on this decision. Authorization by QA is required in several stages when the change is regarded as GxP relevant (see Figure 13.1).

Testing and Implementation of the Change

After evaluation and acceptance, the change can be executed, tested (if applicable) and implemented. In principle, software redevelopment and testing should follow the same procedure as newly developed software. The development and testing of a change must preferably (when possible) be executed in a test/development environment.

Testing is necessary to determine whether a change works properly and has not compromised system functionality. The scope of testing should be based on the impact analysis. Where potential impact on other system functionality or other applications is identified, testing must be extended to include affected areas. Testing should be performed according to a test plan, and all testing should be fully documented (e.g., test description, test items, acceptance criteria, results, date of test, names and signatures of persons who performed the test).

After implementation of the change (in the operation environment), the change should be approved by the system owner (Figure 13.2). This approval can be made based on the test results, or the system owner can decide to perform a separate acceptance test.

Change Completion and Approval

All documentation concerning the change and all documents required for operation with the change need to be completed. The Change Request Form should be completed and passed to the system owner for final review and approval. Depending on the GxP relevance of the system or the impact of the change, QA should review and approve the change. QA should always be informed about the change by sending QA a copy of the completed Change Request Form. The users should be informed (and trained if applicable) about the change. The system owner gives the final approval of the change and releases the system.

System Retirement

General Retirement Issues

Unfortunately, there is no detailed information on the retirement of IT systems available in any of the GxP regulations. The following regulatory issues, however, are considered relevant:

- Raw data include data directly entered into the computer.
- Major requirements related to raw data include retrievability, readability and integrity.
- Although the guidelines do not require that reanalysis of the data should be possible instantly and/or on the same system, original raw data or verified copies thereof as well as the algorithms used have to be available within a reasonable time frame.
- All methods used have to be demonstrably validated.

It is recommended that at the time of system retirement, at least the following system-related documentation be available and archived:

- Retrospective evaluation and/or prospective validation documentation
- Maintenance records/logbooks
- Relevant SOPs
- Information of the access to the system over time
- QA records
- System retirement project documentation

As long as the data can be retrieved and the integrity ensured without using the hardware and software actually used to generate and/or process the data, there is no requirement for archiving that hardware and software.

Management and QA Involvement

Management must be aware of the need to take decisions on system retirement and of their responsibility in this respect. For larger systems, it may be very useful for a project team to supply options, pros and cons and proposals on which management can take their decisions. The following information should be available to management for proper decision making:

- An overview of existing systems, including the life-cycle phase
- All links to other systems
- The validation status of each system or system component
- Costs of upgrading the validation status and new software/hardware purchase and support

- Regulatory impact of each system
- Each system's fitness for purpose

Management and QA should decide on the required QA involvement and document the outcome in a policy and/or procedures. Suggested QA involvement in this respect is to

- advise on needs and requirements,
- assist in a risk/benefit analysis,
- stimulate good validation and archiving practices and involvement in all phases of the system life cycle (as this markedly reduces the retirement effort) and
- perform a preretirement audit.

Planning the Replacement and/or Retirement of IT Systems

Replacement of an IT system is often combined with the retirement of an existing system. Practical issues that should be considered in the discussions on or in the preparation of (possible) replacement of IT systems are as follows:

- Will the transition require parallel running?
- Will the replacement need to be phased, e.g., some functions earlier than others?
- How much training on the new system will be required?
- What, if any, changes in working practices will be required?
- How do we handle any possible user resistance?
- Does the system store any GxP relevant data?
- Do raw data have to be migrated?
- What validation efforts are required?
- What retirement efforts are required?

Chapter 16 discusses the issue of data migration in respect to 21 CFR 11 on electronic record requirements.

One of the things that is generally underestimated is the importance of timely planning of IT system replacement. Usually, it is not an easy decision to retire a system. Too soon means that you will not get the full benefit of the system; too late may result in a crisis situation and can be very risky. Therefore, the advice is to *plan* replacement in time, preferably *NOW*.

There is no best way to retire a system. Depending on the situation, the optimal way of retiring a system can be anything from "pulling out the plug of the old and plug in the new system" up to very complex and expensive activities. For each situation, the optimal solution must be chosen based on a risk assessment and a cost/benefit analysis. The benefit of easy access to a possibly rare chance of occurrence should be assessed.

Raw Data Definition and Archiving

With respect to archiving, there is no one best solution for all systems and situations. For each system, a decision should be taken based on a risk assessment and a cost benefit analysis. There are three main options, each having different pros (+) and cons (–):

1. Leave the data where it is:
   - \+ Data remains accessible.
   - – But for how long?
   - – You are not really replacing a system, but adding one.
   - – Increased support and maintenance overhead.
   - – Old hardware may deteriorate and therefore make data inaccessible.
2. Move the data to a human readable form, off the computer:
   - \+ Human readable (e.g., ASCII [American Standard Code for Information Interchange], paper, microfilm).
   - \+ Portable (not depending on software).
   - \+ Standard archiving techniques.
   - \+ No increased support or maintenance overheads.
   - \+ Reasonable access.
   - – Likely to be reviewer unfriendly.
   - – Not readily available for reuse.
   - – Printouts require space requirements and are not environmentally friendly.
3. Migrate the data to the new system:
   - \+ Access for review and reanalysis.
   - \+ No increased support or maintenance overheads (eventually).
   - \+ Background data pool can grow.
   - – The migration process may require comprehensive validation.
   - – Data structure compatibility, even with a single supplier or system, may be an issue.
   - – A mechanism to ensure data integrity has to be in place.
   - – The amount of data to be migrated can be very high, e.g., in the range of several gigabytes or terabytes for large sets of three-dimensional chromatograms produced over several years.

Whichever method of transfer of data is chosen, it should be properly and demonstrably validated. Paper or microfiche are still the most stable forms. Despite modern technologies, these are still the longest-lasting media.

In many companies, a policy on duration of archiving is still lacking. This is the main reason why, in practice, nothing is destroyed, and voluminous archives exist.

In practice, most questions from regulatory authorities can be answered without any reanalysis. Therefore, depending on the system or data type, in some cases (particularly for numerical data and simple algorithms) the most efficient solution might be storing the data in human readable format. In that case, if reanalysis is required, the data should be reentered.

WHAT IS NEEDED TO SUPPORT VALIDATED SYSTEMS?

Risk Assessment

Because validating computerized systems is time-consuming, expensive and resource intensive, organizations are challenged to identify and prioritize which systems and to what extent will be validated. If time is taken to properly evaluate the risks when assessing the approach to computer validation, the validation efforts may be markedly reduced. There are organizational and system-specific risks to consider. It is highly recommended that each organization establish its own risk assessment process. Some items to consider are discussed below.

Organizational Factors

Each organization should define its policy on computer validation and make sure it is implemented and followed. For regulatory and business purposes, all computer-related systems must be validated which

- are used for the acquisition, generation, measurement or assessment of data intended for regulatory submission;
- are used for the generation, storage and retrieval of data relating to material management, production, quality control (QC), QA and distribution of pharmaceutical products;
- result in GLP (Good Laboratory Practice), GMP (Good Manufacturing Practice) or GCP (Good Clinical Practice) noncompliance in case of failure or latent flaw, and/or
- control information related to the safety and efficacy of the product supplied or the correctness of related information.

Management may also decide that additional systems should be validated to ensure the accuracy and reliability of data for business reasons.

An inventory should be made of all systems. The inventory should indicate whether each system falls within any of the above categoies and if it needs to be validated at all.

System-Specific Factors

The GAMP classification [1](see Appendix 13.2 to this chapter) gives guidance on the possible ways to reduce validation efforts based on risk assessment and the type of system. Another aspect to take into account is the potential risk associated with system failure. In case of low potential risk, validation efforts could be reduced.

The GxP criticality of each module or function within a system should be examined. If they are well defined and documented, the validation efforts could focus mainly on the most critical functions.

Finally, one should consider whether there are any safety nets for errors built into the system or process, e.g., QC measures. If so, it is worth considering a possible reduction of the testing for those aspects.

Risk Reduction

Even more effective than risk assessment is trying to minimize risk in the design and programming stages of system development. Some effective ways to reduce risks and, therefore, also the required validation efforts are as follows:

- Incorporate safety nets or independent controls in the process.
- Use standard pharmaceutical industry methods and configurable off-the-shelf packages as much as possible.
- Select qualified suppliers who have demonstrated to work according to proper SDLC (System Development Life Cycle) procedures and who supply good validation support.
- Take the time to define clear and effective system specifications, which relate to clear, unambiguous URSs.
- Define the basic testing when defining the user requiements. Then at least you know that the system requirements are indeed testable (which often is not the case).

When making use of these suggestions, validation efforts can often be markedly reduced. It should be noted that rationales for skipping activities should be clearly documented and approved by QA.

QA Program

There are several ways to involve QA in order to ensure that computerized systems in GxP environments maintain their validated status. Some important aspects are described below.

QA Involvement in Vendor Reviews and/or Audits

For major updates or new releases of validated systems, the change control process might include a (more or less extended) vendor review, which may need to be followed by an on-site vendor audit. In any vendor review or audit process, there should be QA involvement.

QA Involvement in the Change Control Process

QA involvement is also required in validation and change control processes, as discussed earlier in this chapter.

System-Specific Audits

QA is responsible for auditing computer-related systems that are used in GxP environments in order to assess adherence to policies and procedures. The frequency of the audits should preferably be related to experiences during previous audits (in the same discipline and/or of the same or similar systems). For systems in the operational phase, special attention should be given to the change control process, including the relevant documentation and training.

At least the following needs to be covered by QA procedures:

- System identification (e.g., system components, including version numbers, responsibilities, onset of use of the system or system components)
- Up-to-date documentation on the users (e.g., access rights, training/qualifications)
- Data handling (e.g., access control, data security/integrity, GxP compliance of data entries and edits, data processing procedures, Disaster Recovery/Contingency Plans, archiving and retrieval of data)
- System validation procedures and documentation (e.g., retrospective evaluation documentation, prospective Validation Plans, validation documentation and formal release by the system owner, existence of and adherence to operation and maintenance procedures with emphasis on change control)
- Supplier selection (e.g., selection process, vendor reviews and/or audits [or a documented rationale for the absence thereof], accessibility of the source code, a maintenance contract, proof of application of SDLC principles for internally developed systems)
- System management (e.g., formal contract or SLA, system management documentation, documented responsibilities, overviews of system components, existence of and adherence to change control procedures, backup and restore procedures, physical and logical security, Disaster Recovery Plans)

Audits of the Central Computer Services Facilities

In order to verify the compliance for those parts of systems that are controlled by the central computer services group, facility audits have to be performed of these facilities at regular intervals.

At least the following issues have to be covered in audit procedures:

- Quality policy (e.g., validation policy and/or master plans, SOPs and SOP control, GxP measures and procedures [theory and practice], internal QA and audit procedures)
- Responsibilities (e.g., organograms, job descriptions/tasks and responsibilities, delegation of tasks and responsibilities, training records)
- Physical security (e.g., access to computer rooms, environmental conditions, fire protection)
- Logical security (e.g., user accounts and password policy [theory and practice], access control lists [ACLs], network security)
- Change control (e.g., hardware: servers and clients; system software: operating systems, DBMS, application software)
- Other maintenance issues (e.g., preventative and corrective maintenance, network, maintenance documentation)
- Backup and restoration of data and software (e.g., existence of, adherence to and testing of procedures)
- Problem handling (e.g., communication between helpdesk, problem solving group and maintenance; documentation)
- Archiving of data, software and documentation (e.g., responsibilities, restricted access measures, security, retrievability)

Validation Committee

The best way to assure that all automated systems used for GxP activities are developed, validated, used and maintained in line with GxP regulations and that the support organization implements and follows a GxP–compliant QA system is to have a validation committee. This could be a separate committee dedicated to the validation of automated systems or combined with the central validation committee that often already exists, particularly at manufacturing sites. The major tasks and responsibilites that could be assigned to this committee are presented in Table 13.4.

MANAGEMENT ISSUES

Some key practical issues which affect the successful provision of maintenance and support to IT systems and the ICT infrastructure will be presented here.

Table 13.4. Major Tasks and Responsibilities of the Validation Committee

Responsibilities	Tasks
Keep (local) policy in line with international regulations (and corporate policy if applicable).	• Give advice to management on validation policy. • Keep informed on international regulations on GxP and computer validation. • Initiate updates of the (local) company policy and SOPs. • Generate the required SOPs.
Implement company policy on computer validation.	• Communicate to the company on computer-related validation subjects. • Generate and control templates for computer validation documentation (e.g., Validation Plans, Protocols and Reports) in consultation with users and QA. • Where necessary, actively participate in the preparation of training of staff (general training and/or on-the-job training, e.g., under supervision in a project).
Check adherence to company policy on computer validation.	• Set up a system to generate overviews of GxP–related automated systems and their validation status. • Inform management on the progress and status of validation activities of GxP–related automated systems. • Give advice on validation deliverables (plans and protocols). • Coordinate supplier qualification (audits, etc.) and maintain an overview of suppliers and their qualification status. • Randomly check (audit or review) computer validation practices in order to monitor adherence to company policy (e.g., initial computer validation, security, change control). • Be involved with inspections by compliance monitoring authorities.

Time Synchronization

An important issue for regulatory authorities is the reconstruction of conclusions and reported results. Evidence of data integrity can often be based on the chronological sequence of events. The U.S. Food and Drug Administration (FDA) regulation 21 CFR 11, dealing with electronic records and electronic signatures, mentions in §11.10(e) one of the controls that must be used:

> *Use of secure, computer-generated, time-stamped audit trails to independently record the date and time of operator entries and actions that create, modify, or delete electronic records . . . [2].*

To maintain and support this requirement, the support organisation must properly time all of the components of the ICT infrastructure. Electronic records as defined by the FDA are generated by most computer systems. With client-server applications, there are several sources that could deliver the date/time stamp. There are many applications that rely on the time stamp generated by the client's PC. Usually a client logs into the company's computer network, which can trigger all kinds of log-on activities that the user is unaware of. One such activity must be the synchronization of the workstation's system date and time with that of the server used for logging on. This server is usually not the server where applications store their data. So there is also a need for synchronizing all servers. Several technical solutions are available today. The different servers in a LAN should be synchronized with one computer that receives a time signal from a radio or satellite receiver. Workstations connected via a LAN or WAN can be synchronized within a few milliseconds to the primary server.

The Year 2000 Experience

The changeover to the new millennium caused an almost worldwide awareness of the dependency of society on computerized systems. But keeping in mind that "without problems there is no progress", the Y2k crisis can indeed be seen as the starting point for structuring the maintenance and support of IT systems. The Y2k experience might clarify the need for solving neglected or even denied problems. Four areas are summarized below.

Operational Lifetime

Some computers that were able to handle the millennium changeover correctly however will run into problems at a later date because they calculate the date by counting the days from a certain system start date. At a certain point in time, there will not be enough digital positions available to add more days and, hence, to calculate the correct dates. (For instance, many systems will run into problems in the year 2038.) And there is the problem of leap years which is not handled correctly by all systems.

One might consider these to be design faults (apart from special date values which should be condemned as lazy programming practices). It is obvious that the cost to repair these errors in a later phase of the life cycle has increased immensely. Sometimes replacement rather than repair is the more effective solution.

Testing for date conformity is a rather unique exercise because one needs to design test cases that need to travel in time. This also requires care when deciding on the test data to use. And, of course, system functions that are not date dependent must be tested as well when new versions are delivered.

Contingency Planning

When business continuity is at stake due to the breakdown of the ICT infrastructure, and there is awareness that this is unacceptable, it is time to itemize the dangers that threaten the ICT environment, analyze the risks and propose measures to solve the problem. Contingency planning should consider the entire business process, where ICT–based processes play a supporting role. When evaluating the entire process, a set of measures—not only related to ICT—should be identified and justified against the business value.

Configuration Management

If not already present, the Y2k experience should have produced a complete overview of hardware, software and embedded systems. This should be incorporated in an information system that can serve as a powerful tool for the infrastructure architect and the operational manager.

Software License Management

Knowing what software is used within a company is one thing. Knowing what the software license encompasses is a different story. It is not uncommon for an organisation who uses software to be unaware of issues such as the number of users allowed (named users or concurrent users), the license expiration date, product warranties, (period of) software support, consultancy (tariffs), delivery of upgrades (like Y2k-compliant versions) and so on. A contract management system should track the software (and likewise hardware) contracts.

Test Management

When a new version of the software is delivered, it should be tested. With the number of new versions being delivered, it is not feasible to carry out the test activities manually; the introduction of automated test tools might be considered. This is not a trivial job, and it should be left to a team of dedicated testers skilled in the use of automated test tools. The fastest way to introduce a structured test environment is by hiring system testing expertise that can initiate the creation and operation of a structured testing environment within a company. The benefit from automated testing is in repeating tests of critical functions when only small adjustments are done. The availability of important functions needs to be ascertained when changes are applied. This also makes people less reluctant to ask for functional improvements of validated systems. More and more testing will be required from all participants in the ICT environment, and automated test tools can support this in order to reduce time and money.

CONCLUSION

In order to keep systems in a validated state, at least the following must be arranged properly:

- Roles, tasks and responsibilites, including interdepartmental responsibilites, must be clearly defined and adhered to.
- Clear and proper procedures must be in place for using the system (procedures and/or user manual), backup and restoration, periodic evaluation, maintenance and problem management, safety and security, training, Disaster Recovery and Contingency Plans, business continuity, change control, and system retirement.
- Highest priority should be given to change control and security measures, including access control.
- The role of QA must be properly defined and implemented; QA should at least be involved in vendor reviews and/or audits and the change control process and perform system-specific audits and audits of the central computer services facilities.
- Other relevant management issues that need careful consideration are time synchronization, the Y2k project and test management.

Setting up proper risk assessment procedures and applying them to the change control process markedly reduces the efforts required to keep systems in a validated state. Harmonization, standardization and efficiency can be increased by appointing a company computer validation committee per site and/or at a corporate level. Good cooperation between QA and IT is a prerequisite for keeping systems validated.

REFERENCES

1. UK GAMP Forum. 1998. *Supplier's Guide for Validation of Automated Systems in Pharmaceutical Manufacture,* Version 3. Available from the International Society for Pharmaceutical Engineering.

2. U.S. Code of Federal Regulations, Title 21, Part 11. *Electronic Records; Electronic Signatures* (revised 1 April 1999)

BIBLIOGRAPHY

Huber, L. 1995. Validation of Computerized Analytical Systems. Buffalo Grove, Ill., USA: Interpham Press Inc.

OECD. 1995. *The Application of the Principle of GLP to Computerized Systems*. GLP Consensus Document No. 10. Paris: Organisation for Economic Co-operation and Development.

Stokes, T., R. G. Branning, K. G. Chapman, H. Hambloch, and A. J. Trill. 1994. *Good Computer Validation Practices: Common Sense Implementation*. Buffalo Grove, Ill., USA: Interpharm Press, Inc.

Wingate, G. 1997. *Validating Automated Manufacturing and Laboratory Applications: Putting Principles into Practice*. Buffalo Grove, Ill., USA: Interpham Press Inc.

APPENDIX 13.1. EXAMPLE OF A CHANGE REQUEST FORM

Change Request Form
Name of originator:
Date:
Automated system, name/number/version of the affected item(s):
Description of the proposed change:
Reason for the change:
Reference number:

Change Evaluation and Authorization				
Name	Function or Responsibility	Signature	Date	Approved/Rejected

Change Details (Based on Impact Analysis)
Comments *(Include either reasons for rejection or details of affected items [including documentation] and details of retesting required.)*

Change Completion and Approval		
Change completed:		
Name:	Function:	Signature:
	Date:	
Change approval:		
Name:	Function:	Signature:
	Date:	
Name:	Function:	Signature:
	Date:	

APPENDIX 13.2. GAMP CATEGORIES OF SOFTWARE

Category 1: Operating Systems

Established, commercially available operating systems which are used in pharmaceutical manufacture are considered validated as part of any project in which application software operating on such platforms is part of the validation process. The operating systems themselves are not currently subject to specific validation other than as part of particular applications that run on them. Well-known operating systems should be used, and the name and version number recorded in the hardware acceptance tests or equipment installation qualification (IQ). New versions of operating systems should be reviewed prior to use and consideration given to the impact of new, amended or removed features on the application. This could lead to formal retesting of the application, particularly where a major upgrade of the operating system has occurred.

Category 2: Standard Instruments, Micro Controllers and Smart Instrumentation

Category 2 software is driven by nonuser-programmable firmware. Examples include weigh scales, barcode scanners and three-term controllers. They are configurable, and the configuration should be recorded in the equipment IQ. The unintended and undocumented introduction of new versions of firmware during maintenance must be avoided through the application of rigorous change control. The impact of new versions on the validity of the IQ documentation should be reviewed and appropriate action taken.

Category 3: Standard Software Packages

Standard software packages are called canned or COTS (commercial off-the-shelf) configurable packages in the United States. Examples include Lotus 1–2-3®, Microsoft Excel® and other spreadsheet packages. There is no requirement to validate the software package; however, new versions should be treated with caution. Validation efforts should concentrate on the application, which includes

- system requirements and functionality;
- the high-level language or macros used to build the application;
- critical algorithms and parameters;
- data integrity, security, accuracy and reliability; and
- operational procedures.

The validation process should use the management system defined in the GAMP Guide. Further guidance may be found in ISO/IEC 12119: Information technology—Software packages—Quality requirements and testing.

As for other categories, change control should be applied stringently, since changing these applications is often very easy and with limited security. User training should emphasize the importance of change control and the validated integrity of these systems.

Category 4: Configurable Software Packages

Category 4 software is called custom configurable packages in the United States. Examples include Distributed Control System (DCSs), Supervisory Control and Data Acquisition (SCADA) packages, Manufacturing Execution Systems (MESs) and some LIMS and MRP packages. In these examples, the system and platform should be well known and mature before being considered in category 4; otherwise category 5 should apply.

A typical feature of these systems is that they permit users to develop their own applications by configuring/amending predefined software modules and also developing new application software modules. Each application (of the standard product) is therefore specific to the user process. Maintenance becomes a key issue, particularly when new versions of the standard product are produced.

The GAMP Guide should be used to specify, design, test and maintain the application. Particular attention should be paid to any additional or amended code and to the configuration of standard modules. A software review of the modified code (including any algorithms in the configuration) should be undertaken. Appendix B of the GAMP Guide gives guidance on performing code reviews.

In addition, an audit of the supplier is required to determine the level of quality and structural testing built into the standard product. The audit needs to consider the development of the standard product, which may have followed a prototyping methodology without a user being involved. In such cases, the UK GAMP Forum recommends that suppliers use the Guide to provide visible management structure and documentation during the development of the standard product. European GMPs require that the development process be controlled and documented. A Validation Plan should be prepared to document precisely what activities are necessary to validate an application, based on the results of the audit and on the complexity of the application.

Category 5: Custom-Built or Bespoke Systems

The GAMP Guide provides a life-cycle model approach for custom-built or bespoke systems, and there are many instances of these types of systems in the pharmaceutical industry. The full life cycle should be followed for all parts of the system.

An audit of the supplier is required to examine its existing quality systems, and a Validation Plan should then be prepared to document precisely what activities are necessary, based on the results of the audit and on the complexity of the proposed bespoke system.

Conclusion

It should be noted that complex systems often have layers of software, and one system could exhibit several or even all of the above categories.

14

Auditing Suppliers of Standard Software Packages

Anton Dillinger
Robert Kampfmann
Torsten Wichmann
SAP AG
Walldorf, Germany

GxP regulations governing the pharmaceutical industry require suppliers of standard software products to be audited by the pharmaceutical manufacturers who use them. Supplier audits are used to verify that the supplier has effectively managed the quality of its product. Standard software may require configuration but should require little or no customization to its executable code to enable its application. Significant customization would suggest that the software product is actually a bespoke development. Standard software is sometimes also referred to as shrink-wrapped software or commercial off-the-shelf software. It is not industry practice to conduct audits for established standard software that is widely used, such as operating systems. Supplier audits, however, are necessary if the so-called standard software being used by a pharmaceutical manufacturer has been especially developed for the manufacturer. These Supplier Audits adopt a similar approach to that described in this chapter but with more detailed project audit perspective [1,2]. Supplier audits are also recommended for new releases of standard software that have not been market tested.

This chapter presents the audit experience of SAP, one of the world's leading IT (information technology) software product suppliers. SAP is renowned for its R/3 Enterprise Resource Planning (ERP) software. The R/3 product is typical of what is commonly referred to as standard software.

VALIDATION STRATEGY

Supplier-Customer Relationship

Standard software provides functionality that is not based on the defined requirements of one customer; it is based on the requirements of best industry practice of different industries, represented by top companies, university knowledge or consultant experience.

Standard software is composed of a set of functions that can be configured by logic switches defined in tables to fit individual business scenarios of a defined company. Indeed, a user may even decide to customize its specific application with bespoke programming to meet their particular needs. SAP's R/3 product is a standard software package consisting of several programs.

Pharmaceutical manufacturers are accountable to regulatory authorities like the U.S. Food and Drug Administration (FDA) for the validation of computer systems. Suppliers have a responsibility to ensure that their software, hardware or associated services are fit for purpose. Pharmaceutical regulators, however, do not hold suppliers accountable for validation. The validation of such standard software packages by a pharmaceutical user therefore consists of two aspects:

1. *Qualification* of the package itself by performing a Supplier Audit
2. *Validation* of the individual installation and especially of the changes made in customizing

Qualification of a supplier's standard software package includes ensuring that a full specification of the package exists and that the package is comprehensively tested. For products like R/3, testing must encompass verifying the operability of multiple configuration possibilities. It is not just the pharmaceutical industry that requires documented evidence of a defined software development life cycle. There are other industries, like aerospace and defense, engineering and construction or high tech that want to see a comparable effort in quality activities. SAP installations are typically considered by their users to be mission critical systems. These customers, therefore, need confidence that their business system is able to perform as expected.

For SAP, it became obvious that all of these requirements are more or less the same, and therefore treating Good Manufacturing Practice (GMP) requirements for some modules of the SAP system as standard practice in every area will result in less absence of quality and thus less problems in installations after release of the software. Fifteen thousand customers may mean that a bug in a critical program has to be fixed 15,000 times in the worst case. Thus, doing best practice qualification helps to save money.

SAP's Experience of Customers' Audits

Like many other IT software suppliers, SAP has only recently begun to receive audits by pharmaceutical industry customers concerning GxP compliance. This is because pharmaceutical manufacturing companies are only just becoming aware of the the impact of GxP on ERP systems. Pharmaceutical regulators such as the FDA and UK Medicines Control Agency (MCA) have also started inspecting the validation of these systems.

For SAP, the first Supplier Audit mentioning validation for GMP regulations occurred in 1993. It was only at this point that SAP was made aware that their pharmaceutical customers had to validate their products and their production processes, including the software products that are part of production processes. In prior years, SAP received only poor feedback from pharmaceutical customers. There had been some rumors and complaints about necessary extensive tests of a new release or version. None of these customers explained to SAP why that had to be done.

SAP had already started to develop a Quality Management System (QMS) according to ISO 9001. The expectation was that having an ISO 9001 certificate would also fulfill GMP. The audit report stated that SAP was on the right way, but that there were some gaps in the defined software development life cycle. ISO 9001 is a basic requirement but not sufficient for GMP.

SAP founded a so-called "FDA work group" with interested customers and consultants to define a common idea about GMP compliance for a software supplier. This group met several times, and SAP learned a lot about validation and qualification and changed its quality management to fulfill these legal requirements. Comparing GMP with other industry-specific quality rules like AQAP (Association of Quality Assurance Professionals) or QS 9000, SAP learned to see "validation as common sense", as Pfizer's Ken Chapman once stated in an article. SAP now has quality systems and a qualification strategy that satisfies the expectations of different industries.

In the first session of the FDA work group, it became obvious that the industry that has to fulfill these requirements had no common understanding about what is really necessary, what is helpful and when is it more than expected. As a spin-off initiative, six companies came together to define an audit group. They established a common audit standard for their companies and performed two Supplier Audits as joint audits. This group is now developing a common validation strategy to validate SAP installations in their companies.

Parallel to the activities of the FDA group in Germany, the number of Supplier Audits within SAP increased dramatically. While only 4 customers conducted a Supplier Audit at SAP AG in 1994, this number increased to more than 15 Supplier Audits per year since 1996.

All of these audits brought some new experiences because customers often focused on different points. When SAP was confronted with an exploding number of companies that wanted to perform audits, it became obvious that SAP needed to develop a strategy to organize these audits in a more efficient way because the efforts for accommodating the auditors and the loss of time in the different departments were too cost intensive.

SAP developed a standard procedure for these Supplier Audits as an offer to customers and prospects coming in to perform an audit. SAP now offers two different events per month to everyone asking for an audit and organizes a joint audit:

- A two-day event for customers/prospects coming for the first time
- A one-day event for customers coming in for a follow-up audit

SAP now wants to go one step further. In the past, SAP experienced that some prospects only sent a consultant to perform the audit; others sent questionnaires, and some tried to buy audit reports performed by someone else. At present, the U.S. PDA (Parenteral Drug Association) is discussing an audit repository that contains audit reports performed by different companies, at least one per software supplier [3]. These reports only contain findings, not ratings. The evaluation of the audit report has to be performed by the individual company.

SAP is working to support such an audit repository. The Industry Business Unit Pharmaceuticals (IBU Pharma) has ordered a third-party audit by KMI (Kemper–Masterson) and will provide this audit report to any prospect or customer as a first step. SAP expects this to be sufficient for a large number of customers, especially small companies. Others can base their audit activities on the results of the public audit report and investigate details that may not be covered.

The mentioned audit report will be a baseline for further releases. IBU Pharma will charge KMI or other companies with a follow-up audit of future releases. These audit reports analyze whether SAP has followed its own strategy to qualify a new release.

ORGANIZATION OF A SUPPLIER AUDIT

Objective of Supplier Audits

Suppliers should consider developing a qualification strategy so that they can demonstrate to a customer how they manage the development and integral quality of their product. SAP's own qualification strategy is based on

- its QMS (certified according to ISO 9001);
- very extensive discussions in workgroups with customers;
- advice by several GMP–experienced consultants like Weinberg, Hambloch, and KMI;
- feedback from well-known auditors like Guy Wingate (now with GlaxoWellcome), Martin Browning, and Novo Engineering;
- the definitions of the GAMP (Good Automated Manufacturing Practice) guideline developed to describe best practice for the mentioned development; and
- the business needs and interests of SAP as a supplier of standard software.

During an audit, the supplier aims to present its quality management system, demonstrating the qualification strategy ("say what you do") and the documented evidence that everything is followed during development of a new release ("do what you say").

The efficient performance of customer audits requires careful planning between the vendor and the customer. The points given below should be taken into account and passed on to the customer during the planning of a customer audit.

Date, Participants

The date of the audit should be determined at least four weeks in advance. The areas to be audited should be agreed on three weeks before the date at the latest. If a checklist is to be used during the audit, it should also be made available to the supplier sometime before that date. SAP requires an advance copy of the auditor's checklist three weeks before the audit. It is in the customer's interest to share its checklist with the supplier so that the supplier can ensure relevant information to checklist items are readily available during the audit. The duration of the audit should be limited to a maximum of two days.

Suppliers should expect an overage of skills from the customer's audit team:

- One customer representative with knowledge of ISO 9000 or industry-specific field assignments such as GMP and AQAP
- One customer representative with knowledge of software quality management
- One customer representative with experience in using the supplier's products and/or service or a comparable solution

This combination is similar to the definition by Dr. Heinrich Hambloch in chapter 8 of *Good Computer Validation Practices* [4].

Audit Preparation

The supplier should prepare presentation material to introduce the QMS, the quality management organization and an overview of the main procedures. The supplier may well be asked to respond to a postal questionnaire prior to an audit on the supplier's premises. This response may even negate the need for an audit of the supplier's premises if information returned fully answers audit questions. Even if an audit visit is still required, a good postal questionnaire response will enable the customer to focus on open topics during the audit, so it is in the supplier's interest to put sufficient effort into the postal exercise.

Nondisclosure Agreement

The customer agrees in writing not to make the results available to third parties without the agreement of the supplier (Nondisclosure Agreement). Written agreement is also required before the audit that the customer will make available to the supplier the results of the audit and therefore will give the supplier the opportunity to make a statement with regard to incorrect or misleading changes before the official distribution. In addition, suppliers should consider agreeing with their customers to receive the final version of the audit report. This enables the supplier to refer back directly to audit observations and recommendations in the future when conducting their own QMS improvement reviews.

Audit Report and Supplier Feedback

The customer should provide the supplier with a draft of the audit report. Any misleading or incorrect information regarding the supplier and its development and service processes will be corrected by the supplier or explained to the customer. If required, the supplier can respond in writing. SAP normally agrees to comment on the draft report within four weeks of receiving the draft. It is recommended that the supplier agrees in advance on a response time when reviewing the customer's draft audit report so that the customer does not mistakenly believe after a waiting period that the supplier has no comments and is happy with the report. Suppliers should endeavor to give prompt feedback.

THE AUDIT—DEMONSTRATING VALIDATION CAPABILITY

Audit Agenda

Based on the experience of more than 50 Supplier Audits, the following agenda for a software Supplier Audit is suggested:

- Introduction to the supplier's QMS and qualification strategy
- Detailed analysis of the quality management documents and the qualification strategy (qualification of the QMS in the sense of certification)
- Detailed analysis of development projects and maintenance activities by documented evidence showing that the defined processes of the QMS are known and adhered to
- Roundtable discussion with quality responsible persons of different groups
- Visit to the IT center (based on risk analysis of the audit team)
- Q&A session for remaining questions, open problems
- Audit conclusion

Audit Topics

In the course of a two-day audit, it is not possible to get a detailed picture of every area to be audited. However, examples of documented evidence that is well defined and required from the audit team should give a sufficient impression of the supplier's quality reliability. The following topics are usually covered in the course of the Supplier Audit.

Quality Organization

- How is quality personnel organized?
- Who is the quality responsible person?
- What are the roles, tasks and responsibilities of the employees working in internal quality management?

Quality Management System

Quality management defines the quality policy for quality planning, quality assurance (QA) and quality control (QC). The corresponding processes should be appropriate to meet the supplier's business goals. In particular, the following issues are relevant:

- Quality policy, quality goals
- What is the value of quality in the company (commitment of the board)?
- Quality management documentation
  - Types of documents (written procedures, internal standards or Standard Operating Procedures [SOPs], guidelines, templates)
  - Configuration of documentation (versioning, archiving, change control)

- Quality management training of personnel
- Basis and further training of employees
- Internal audits (organization and performance, management review)

Production Life Cycle

- Software production should be based on a well-structured development life cycle
- Collecting requirements:
  - How are requirements identified?
  - Who decides about new requirements?
- Project planning and quality planning
- Product specifications:
  - Who is responsible?
  - Are formal reviews conducted?
  - How is document control organized?
- Design documentation:
  - Who is responsible?
  - Are documents reviewed?
  - Who controls documentation?
- Implementation:
  - Tools used
  - Source code reviews
  - Configuration control (authorizations, versions, release of objects)
- Testing:
  - Test plan
  - Test plan reviews
  - Manual tests versus automated tests
  - Black box versus white box tests
  - Can tests be reproduced?
  - Test documentation
- Release decision:
  - Who is responsible?
  - Quality report
  - Criteria for release

Maintenance Phase

- What does the maintenance life cycle look like?
  - Recording of problems
  - Creation and delivery of corrections to customers (single corrections, packages of corrections, correction releases)
  - Change control (authorizations, traceability of changes, versioning, change documentation [author, date], test of correction)

Subcontractors

If third-party products are integrated or interfaces to third-party products are offered, a corresponding procedure should be in place.

After Release Support

What support does the supplier provide once the product is shipped to the market? Are there procedures in place?

- Implementation
  - Implementation roadmap
  - Implementation validation support
  - Testing the implementation
  - Documenting the implementation/implementation testing
- Software problems/support requests
  - Fault logging procedure
  - Problem recording
  - Hot news for customers
  - Providing short-term bug fixes

Other Topics

- Backup procedures
- Archiving of delivered versions of the product
- Security (authorization control, access permissions, virus checks, network security)
- Disaster recovery procedures for development and support systems

MATCHING THE AUDITORS' EXPECTATIONS

Quality Management System

All suppliers should be able to describe their QMS to the auditor. Indeed, the auditor will expect the supplier's QMS to be formalized and documented. The QMS should contain all procedures, internal standards and guidelines that are required when developing, maintaining and qualifying the product.

SAP's *HORIZON* describes the entire software life cycle involved in developing and maintaining software products. This cycle includes the following phases: product and project planning, specification and design, actual software implementation, testing and maintenance. In the QMS, particular emphasis is put on a well-structured Software Development Life Cycle (SDLC). All releases are developed in accordance with the quality management documents valid at the time and with the qualification strategy.

Quality Management Documentation

The supplier must not only be able to show that there is a QMS but also that it is actively used. SAP manages quality management documents and quality item lists electronically and thus ensures that employees have constant access to the most up-to-date information. Because it is managed electronically, it is possible to construct automatic workflows to ensure that the process steps prescribed are followed. It is important to note that suppliers do not have to conform with U.S. 21 CFR 11 governing the use of electronic records and signatures. Customers should nevertheless expect suppliers to ensure that electronic records are secure and that their integrity cannot be compromised by data corruption or unauthorized manipulation.

Quality Assurance

The supplier should have a nominated quality manager who is supported by a quality department or team. In larger supplier organizations, a central quality management department is normally responsible for organizing and coordinating all quality management activities company wide, including:

- maintenance and enhancement of the QMS,
- central organization of tests,
- internal quality audits and preparation of the management reviews,
- monitoring of customer quality audits,
- process improvement and
- quality key performance indicators (KPIs).

This is essentially a facilitating role in many supplier organizations and is supported by development team members with specific project quality responsibilities.

Quality Control

In addition to the central quality group, each development department headed by a program director has an assigned quality management team. The quality of the product (or certain parts of the product) should be the top priority for the quality management team. The main tasks of these decentralized quality management teams are as follows:

- Prepare quality plans for release by a specific development team.
- Support and monitor the preparation of development documentation.
- Coordinate and lead reviews in each development phase.
- Plan, support and analyze development tests.
- Plan, create and support automatic tests.
- Decentrally plan, support and analyze the integration test.
- Prepare a concluding quality report for the specific development area.
- Support the correction cycle during the maintenance phase.
- Release the product together with the program director and other parties with decision-making authority.

Software Development Life Cycle

In addition to a defined QMS, a supplier should have a defined SDLC. SAP's SDLC, shown in Figure 14.1, is consistent with current industry practice and has the following phases:

- Planning
- Specification
- Design
- Implementation
- Test
- Maintenance

For each of these phases, the required tasks and results are described.

Qualification Strategy

Qualification of a New Release

A key aspect to a supplier's QMS for standard software products is its application to new product release. Without appropriate qualification of new developments, the quality attributes of the original system may be lost. Pharmaceutical industry customers are looking for consistent, if not improving, quality compliance, and this is especially so with product upgrades.

Figure 14.1. SAP Software Development Life Cycle

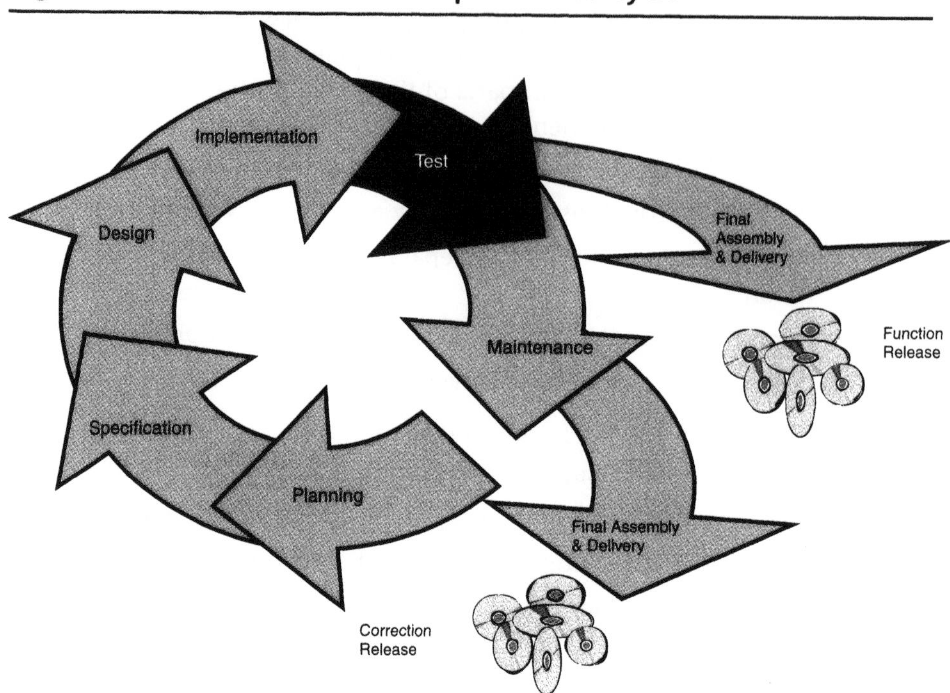

The SAP qualification strategy for a new development is shown in Figure 14.2 and involves taking a qualified release as the basis and describing the transition to a new release that still has to be qualified. Important factors considered during software qualification are compliance with a structured SDLC and verification of the outcome of the individual development phases. It is important that planning and verification are logical and traceable when moving from one qualified release to the next.

The documentation of new developments includes process models, data models, programs (including programmer's comments, tools such as "where-used" lists and links between the models and objects) and user documentation. In addition, the entire test catalog contains the entire range of manual and automatic test cases and test procedures.

Development Documentation

The difference in functions between a new release and the previous release should be described. SAP includes development documentation with the issue of a new release, including, for example, detailed specifications of the product's new features. These features are then turned into "real" functions as part of the SDLC. A project plan specifies which functions are to be developed

Figure 14.2. Qualifying New Developments

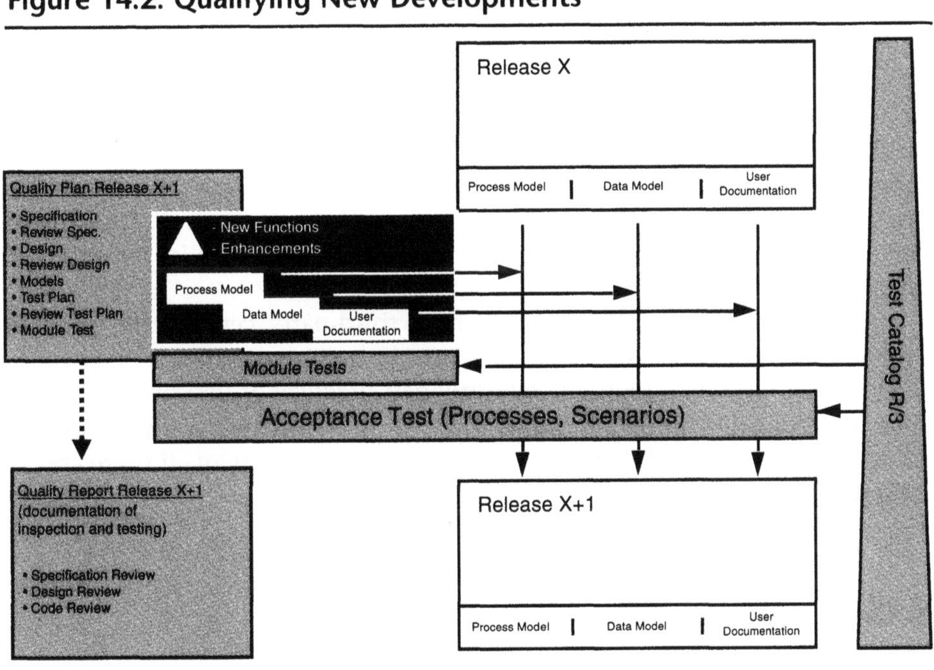

as "work packages" and at what expense. The project plan also specifies the necessary "project milestones" and which groups are responsible for which tasks. The quality plans of the individual areas describe the QA measures of the various work groups. Above all, it specifies criteria for testing and releasing the release.

Test plans document the planned scope of the verification. The test cases are based on developers' and quality management team members' expertise. Test packages document the implementation of the tests.

The new functions a release is to contain are incorporated in the existing product during the SAP SDLC. This involves extending the existing process and data models and adding to the application documentation and test catalog. New test case descriptions for new functions are added to the test catalog. New process chains that link the old and new functions in test scenarios are included as test cases.

Model for the Future

SAP's experience to date has highlighted tremendous differences within the pharmaceutical industry about how to fulfill GMP requirements. SAP therefore welcomes initiatives like GAMP, which establish a common understanding about best practices in the validation of automation and business systems.

Unfortunately, there still is a lack of advice for the validation of standardized software packages. Due to this, SAP has defined its own way in a qualification strategy. This strategy describes what SAP does during development of a new release to ensure that the outcome is qualified in the sense of GMP. SAP is working on the so-called *Supplier Forum*, established in the United Kingdom in 1998 by Dr. Guy Wingate (then validation manager at ICI Eutech) with sponsorship from the U.K. Department of Trade and Industry. In particular, the Supplier Forum is preparing a *Supplier Guide to Customer Audits* [5]. The expectation is that GAMP version 4 will include the best practice for any type of software solution.

In an ideal world, there would be a third-party certification program similar to ISO 9001 certification that gives documented evidence that a company fulfills GMP requirements by using best practices as described in GAMP. Such a certification program would then be in accordance with GAMP, and GAMP would include ISO 9001 and special GMP requirements if there are still differences in the future.

BACKGROUND INFORMATION

Who Is SAP?

Founded in 1972, SAP AG (Systems, Applications and Products in Data Processing), based in Walldorf, Germany, is the leading global provider of client-server business application solutions. As of the end of 1998, more than 10,000 customers in over 90 countries have chosen SAP client-server and mainframe business applications to manage comprehensive financial, manufacturing, sales and distribution and human resources functions essential to their operations. For more than 25 years, SAP has been recognized by its customers for excellence in software technology.

Nearly all customers who chose SAP products at the beginning of the 1970s (starting point for commercial data processing to a certain extent) are still using these products today. The customers underwent a lot of changes regarding IT infrastructure (hardware and software) and their way of doing business, but they kept SAP software as their strategic business solution.

SAP always provided these customers with a state-of-the-art software solution.

Most prominent in SAP's product range are enterprise applications R/2 and R/3: R/2 applications are for mainframe environments, and the more recent R/3 applications are for open client-server systems. With both R/2 and R/3, customers can opt to install the core system and one or more of the functional components or purchase the software as a complete package.

SAP customers have chosen to install R/3 in more than 15,000 sites worldwide. R/3 is accepted as the standard in key industries such as oil, pharmaceuticals, chemicals, consumer products, high technology and electronics. SAP AG employs a workforce of over 15,000 and has offices in more than 50 countries worldwide. SAP is the number one vendor of standard business application software and is the fourth largest independent software supplier in the world. In its most recent fiscal year, ending Dec. 31, 1997, SAP AG reported revenues of DM 6 billion, a 62 percent increase over 1996 revenues. In the same period, sales of R/3 rose by 63 percent.

SAP is a market and technology leader in client-server enterprise application software, providing comprehensive solutions for companies of all sizes and from all industry sectors. Cultivating innovative technologies based on solid business experience, SAP delivers scaleable solutions that enable its customers to continuously improve upon best business practices. SAP products empower people to respond quickly and decisively to dynamic market conditions, helping businesses achieve and maintain a competitive advantage.

SAP Pharmaceuticals/SAP Chemicals

SAP has long-lasting experience in the pharmaceutical and chemical sector. Some of SAP's first customers were from these industries. The very first customer was ICI Fibres. More or less, this was the cradle of SAP software. At a very early stage, the IBM–based (International Business Machines) solution was adopted to SIEMENS platforms at Schulze, a pharmaceuticals wholesaler. Approximately 15 percent of SAP's business is made in the chemical and phamaceutical sector. These industries have been and continue to be strategic for SAP. In the past, SAP placed substantial investments to keep the product best suited for these industries, developing solutions according to best business practices and meeting the legal requirements of GxP–regulated companies. In addition to these functional requirements, SAP also fulfills the GxP requirements of a model-based SDLC. All activities have always been performed in close contact with customers and prospects.

History of Quality Management at SAP

SAP was founded at the beginning of commercial data processing. At the time, almost nobody thought about developing standard software for commercial usage. Each program was project-specific software developed for very special business needs. Every project was a pioneer's work. Users had no high expectations regarding the quality of software. The most important objective was to have the required functions nearly in time. Users were used to a lack of quality in an early version. Normally, software bugs were only detected and fixed once. Over the years, software increasingly became a normal business tool, and the expectations of users about quality changed. SAP's first quality initiative started in the early 1980s. At that time, R/2 was used by a larger number of customers. SAP found that every bug shipped with the software came back as a problem reported by several customers at once. Four key developers were announced to form a quality circle. Their task was to organize software tests and to define rules of good programming. Over the years, this quality circle became a quality department focusing on the quality activities at the end of the process, i.e., organizing

- acceptance tests at the end by hiring test professionals,
- the developing phase of a new software version and
- problem handling and solution to support the customers.

This process changed dramatically in 1993. SAP became aware that quality has to be incorporated in the software and cannot be tested in only at the end of development. Quality management became a major task of four special quality teams in the different development groups. At the same time, a special ISO 9000 initiative was launched. In a pilot project, a special development team responsible for developing software for quality management was certified. In a second step, the entire development department of SAP Walldorf was certified according to ISO 9001 in December 1994. Today, SAP has quality management teams in every development group around the world responsible for the quality of their products (approximately 90 quality managers with their teams) and a central QA group based in the Walldorf headquarters. The leader of this team reports directly to an SAP board member responsible for quality. The task of this QA function is the maintenance of the QMS and the organization of internal or external audits as well as the organization of Supplier Audits by SAP customers or prospects.

CONCLUSION

This chapter has given insight into the process by which pharmaceutical manufacturers audit suppliers of standard software packages for GxP compliance. It is vital that suppliers manage these audits to ensure that their customers get all the information they need but also that the supplier puts this information in the right context so that the customer does not unknowningly take away any misunderstandings. This is very important when a Supplier Audit is being used as a supplier selection activity. Those preparing to conduct or to receive such audits are recommended to read other useful information identified in the References.

REFERENCES

1. UK GAMP Forum. 1998. *Supplier's Guide for Validation of Automated Systems in Pharmaceutical Manufacture,* version 3. Available from the International Society for Pharmaceutical Engineering.

2. Wingate, G. A. S. 1997. Validating Automated Manufacturing and Laboratory Applications: Putting Principles into Practice. Buffalo Grove, Ill., USA: Interpharm Press, Inc.

3. John, J. 1999. *Supplier Evaluations in the Regulated Pharmaceutical Industry,* incorporating PDA Task Group Draft Supplier Audit Guide and Checklist. Presentation for Achieving Cost-Effective Computer Systems Validation for cGMP in Pharmceuticals: Business Intelligence, 27–28 April in London.

4. Hambloch, H. 1994. Audit of External Software Vendors. In *Good Computer Validation Practices Common Sense Implementation,* edited by T. Stokes, R. C. Branning, K. G. Chapman, H. Hambloch, and A. J. Trill. Buffalo Grove, Ill., USA: Interpharm Press, Inc., pp. 141–156.

5. Supplier Forum. 1999. *Guidance Note on Supplier Audits Conducted by Customers,* available on the Internet at http:/www.eutech.com/forum.

15

Auditing Software Integrators and Hardware Manufacturers

Howard Garston Smith
Pfizer Limited
Sandwich, Kent, United Kingdom

Software lies at the heart of all information technology (IT) systems. Such a statement of the (apparently) obvious may seem superfluous in a book on IT systems validation, but the need to reemphasize it is pressing. Because of its complexity and the fact that its structure and detailed functioning are understood only by a small eclectic group of individuals, the importance of ensuring its validatability at the outset of an IT system validation project is frequently underestimated. Indeed, there have been many instances in the history of regulated industries when the issue has been ignored altogether.

Validation has already been defined in the GAMP Guide [1]:

> *To produce documented evidence which provides a high degree of assurance that all parts of the facility will consistently work correctly when brought into use.*

As understood and interpreted by the pharmaceutical industry, validation is concerned with gaining assurance that the software will do what it purports to do, and will **not** do what it purports **not** to do. The term *validation* is usually used to mean the practical activities aimed at ensuring this state of affairs. These activities are essentially systematic structural and functional testing, supplemented and buttressed by the acquisition of supporting process documentation.

What is seldom realized, however, is that there is a problem here in connection with computer software. It is generally assumed that validation testing will exercise all of the pathways in the code to such a widespread extent that the passing of these tests will thereby deliver the high degree of assurance required. Regrettably, this is seldom the case with modern software. This arises from the large number of possible pathways found in most of today's software, and the sheer impossibility not only of testing every one but even testing any more than a small proportion. Back in 1979, Myers [2] succinctly explained this inescapable reality. It follows, therefore, that some assurance is needed regarding the behavior of the software when executed along all those unseen pathways—usually the overwhelming majority that are rarely followed and, due to the constraints of time and money, cannot be tested functionally. If such assurance can be obtained, we may then be entitled to declare the software *validatable.* In other words, we must obtain an insight into the fitness for use and behavior of that part of the code that the validation activity, applied in isolation, cannot reach.

Two conclusions are therefore inescapable:

1. Ensuring the validatability of software is one of the principal foundation stones and an essential precursor for the success of any such validation exercise.
2. The widespread belief that all software is somehow validatable, providing one makes enough effort to test the code and document it, betrays a complete misunderstanding of the realities applicable to all but the simplest and most rudimentary programs.

For example, if the scope of a program's validation testing exercises 10 percent of the program's pathways, 90 percent of the code remains untested, and a large proportion of defects remain concealed. In the absence of an inspection of the development process, practically nothing whatsoever is known about 90 percent of the program's potential behavior. The indispensability of such an inspection and its complementary relationship to validation testing is thus clear.

Furthermore, it is also plain from the definition of validation above that documentary evidence is the lodestar of validation and the tool by which the assurance is derived. In respect of software, the assurance needed will be in the form of documentary evidence, not of the product directly (though this may be included) but, as we have shown, from an examination of the development process that produced it. This strongly implies that the product is inseparable from the process, and the fitness for use of the product can only be secured through the rigorous control of the process that produced it. This ought to come as no surprise, since it is precisely this principle, the essence of the science of *statistical process control,* that determines the quality of any manufactured or fabricated product. In other words, the quality (of fitness for use) of an item is entirely determined by the process that produced it, where quality is defined according to the ISO 8402 [3]:

The totality of characteristics of a product or process that bear on its ability to satisfy stated and implied needs.

The inspection of the capability of the process in order to gain the required assurance as to the fitness for purpose of the developed software—a measure of its validatability—is referred to here as a *software quality assurance audit*. The software quality assurance audit is occasionally referred to as a Supplier Audit (e.g., in the GAMP Guide). However, a Supplier Audit is also used to audit original equipment manufacturers, hardware suppliers, independent contractors and even an internal department with a pharmaceutical organization. There may be no "software" involved at all. For that reason, the term *software quality assurance audit* is more preferable than *Supplier Audit*.

THE SOFTWARE QUALITY ASSURANCE AUDIT

According to ISO 10011, a quality audit is defined as follows [4]:

A systematic and independent examination to determine whether quality activities and related results comply with planned arrangements and whether these arrangements are implemented effectively, and are suitable to achieve objectives.

The planned arrangements in this case must be the standards, procedures and methodologies in place to supervise and control the development process in order to deliver a product of the required quality. These standards and procedures are colloquially known as a Quality Management System (QMS) or Quality System (QS). ISO 10011 defines a QMS as follows [4]:

The organizational structure, responsibilities, procedures, processes and resources for implementing quality management.

The audit has two components: (1) ensure that adequate controls are in place to ensure a satisfactory product, and (2) to measure the extent to which these controls are followed in actual practice. The second aspect is crucial, for there have been numerous cases of comprehensive and sophisticated QMSs being set up that have subsequently fallen into disuse for a variety of reasons. This situation tends to remain concealed from customers in the normal course of events, until some unexpected event led to the quality of the product being questioned.

This chapter will provide a summary of the essential standards by which software development may be controlled. These standards are documented in the TickIT Guide [5], which is based on ISO 9000–3 [6]. These are consistent with the standards featured in the GAMP Guide [1], and a process of convergence between them has been apparent for some time.

This chapter does not attempt to explore the software quality assurance audit process or the nature of currently accepted standards of good software development in exhaustive detail. Other works in print explore this area in the context of both ISO 9000–3 and GAMP (Good Automated Manufacturing Practice), some of them broadening the application of software quality assurance (QA) to inspection of the process relevant to the development of software of *all* kinds, not just that intended for use in regulated areas. We will concentrate here only on the leading aspects of the software quality assurance audit, as they impinge on the particular issues faced by hardware manufacturers and systems integrators.

Systems Integrators—Special Issues

Systems integrators provide IT solutions by combining software and hardware components from more than one source into a consolidated product, not only adding value but also producing final products that meet the needs of niche markets. In the course of such activity, integrators may contribute bespoke code to supplement the code of standard packages. This may involve either a configuration of the latter (in which case the packaged native code remains unchanged) or a modification of the packaged source code itself. Frequently, independent subcontractors provide code, sometimes to several systems integration houses.

Whatever the commercial arrangements, an effective quality-assured development process for in-house code and integration activity is essential. However, the development processes of all the software components being used must also be under effective control, and this is rarely the case. Few systems integrators are willing to impose adequate controls on their suppliers, and pressure is usually necessary to secure such controls for a specific systems project. While subcontractors may remain commercially independent, they must be willing to be regarded from a quality standpoint as *employees* of the systems integration enterprise. No unilateral declarations of independence in standards and methodologies are acceptable, whatever the private commercial arrangements for the subcontractors may be. If any doubt remains over the effectiveness of the standards superintending the work in any of the anticipated subcontractors, customer organizations must not hesitate to insist that the systems integrator's own QMS be formally applied to all software suppliers involved on a given project. Indeed, this stipulation should be contractual.

Hardware Manufacturers—Special Issues

The days when all precision engineering equipment consisted exclusively of mechanical and electrical components are now passed, at least for all except at the most elementary level of complexity. Since the Intel® 8080® family of microprocessor chips emerged in the middle 1970s, manufacturing equipment and instrumentation of all kinds have increasingly adopted this technology for control purposes. Such in-built intelligence has not only relieved the operator or pharmaceutical manufacturer of routine operational functions but also blazed a trail of enhanced internal complexity, sophistication and automation across the entire engineering spectrum. Among the most obvious examples of this extraordinary revolution are the sophisticated instrumentation found in today's analytical laboratories and hardware platforms supporting corporate computer systems.

Because of the sheer ubiquity of the microprocessor in every aspect of daily life, it is easy to take it for granted and overlook the fact that *software* (in the form of code burnt onto EPROM [electronically programmable read-only memory] chips and generally known as *firmware*) is also involved, with all that this implies. These implications have escaped not only the layman, for whom such lack of awareness is quite forgivable, but also the enterprises that deploy these devices in their products. At present, most of these equipment suppliers continue to see themselves, quite rightly, as engineering companies but do not *also* regard themselves as software houses.

This blind spot leads directly to glaring double standards in the engineering approaches adopted in mechanical and electrical engineering on the one hand and software on the other. In the case of the former, every aspect of the work, from the cutting and machining of steel to the design and fabrication of printed circuit boards, is subject to the most exhaustive design methodology and documentation. This is designed with one supreme objective in view: to eliminate every source of error in the manufacturing process and ensure that every mechanical and electrical characteristic of the final product conforms exactly to what was originally specified, down to the last detail. Engineers have long accepted that even if it were possible to rectify defects in the final product after the event, the cost is so prohibitive that the future of the company depends on getting it right the first time. Even questioning this tacit assumption in such companies is greeted with astonished amazement—the principle is so universally accepted as obvious. The implementation of this principle gives rise to draughtsmen assembling the most exhaustively detailed drawings in advance of construction work, buttressed today by Computer-Aided Design (CAD) tools to eliminate every possible source of defects at the earliest stage of development. No work commences until these documents have been subjected to the most rigorous reviews and approvals, in order to eliminate errors when, by overwhelming consensus, it is relatively inexpensive to do so.

Now suppose that someone had the temerity to suggest that these approaches are needlessly bureaucratic and that sole reliance on the skill of toolmakers or electricians could save money. Just a simple drawing on the back of an envelope is all that is required, since skilled machinists at their lathes (for example) are quite capable of filling in the details themselves—they are very experienced and have done many similar jobs previously. *What a ludicrous idea—are you out of your mind? We would be out of business instantly!* Yet this is precisely the approach adopted for software development! Seldom can one find a Functional Specification or design specification for software to complement those for hardware (mechanical and electrical). Reliance is placed solely on the skill and professionalism of programmers, working in conjunction with other disciplines but in the absence of documentation and standard methodologies adopted instinctively elsewhere. After all, it is claimed, they are very skilled and have developed code for similar jobs in the past! It is therefore not surprising that the majority of rework effort in such companies arises from the need to rectify defects in software, which in turn controls flawless (at least relatively speaking) electrical and mechanical components. Not at all! If this were not so, it would be as miraculous as the spontaneous appearance of a Rolls Royce at the end of an automotive production line designed to produce Minis®.

The revelation and identification of this dichotomy of standards and its inevitable outcome is greeted with relief in such companies. Someone has at last pinpointed the cause of what hitherto had remained something of a puzzle. In addition, bringing the approach to software development into line with those of the other two main branches of engineering actually pays for itself in the medium term, as the classic TickIT diagram in Figure 15.1 shows. The relationship between the initial investment required to establish effective QA in an organization and the much larger benefits accrued in the medium term directly attributable to getting the job right the first time have led to the notion that quality really is free. Among the best explorations of this fundamental truth underlying all forms of processed-based engineering is that by Crosby [7].

Audit Versus Assessment

In spite of the consistent pattern of commercial success experienced by software companies of all shapes and sizes through the adoption of ISO 9000-3 standards, quality improvement and the cost savings to be reaped is not generally high on the scale of priorities across the software industry. Though regrettable, this is a somber fact of life. Many software houses, for whom regulated industries such as pharmaceutical and nuclear represent no more than a small proportion of their total market, tend to dismiss documented standards and methodologies as a bureaucratic distraction. This culture has taken root in an environment dominated by the inexpensive personal computer (PC), a landscape where defect-ridden software developed by amateurish, craft-based, labor-intensive methods has always been the norm. Markets have acquiesced in the forswearing by software houses (in the text of their licenses) of all responsibility for the quality of their products and the impact of defects. This situation would never be tolerated with any other commodity. The evasive wording currently being adopted in statements of Year 2000 compliance by software houses further exemplifies and compounds this

Figure 15.1. Cost Versus Quality

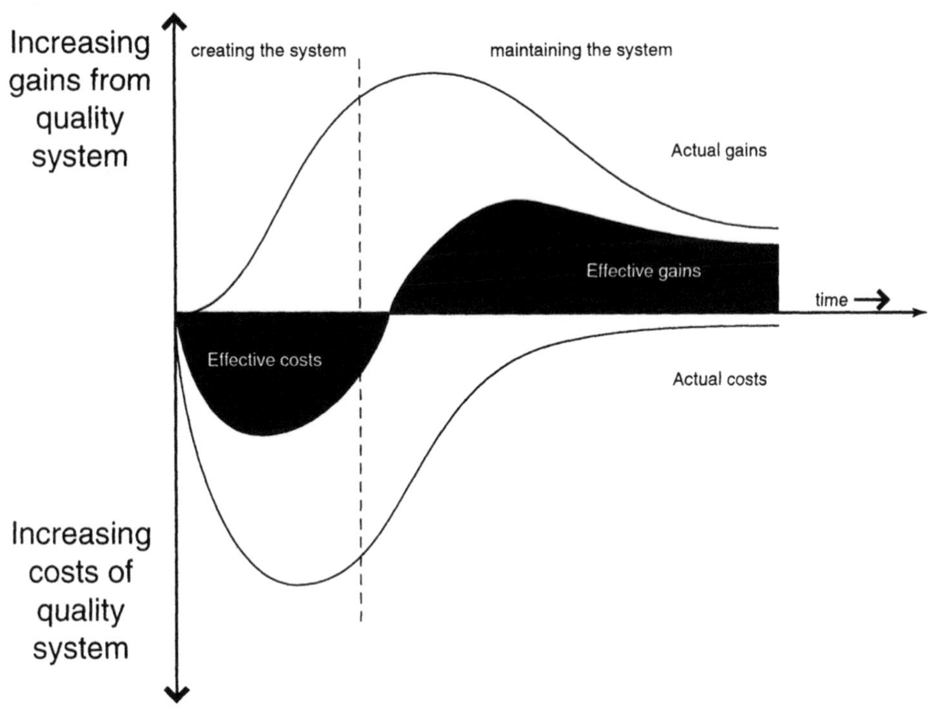

irresponsible attitude. Firms that have been obliged to face up to the rigorous quality requirements demanded by regulated industries have discovered, sometimes to their surprise but always to their delight, that getting it right the first time really is in their best commercial interest.

In light of the prevailing culture, therefore, the initial approach adopted by auditors to software houses is critical. Companies currently making heavy weather under craft-based, uncontrolled development methods need to be helped to put their house in order and motivated to do so by enlightened self-interest. This can only be achieved if auditors adopt a positive *consultancy* approach to firms, offering sound practical advice based on hard software development experience to bring the development process under control in a cost-effective way. The word *audit* is itself a stumbling block. It conveys negative overtones of inspection, perhaps evoking unhappy memories of the grudging compliance of reluctant school pupils under a strict schoolmaster—it may even raise the specter of the police raid! We substituted the word *assessment* in this context long ago and qualified it by the term *team* (i.e., *team assessment*), thus emphasizing the need for developer firm staff to examine themselves. Once this spirit (facilitated by the auditor) is established, the case for improvements tends to arise naturally, since in all firms bad practice that needs amendment is already intuitively recognized by most, though seldom confronted. Remedial action consistent with the prevailing culture, suggested and therefore owned by the staff themselves (rather than being imposed externally), emerges into the open as public property and becomes politically acceptable. Such wholehearted commitment to change within the firm is the only basis for a sense of optimism that improvements can be achieved and institutionalized into permanence.

PROJECT MANAGEMENT

Project management is an area of critical importance for all groups involved in software, as it is for any extended team activity of technical or logistical complexity. This almost certainly explains the fact that failure to control projects effectively, from inception of an idea through to the successful market launch of the product, is the principal cause of insolvency in the software industry. Coopers and Lybrand have reported on a study revealing that two-thirds of IT firms in the United Kingdom had suffered the failure of at least one project, in terms of delivering the expected benefits, while one-fifth of these firms had suffered actual financial losses. It has been reported that in the United States in 1995, one-third of an estimated quarter of a million projects were cancelled, at a total overall cost of $80 billion [8]. No assessment exercise can therefore afford to overlook this strategic area.

The challenges posed by the management and coordination of the interrelated activities of a team of individuals engaged in a task of technical complexity over an extended period of time are formidable. This fact is acknowledged everywhere. Indeed, anyone daring to suggest that project management is a trivial exercise that might readily be handled informally, while retaining a good chance of successful completion on time and within budget, can expect to be dismissed as either quite out of touch with reality or hopelessly naïve. How extraordinary then that the majority of software houses still approach projects in precisely this manner! The difficulties, the setbacks, the cost-overruns, the late deliveries and the uneasy perception of great risk are ubiquitously berated. Yet little is done to alter the approach that is adopted from one project to the next.

The solution is well known, easy to state and has been used to transform thousands of firms. It is the deployment of *standards, procedures* and a *defined methodology*—in fact, the principles that undergird ISO 9000. Unfortunately, this is far easier to state than to implement. Surprisingly, the main difficulties lie not in the technical tasks of defining and documenting best practice but of overturning the entrenched conservatism of staff at all levels to altering the *status quo* in what appears to be something akin to a revolution. To be fair, this is true up to a point, although the universal experience of firms that have embarked on transforming themselves into quality assured organizations has been to ask themselves, at quite an early stage, why the old chaotic work habits were tolerated for so long. The usual underlying reason for the long delay in implementing a standardized approach is that the true costs of working without standards, procedures and methodologies, while huge, is unknown and largely unknowable. Regrettably, it is extremely difficult to reliably estimate the long-term cost of loss of new potential customers or the leakage of repeat business from existing clients. What remains hidden rarely becomes politically controversial, thus change is hard to justify.

The first document fundamental to mitigating risk in the project process is a written description, ideally illustrated with a diagram, of the *project life cycle*. This is a description of the discrete sequential stages through which a project passes, from the incipient bright idea to the market launch of a new product. The emphasis in the project life cycle is *not* on the quality of the intermediate deliverables or the final product, which will be addressed separately. Its focus is on the monitoring of activities, timetables, deliverables and possibly also the budget. In most software firms, especially hardware manufacturers using software to control their machine or instrument products, these stages are already well defined even if undocumented. However, the lack of documentation leads to differences of perception between equally professional and skilled staff of what constitutes best practice. Documentation of the process exposes these differences, often for the first time, allowing the firm to address deep-seated causes of waste and poor quality.

Once the process is agreed on and documented, the procedures to be followed in executing the process by the various players become much clearer. They bring to light the interrelationships and dependencies not only of the activities of each individual player but also those within the project itself. Procedures should be drafted in the simplest manner possible, intended only to constrain in respect of those practices that lead to waste or lack of control. They should outline basic best practice but at the same time avoid becoming unnecessarily rigid or prescriptive, being sufficiently flexible to accommodate the varying requirements of individual projects.

Central to the control of the process are two key documents that ought to be drafted at the outset of all projects: the *project plan* and the *quality plan*. The GAMP Guide [1] combines these two into one, defining the Project Quality Plan as defining actions, deliverables, responsibilities and procedures to satisfy customer quality and validation requirements. Whilst there is nothing in principle wrong with this, most firms have found it helpful in practice to separate the two. In this case, the project plan concentrates on the activities, deliverables and also perhaps the budget control measures. The GAMP Guide suggests that the core content of a project plan includes, in addition to a description of the organizational arrangements, the following specific activities:

- Project milestones (predetermined, clearly identifiable project events)

- Project activities, e.g., design reviews, code walk-throughs (or for larger software teams, formal software inspection—both forms of *structural testing,* described below)
- Personnel assigned to each activity
- Planned start date and duration of each activity
- Deliverables expected

Control is most successfully exerted by means of a control chart known as a Gantt chart, for whose drafting and maintenance a variety of labor-saving tools are now available. Two well-known and easy-to-use examples of project management software packages are Microsoft Project® (Microsoft Corporation) and Schedule Publisher® (AMS Software Inc.)

The quality plan, on the other hand, concentrates on defining the *documentary evidence* that must materialize at given stages in the project that will **prove** that the product under development is acquiring the necessary quality characteristics that will fit it for its intended purpose. Examples of such quality controls might be as follows:

- Specific tests within a structured test plan that demonstrate that particular functions work correctly *(evidence: the test record)*
- A review aimed at ensuring an application's compatibility with the UNIX environment *(evidence: the minutes of the review meeting)*
- The running of a CASE (computer-aided software engineering) tool that looks at source code to see that it kept to the specified programming standards *(evidence: the tool's hard copy report showing the absence of any violations)*

The plan thus defines the QA monitoring of the project and thus provides a safeguard against a key element being missed. In this respect, quality controls are an important risk-reduction device.

Typically, the project manager would own the project plan, while the QA manager or team would use the quality plan to monitor the incipient product. Notwithstanding this distinction and separation, the link between the two plans and their patrons must be extremely close if the project is to be prosecuted successfully and harmoniously. A lack of procedures defining the use of such plans is a certain pointer to a project process out of effective control.

Another critically important procedure in relation to projects is the management of change. Few projects ever proceed as originally intended, even for a week, due to the imposition of changes from every quarter, both internally and externally. The threat this poses to software-related projects is particularly serious because of the complexity of much modern software and its sensitivity to even the smallest change in requirements or characteristics of hardware to which it may be intimately linked. Examples of such situations abound in the computer press. To avoid frustration, disillusionment and chaotic slippage of both timetables and budgets, a process that is documented in specific procedures for maintaining control of software-related projects in the face of such contingencies is vital.

In most hardware manufacturers, the project management procedures are reasonably well documented, since the needs of the engineering processes have determined this. Since in these cases the software is intimately associated with the electrical aspects of the product, software project management is automatically catered for. Unfortunately, the situation in relation to project management and control in software integrators is seldom as healthy, and the auditor is usually obliged to urge the standardization and documentation of the basic elements of the process.

QUALITY ASSURANCE

In general, hardware manufacturers and software integrators, looked upon as groups, provide an instructive contrast in their attitudes to formal QA and in the degree to which a quality culture has taken root. Steeped in the disciplines of high precision electrical and mechanical engineering, hardware manufacturers without exception have some kind of formal QA presence in place. The sheer demands of ensuring the successful fabrication and machining of complex assemblies, the prohibitive expense of abortive effort and the futility of attempts at rework have driven such firms to adopt strict standards and procedures. This industrial sector was among the first to embrace the use of computers (CAD/CAE [Computer-Aided Engineering]), whose computational and graphical tools had been found capable of radically reducing the huge effort hitherto required in the manual development and maintenance of detailed drawings. These examples of design documentation had long ago been found critical to the successful development of even the simplest mechanical or electrical product. In view of the tumbling cost of such equipment and the ingrained quality culture bred through years of practice, it is now inconceivable not to base an engineering business on the use of such methodologies. Personnel in these enterprises are therefore open and receptive to suggestions of improvements to the level of control in the process, especially in relation to software that, because of its intangibility and relative novelty, has tended to slip between the vice-jaws of existing documentation disciplines.

Software integrators, on the other hand, are a very different kettle of fish. In the early days of computing, the exorbitant expense of computer hardware necessarily confined programming to a very select community within academia and a handful of large, commercial computer manufacturers. Most of the code was constructed in the early process-related languages such as COBOL (Common Business-Oriented Language), ALGOL (Algorithmic-Oriented Language) and FORTRAN (Formula Translator), the algorithms on which such programs were based lending themselves to standard and relatively straightforward documented methodologies. While it must be conceded that not all of the software produced in that period was well designed or of high quality, much of it certainly was. The success of IT in those decades owes much to the robustness of the major operating systems of the day and the core application software on which much of the corporate West came to depend.

The advent of the PC in 1982 was to wreak profound changes on the IT scene that even IBM (International Business Machines), the makers, did not anticipate. Indeed, the pace of technical advance in under 20 years is scarcely believable. The first PC, the *Sirius,* was launched in 1981 and followed by the better known IBM model, also based on the Intel® 8088® microprocessor. This machine ran at just 8 MHz, had only 64 kbytes of memory and used a 160 kbytes 5.25 in. floppy disk drive. Far from remaining the exclusive province of the hobbyist or the small business, this small device has transformed the whole basis upon which business and commerce operate. From a software perspective, it opened floodgates to a vast community of untrained *amateur* (descriptive of someone who pursues an interest purely for the love of it) programming enthusiasts, able to exploit the power of a computer now unprecedently within their sole control. The software industry of today has grown directly from this seedbed, fertile with ideas but barren of the disciplines and training in good software development practices that were the lifeblood of the early professionals. Most of the contemporary software industry still needs to learn these principles and take QA seriously, an essential prerequisite to the development of software demonstrably fit for purpose.

Two of the most important indicators of a quality culture in a firm, in the absence of official accreditation to a public standard like ISO 9000, are as follows:

1. The presence of a person or group, mainly or exclusively responsible for quality assurance
2. The existence of a documented set of procedures defining how things are done, known as a QMS or just QS

Of course, if a firm is ISO 9000 accredited, this already speaks volumes about its commitment to quality, although the limitations of ISO 9001 to software development need to be clearly understood. These shortcomings have given rise to ISO 9000–3 (TickIT), to which over 1,000 firms have now been accredited; however, this is a drop in the ocean when viewed against the industry as a whole. The consistent pattern of commercial success enjoyed by TickIT-accredited software houses is no coincidence and should be used by the auditor as a powerful endorsement of the benefits of a quality-assured process in securing the future prosperity of a firm involved in any way with software development.

Most hardware manufacturers already have a QS in place, although this is usually heavily or exclusively focused on electrical and mechanical engineering. Clear guidance will be needed with regard to the extension of these principles to the control of the software development process. Some of the central details of this will clarify as this chapter progresses. For software integrators, the auditor will often find that an understanding and acceptance of QA is far less mature, a situation exacerbated if one or more independent subcontracting software firms have been engaged. Getting a commitment to quality in the software integrator firm is one thing. Getting them to impose this on their subcontractors is quite another.

Even in firms with a formal QA program, QA's role is not always clearly understood. In many firms, the role remains as *quality control (QC),* a title emphasizing *testing action* at the end of the process, when the product and its quality are fixed, cannot be changed and only attempts are made to measure it. An infinitely better approach is that of QA where the focus of activity is shifted from testing to the *process itself,* seeking to root out the causes of failure and to improve the capability of the overall process. The product will then become a more predictable outcome and to some extent look after itself, allowing the testing effort to be scaled down. If the level of defects hidden in the product is less, we can spend less effort seeking them! This is one of the hallmarks of a mature organization. The role of quality assurance is principally one of ensuring compliance with in-house established standards of good practice, following up noncompliance and fostering a culture of continuous improvement. Such a role is exemplified by the classical Shewart cycle, illustrated in Figure 15.2.

THE SOFTWARE DEVELOPMENT LIFE CYCLE

Most of the contemporary software industry continues to approach the development of software packages on the tacit assumption that the elements of success, in essence, are the creativity and flair of programmers. This belief has been greatly reinforced in recent years by the appearance of modern graphical tools, whose capability now enables programmers to deliver working prototypes much more rapidly than hitherto. Unfortunately, because prototypes of this kind are usually bereft of any underlying control on what they will ultimately be expected to do, this results in the delivery of final products that are teeming with defects. The fundamental reason for the poor quality of such software continues to mystify most of those involved in delivering it, especially since the tools used to assemble it are so visually attractive and most impressive on their own terms. The possession of sophisticated tools does not automatically confer upon the pharmaceutical manufacturer the ability to use them to their full potential.[1]

[1]This principle was powerfully illustrated recently in the author's own experience in the purchase of a professional spray gun and compressor to respray his car. Without the skill to use these tools, the final result fell far short both of his expectations and of commercial standards!

Figure 15.2. The Shewhart Cycle

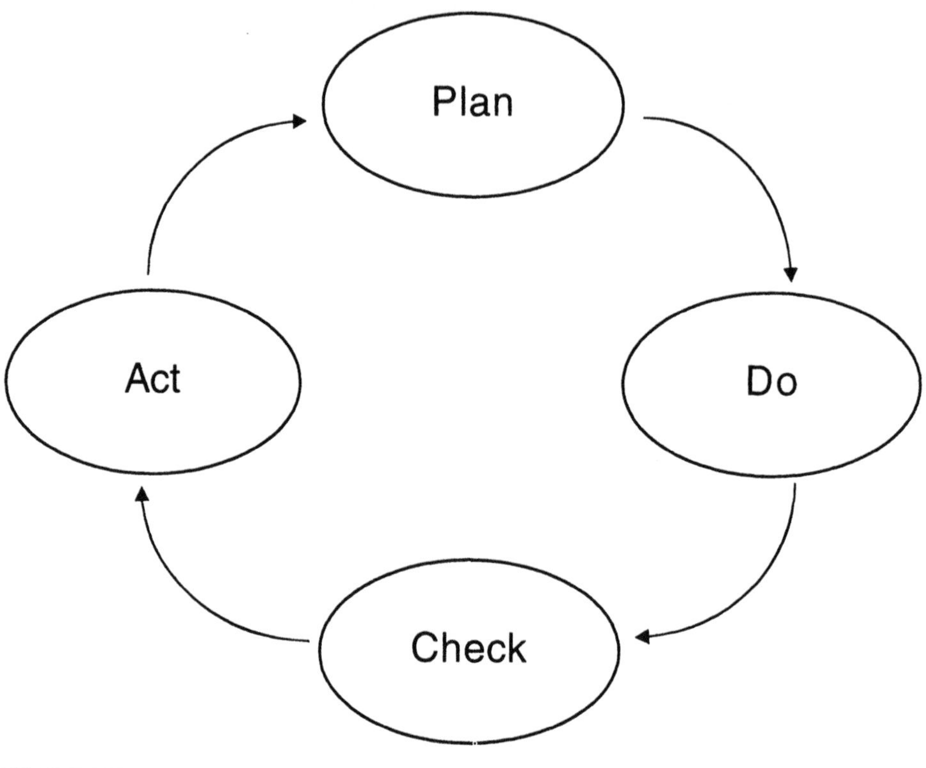

The disappointed expectations both of software developers and pharmaceutical manufacturers (and indeed users of software in general) arises from a misconception of the formidable intellectual challenge posed by the requirements for modern software, whose complexity on the fast and inexpensive machines of today far surpasses that in the early days of commercial computers. However, even then the difficulty of delivering software that did what it was intended to do, and perhaps more importantly did not do what it was not supposed to do, was driving software houses to recognize the weakness of their craft-based methods. These approaches dominate the industry of today, which generally has learned little from the huge improvements in software quality and development capability that can be effected by introducing formal controls into the process. Such formal controls are founded on the recognition that development is a stage-wise process, each stage can be defined and can also be a wellspring of errors in the final product. Echoing the principles of statistical process control, these errors can only be avoided by addressing the causes of their ingress at each stage.

This stage-wise approach to software development is known as a *Software Development Life Cycle* (SDLC); when implemented, it formally introduces control into the process and causes the final product to become far more predictable in both functionality and innate quality. Because of its similarity with the conventional approaches adopted in mechanical and electrical engineering manufacture, the term *software engineering* was coined (this term actually seems to have originated at a scientific conference organized under the auspices of the North Atlantic Treaty Organization in 1968). The link with the project life cycle discussed above is intimate, though this recognition often leads to the confusion of one with the other. The distinction here is that the SDLC addresses the *quality characteristics* of the product and extends beyond the launch of the product (which is where the project life cycle terminates), into maintenance and support, concluding only in product retirement.

There are a variety of life cycle models in use today, some of which are described in the STARTS Guide [9]. One of the oldest, the *Waterfall* model, first recognized and named by Royce [10], is perhaps the most commonly used model. This simple but elegant model defines the basic stages of development (Figure 15.3). The life cycle introduces control in two ways. First, it separates development into discrete stages, each of which must be satisfactorily completed to defined quality standards before it can constitute a sound base on which to safely base subsequent work. Of course, in real life, this one-way traffic flow is not absolute; there is usually some measure of interaction between successive stages. However, the principle is paramount. The large arrow represents a controlled change process after release. Second, each stage delivers documentation. This constitutes evidence to everyone involved that the required quality characteristics are being built into the product *as development proceeds*. Should this begin to break down, the developers are warned at the earliest moment, when the deviation (incurred errors) can be rectified at minimum cost. It is these features that explain why working to a life cycle is the most cost-effective and surest way of delivering a product with the least defects, ensuring customer satisfaction, repeat business and hopefully the long-term prosperity of the firm.

Another life cycle often used in today's Windows® oriented graphical environment is the *spiral model,* illustrated in Figure 15.4. This model, appropriate to rapid prototyping or other forms of incremental development, acknowledges the use of the graphical Rapid Application Development (RAD) tools mentioned above. However, it brings them into a controlled framework so that the traditional benefits of the life cycle can be combined with the speed of development—the best of both worlds.

The auditor has to ensure that the hardware manufacturer or software integrator is working to a life cycle and has documented this work. Most hardware manufacturers already have a stage-wise, documented approach to engineering operations and are therefore usually receptive to the need for

Figure 15.3. The Waterfall Diagram of Software Development

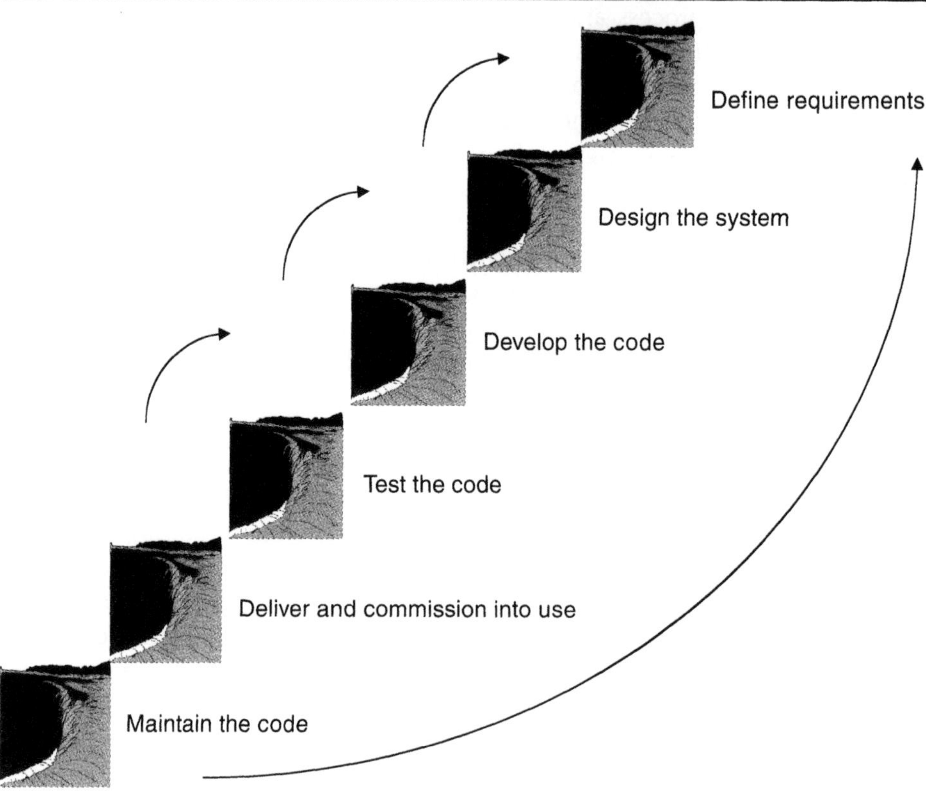

a documented life cycle for software. Auditors frequently find software integrators with little knowledge of the benefits of a life cycle, and that no effort has been made to require contractors to follow it. Often it will be claimed that since the subcontractor is a one- or two-man team, a documented life cycle is unnecessary and inappropriate. Yet to rely solely on verbal communication, with all its accompanying vagueness and ambiguity, to govern such an intellectually demanding and complex activity as software construction is to guarantee that hundreds of errors will ultimately be passed on to the customer, bundled into the product. Needless to say, this characteristic of the product is never featured in the brochure!

The life cycle description should be as simple as it can be while still being adequate as a working description. It should describe the contents of the document delivered at each stage. Details of the suggested contents of the *User Requirements Specification* (URS; rarely used by hardware manufacturers), the *Functional Specification* and the *Software Design Specification* (SDS) are

Figure 15.4. The Spiral Model of Software Development

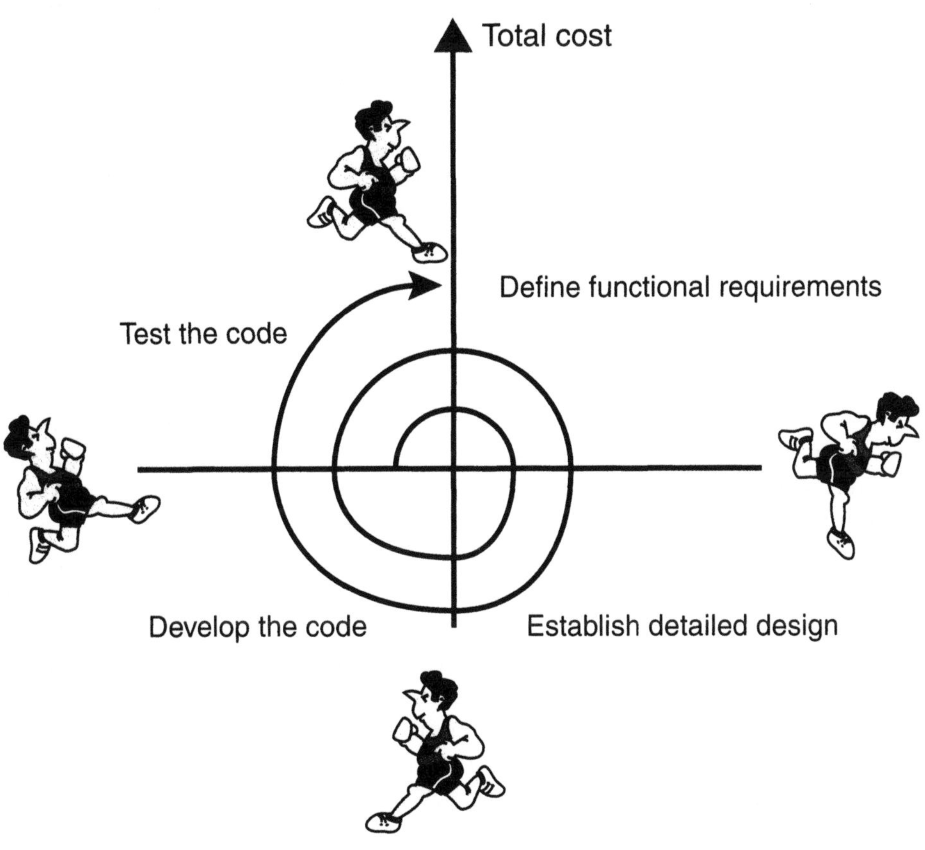

found in the GAMP Guide [1]. The most successful software integrators use a single, central documented life cycle to control all the activities of their constituent suppliers.

The design specification may be regarded as the center of gravity of the entire life cycle, around which both the definition of the requirements and the coding/testing processes revolve. The SDS is the document on which programmers will base their work and flows out of the functional requirements on which it is based. Hardware integrators sometimes justify the omission of this document on the basis that it is subsumed in the electrical drawings. However, closer inspection usually reveals that this shortcut exacts its own penalty, since the software must have some kind of design that, if undocumented, will be perceived differently throughout the team. Errors can often be tracked back to such deviations in perception.

We have implied that the design specification is based on the functional requirements. However, both the importance of ensuring this and the way to achieve it simply are seldom understood. All of the functional requirements must somehow be catered for in the design. This applies no less to what the system must do as to what the system should not do. Failure to achieve this is one of the most common causes of the most fundamental defects of delivered systems, those errors that are, in general, the most expensive to rectify. Ensuring such comprehensive scope in designs is one of the responsibilities of QA, of which more will be discussed later. In the meanwhile, a simple but reliable system of *document traceability* must be in place to enable QA to sign off the document for comprehensiveness.

One of the easiest ways to achieve this in practice is to implement a standard for traceability in the form of paragraph numbering. Under this standard, Functional Specifications are drafted by uniquely numbering each individual requirement. In the design specification built upon it, each numbered design feature references the functional requirements paragraph covered by that aspect of the design. QA personnel, while not necessarily possessing the technical expertise to review the design for technical merit, are certainly able to ensure, as part of their approval, that every functional requirements paragraph is mentioned in the design. A mechanism is then in place to ensure the comprehensiveness of the design. The same traceability principle may operate between the design specification as the source document and the structured test plan as the derived document. Auditors should seek evidence of traceability and, if absent, recommend a simple system such as this as a powerful tool for the early elimination of errors.

Any life cycle worth its salt should deliver at least the following foundational documents:

- Functional Specification
- Design specification
- Source code
- Structured functional test plan
- Installation instructions (if applicable, but usually not for hardware)
- User manual

At this point, it is worth noting that the source code is just a life-cycle document. Often this comes as a surprise, as it is so closely identified with the product and often thought of as the product itself. Of course, the product is really the executable code that makes the computer useful, while the source code is simply a documented listing of the instructions in human readable form—in other words, just another item of life-cycle documentation!

Auditors will often find that an intuitive life cycle is in place but not documented. While this situation is preferable to creative chaos, it falls far short of the standard for validatability, which is defined as in terms of documentary evidence. However, the benefits of documenting the life cycle, even in very small firms of two or three people, quickly become apparent, a fact that should be of much encouragement to the novice auditor.

Programming Activity

Although we have asserted that the design phase is the center of gravity of the life cycle, programming activity lies at the very heart of software development. For this reason, it is important that the auditor pays attention to the individuals involved and the standards by which they work, since both have an important bearing on the quality of the code produced.

A key feature of the ISO 9000 suite of standards is the weight accorded to staff training, in terms of the acquisition of skills and competencies needed to perform their function in the business. There are two factors, apposite to computer programming, that powerfully reinforce the importance of training in this sphere. The first is the degree of intellectual rigor demanded by the remorseless logic of computer programming in any language. This on its own carries enough weight to justify a software company investing in training to adequate levels. However, an additional factor, which shows no sign of abating, is the frenetic pace of change in IT, imposing ever shortening lifetimes on many particular skills, competencies and areas of specialist knowledge. Software houses seeking to save on training quickly discover how false such economies are, as staff perceive the managerial myopia implied by such attitudes and leave for greener pastures. The economic loss suffered by such hemorrhage of staff is incalculable.

Auditors should therefore seek documentary evidence that effective training of programmers, both on and off the job, is taking place. Such evidence could be the ubiquitous certificates handed out after the completion of formal training courses or internal records of new skills acquired on the job, for example, the mastery of a new tool or language. In hardware companies, there is normally an established training scheme for engineering apprentices, of which software skills training would be a natural extension. For software integration firms, this might be a new concept. In any event, there should be a periodic development interview, conducted at least annually, at which staff training requirements are reviewed against the shifting background of IT. This can be quite simple in small firms, though records are essential. However, more formal appraisal systems are necessary and usual in larger firms.

Certain companies in the United Kingdom are participating in a scheme known as *Investors in People* (IIP), a systematic approach to staff training based on formal methods that mirrors ISO 9000. Its intention is to provide a framework of standards that ensure that the training provided for each member of staff is formally geared to the needs of the business. In this way, it becomes possible to demonstrate that the firm is deriving the maximum benefit from all training investment.

Auditors should record the type of staff training taking place, the method of recording this and the way in which the needs of the business are considered in relation to it. Furthermore, the average longevity of programming staff is an important indicator not only of morale but also of the longer-term effectiveness of the training methods employed.

So much for the individual programmers; let us now consider the standards by which they work. Only about half the firms involved in software today have documented the programming standards for the languages in use. Those without programming standards continue to rely solely on the handbooks provided by the manufacturer of the computer language or tool in use, claiming this to be adequate on its own. It is not difficult to show how disingenuous this claim is. In every programming environment, in-house conventions in style and approach quickly arise that determine the layout and content of code. Examples of these are the naming of variables, indenting branches and loops in code (if not imposed by the language itself), using of header and paragraph comments and error handling.

While the programming team remains small and tightly knit, the absence of documented standards is often no more than a minor handicap, since the conventions are well understood and instinctively followed by consensus. Thus code maintenance takes place safely, in a manner entirely consistent with the original patterns of development and the agreed conventions. Providing the authors remain with the firm to develop the code in line with changing requirements, the weakness remains hidden. It is only fully perceived, however, when the organization enters a phase of rapid expansion, and novice programmers are recruited. In the absence of documented standards, the incumbents are then faced with the task of verbally imparting to the newcomers all the detailed assumptions adopted instinctively within the group. The challenge presented to them is not only to recognize, let alone to remember to declare everything of importance, but also to ensure that this information has been heard, fully understood and digested by their audience. The exponential rise in defects appearing in software maintained by newcomers in these circumstances demonstrates that this communication process is always to a greater or lesser degree unsuccessful.

How much better, then, for the firm to insist that all of the accepted in-house conventions for programming are exhaustively documented in a set of standards for each language or tool. These should be supplemented with examples not only of Good Programming Practice but also of prohibited practices, the latter often being far more telling in constraining coding to the righteous paths of best practice than the former. The investment involved is quite small, since the standards once drafted can quickly be supplemented over the course of a few weeks or months to a comprehensive level.

The auditor should review these standards for adequacy, based either on his or her own knowledge of the language in use or against a published checklist such as the one in GAMP Guide [1] and in Appendix 15.1 of this chapter.

A much fuller treatment of the significance of such checklists and their use by both software development groups and software user companies (whether within the pharmaceutical sector or not) in fostering software quality improvement has been published by Garston Smith [11]. An absence of such standards is a serious weakness in any company, such as hardware manufacturers or software integrators, where software forms a key part of the product. In the latter case, the standards used by all of the subcontractors should be similarly reviewed, and the reason for the lack of adoption of a single set of standards explored.

Structural Testing

We have already referred to the complementary nature of structural testing to functional testing in the context of a validation exercise. This relationship is, of course, equally true outside the scope of the pharmaceutical validation sphere and applies to software of all kinds. The fundamental limitations of functional testing in exposing the true level of defects in software, in other words the innate level of quality, is still not understood throughout large parts of the software industry. While it may be argued by hardware manufacturers that the embedded code is often structurally less complex than many other types of software, the limited usefulness of functional testing as a quality measure remains.

Because structural testing seeks to "test" those pathways that other types of testing cannot reach and is relatively inexpensive to conduct, it is one of the most efficient ways of building quality into software. It can take one of two forms. The simplest form consists of one programmer *walking through* a piece of code that he or she has developed in the presence of one or two colleagues, who are equally technically proficient and therefore able to identify defects, especially in the logical structure. These scrutineers are responsible for raising issues relating to the software (the word *defect* is assiduously avoided in this context in the interests of frankness), which the author is committed to investigating and resolving. The number of such issues raised closely correlates with the innate level of *structural integrity* of the code. More mature organizations, experienced in the practice, will preset an *exit quality level* of the maximum number of issues permissible prior to the code being released into a project. A given piece of code may require several walk-throughs before reaching the company quality level. The reference document under which the inspection is conducted is the design specification.

In a larger organization, the process is fundamentally identical, except that the number of scrutineers is greater, and the session is moderated by an independent moderator, perhaps the QA manager or some other impeccably impartial figure! In both scenarios, a record should be kept of the date of the event, the number of issues raised and the date of their resolution. These records are fertile soil in which process improvement initiatives can grow.

The auditor should seek evidence of the conduct of structural testing as complementary to functional testing. There should be a procedure within the QMS describing the practice in detail, including record keeping. In the absence of such a process, the auditor is obliged to explain its virtues and advocate its adoption, in the interests of eliminating defects at the earliest (cheapest) phase of the process and minimizing the overall cost of software development and support.

It should be noted that this technique may be applied with equal benefit to other types of life-cycle documentation. Indeed, the most mature software houses do precisely this (usually within one of the classical life-cycle models) in order to remove defects at a minimum cost.

Functional Testing

A feature of structural testing that in no way undermines its effectiveness is the lack of independence. The scrutineers may be part of the same development team as the author of the code, only the moderator (if there is one) being independent (perhaps to keep order). Expertise is more important than impartiality at this early stage of the process.

Contrast the situation in functional testing, which is very different. Here the behavior of the finished product is being compared to that described in the Functional Specification. Because of this, those conducting the tests do not need any detailed knowledge of how the software works; hence the term *black box testing* is used to describe this activity (a black box being one whose contents remain hidden). Nevertheless, it must not be assumed that the testing need not be *structured* at all. Indeed, it is lack of structure in functional testing that lies at the heart of the ineffectiveness of much functional testing conducted today.

The remedy here is for functional testing to be conducted under a *structured test plan*. Hardware manufacturers commonly used such plans as a direct outcome of the electrical engineering methodology in use, where the testing of the fabricated machine (operated by firmware that is also necessarily tested) is organized by functional area. In the software industry at large, however, the situation is less secure. Functional testing is often based on the ability of QA to formulate a plan on the basis of intuition, rather than with any formal methodology or tools. The chance of achieving comprehensiveness in testing positive product behavior (what the system is supposed to do) is quite low, while the situation in relation to negative behavior (what the system is not supposed to do) borders on the superficial.

Both these evils can be avoided with structured test plans based on the design specification and formally traceable to it. Few companies have reaped the benefits of increased functional testing efficiency and additional QA that formal traceability delivers. By this means, the structured test plan can be formally checked by QA, prior to use, as capable of exercising every feature of the design (positive and negative) that is amenable to independent testing.

The auditor should seek a procedure defining the assembly of structured test plans. Ideally, actual test plans should be traceable to the design and include a formal review for comprehensiveness before any testing commences. Test reports summarizing the outcome of functional testing would normally be expected as evidence that the testing was actually carried out and that the product met its defined requirements.

Documentation

We have already established that documentary evidence is central to validation. However, the same principle applies to the control of any industrial or manufacturing process; software development is no exception. Unfortunately, documentation can quickly degenerate into a bureaucratic nightmare and thus justly attract the contempt of everyone, unless it is conducted in accordance with standards of good practice. Otherwise, far from being recognized as a crucial asset on which the success of the operation depends, it serves to undermine the very efforts to achieve the effective QA that it purports to support.

Documentation standards must apply not only to the life-cycle documents related to a particular product but also to the Standard Operating Procedures (SOPs) that constitute the QMS. In respect of the latter, the fundamental link between these, and the skills and competencies training program within the firm, is often overlooked. When personnel are competent to perform their assigned tasks, the detail included in SOPs need only be that essential to constrain bad practice. It should be sufficient to enable a trained, competent person to execute the process securely in accordance with the prevailing standards to ensure final product quality. It should not attempt to teach the person how to do the job. Many documentation systems fall into disrepute for failing to abide by this simple rule.

In most hardware manufacturing firms, the challenges posed in successfully manufacturing a product using precision mechanical and electrical engineering will have long since brought into being a basic documentation system. The auditor needs to ensure that this possesses the key elements of document issue and control. In other words, each specification or SOP should bear the name of the author, a second reviewer capable of finding defects, a QA review (highly desirable but not essential), date of issue, revision number, superseded revision number and the reason for change from its predecessor (the audit trail). In the software integration world, however, the situation is less satisfactory. While many firms have highly effective QMSs supported by good document control, a substantial proportion have yet to reap the benefit of such a transformation. The charge of bureaucracy leveled at process documentation is heard far more often in this arena.

Although the QMS may be adequately controlled, it is important to gauge whether it is actually being followed in practice. The rate of change of procedures is a good indicator in this respect; moribund systems that have fallen into disuse and which are out of step with actual practice readily betray themselves through SOPs remaining unchanged for years, despite the obvious evolution of the business. An alert auditor can quickly expose this situation by asking the key players to discuss the usefulness of the procedures as they are presented for inspection.

Finally, there is an enormous advantage of using Electronic Document Management Systems (EDMSs) over paper as the message medium. Once prohibitively expensive, such systems can now be acquired relatively cheaply and within the budget of even small companies. The availability of inexpensive color scanners, sold almost universally with basic optical character recognition software, means there is really no excuse for any firm not to reap the huge advantages of a central electronic document repository. Conversion of earlier typed paper documents can be implemented by secretarial staff, thus making the establishment of a simple electronic library in a Windows 95® folder set a rather straightforward (though perhaps a little laborious) task. Auditors should make sure that firms still using entirely paper-based QSs are aware of the savings to be reaped by migration into the electronic age!

Change Control and Version Control

By its very nature, most software today is in a continual state of flux, reflecting not only the steady expansion of functionality driven by the fertile imagination of current users but also by the frenetic pace of improvements in IT hardware. This lack of stability enhances the importance of *change control* in any organization involved in the development and maintenance of software. Without formal change control, as many new errors are liable to be introduced into the software as the number of existing ones the maintenance effort intended to remove.

Briefly, change control refers to the formal procedures by which changes made to software are defined, evaluated for legitimacy and approved or forbidden, implemented, tested and signed off. The entire process must be documented in a manner permitting the implemented changes to be traced backward from the implemented change (to code or to a document) to the original request or definition of need that initiated it. Usually, this includes some kind of Change Request or Engineering Change Form, uniquely numbered and logged in a book or a database, often of commercial origin but sometimes one written in-house using a commercial package such as Microsoft Access®. The practice has two elements: (1) a procedure or set of procedures describing the activities that implement the process, and (2) the capability of providing a visible trail that can be followed subsequently.

In auditing the effectiveness of change control, the auditor should be aware of the need for and seek evidence of the actual presence of both components. In Supplier Audits focusing on a particular software product and version, the obvious framework for such inspection is the source code suite of the product, together with the change control records. Selecting a particular change and working forward to the item changed is followed by a similar exercise in the reverse direction. Any weaknesses in the procedures soon become painfully apparent!

Where the software suite consists of a number of discrete component items (whether code of various kinds or documentation), developed separately but combined to form a unitary whole, the complementary discipline of configuration management also assumes great importance. This discipline seeks to trace the relationship between the various versions of each component item and the version number allocated to the total product. At the component level, intermediate changes made to these items give rise to *variants*. The changes expressed by several variants may then be consolidated into an incremental stage, called a *version*. From time to time, attention will be switched from the component level to the final product level by creating a snapshot image of the total product, in terms of the variant/version numbers possessed by each component at that moment. This stage is known as a *baseline*. The development of the overall product may traverse several baselines before reaching its first internal release stage. Configuration management is thus a formal means of tracking the development of the overall product in terms of the version numbers of each individual component, of which the total product is comprised. The principle is illustrated in Figure 15.5. It is not difficult to appreciate that for the creation of a software product with even a modest degree of complexity (by today's standards), the lack of formal configuration management is a recipe for chaos. In the event of trouble, there is no means of returning to a defined *status quo ante*.

Secure configuration management is especially important to software integrators, where by the nature of the work multiple components are being incorporated (integrated) into a unique whole. It might be assumed that the discipline is of less importance for hardware manufacturers. While this is undoubtedly true where only one or two software components are involved, there are pitfalls to be avoided through ignoring it, a fact that may be more clearly illustrated by a commercial example.

In one of the leading UK pharmaceutical companies, a printed circuit board assembly in a plant controller was replaced after a failure with an ostensibly identical unit. To everyone's complete surprise, this simple action, regarded as relatively routine, led to a major operational incident. The cause was found to be an EPROM chip on the replacement board. This chip not only carried an upgraded version of the software on the original but also possessed different operational parameters. Despite the fact that the replacement printed circuit assembly had been successfully tested (in accordance with quality procedures), it was the lack of effective configuration management that allowed the change in the software to elude detection, which led directly to the failure.

Auditors should therefore seek evidence of the use of procedures that exert control over version numbering and configuration management, which

Figure 15.5. Configuration Management in Software Development

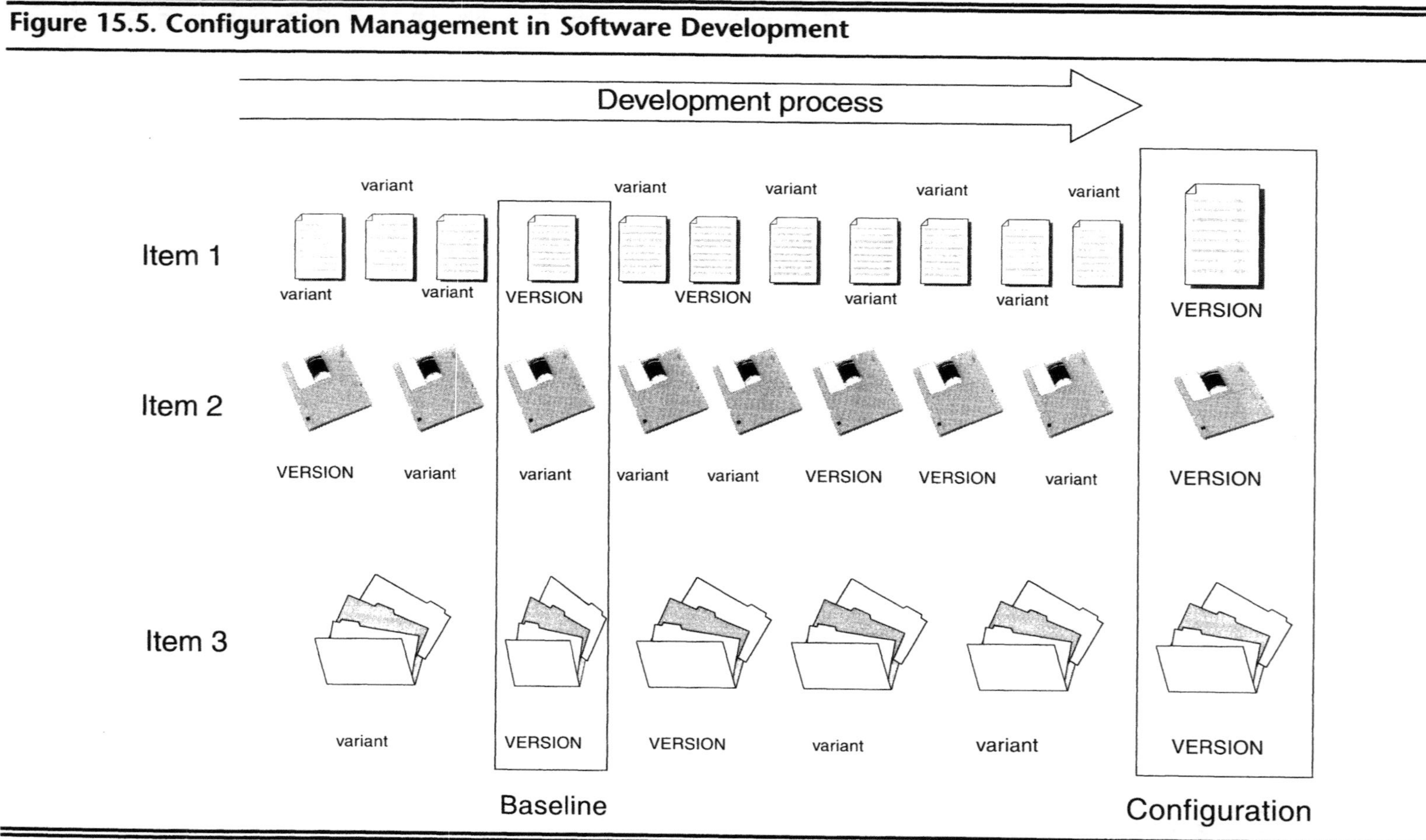

has been the subject of its own ISO standard since 1995 [12]. In respect of the former, there should be a description of the version nomenclature and its meaning, and the criteria dictating the issue of major and minor releases.

Managing the Environment

Although the environment under which software is developed may not seem at first to have a major bearing on the quality of the development process, it cannot be entirely divorced from it; in some respects, it is of major importance. An auditor should not overlook the environment in the assessment of the process and the software product resulting from it.

Data Security

Unauthorized access to source code files is rarely a matter of much concern to small-and medium-sized software companies, but it occasionally becomes a problem in larger companies. Nonetheless, regardless of its size, every software house must preserve the integrity of such a critical resource. Network access controls in the form of user identification and passwords are used for the ubiquitous client-server architectural model, with source code files being held on a central server. For protection against hard disk crashes, perhaps the greatest threat to the data in terms of probability and seriousness weighed together, the use of RAID (redundant array of inexpensive disks) technologies, where multiple copies of a file are simultaneously held on separate drives, mitigates this risk at very low cost.

Although user identifications and passwords usually appear to work reasonably well, there should be a procedure in place for ensuring the regular and frequent changes in passwords, an issue that assumes greater importance with increasing company size and organizational complexity. It becomes an even more serious issue if the QMS, already implemented in electronic form, is fortified with workflow. Here the business processes are managed through the transfer of electronic documents from one person to another, approvals within such electronic records being recorded by means of electronic signatures. These signatures must be created in a manner of sufficient integrity that they cannot easily be repudiated as genuine. Coincidentally, the principles and issues underpinning them have been the subject of intense study in recent years by the pharmaceutical industry and its regulators. Both have been grappling with these very issues as GMP regulations strive to mature and retain their relevance in the face of technological advance (particularly in IT) at a breathtakingly frenetic pace; indeed, the rules have only recently been agreed on and finalized.[2] Of course, software houses are in no way obliged

[2]The Code of Federal Regulations may be found at http:/www.access.gpo.gov/cgi-bin/cfrassemble.cgi? title-199821 or using the index feature at http:/www.fda.gov. Type "electronic records" and "electronic signatures" in the search box (exactly as shown) and click SUBMIT. The search results will reveal 21 CFR 11: *Electronic Records; Electronic Signatures*.

to comply with such regulations incumbent on pharmaceutical manufacturers in this regard, however relevant. Nonetheless, the study of (and even compliance with) them has much to commend it. Most software houses, having *closed systems* as defined in the regulations, already comply with some of the spirit, if not much of the letter, of these regulations.

Data Backup

The second area of interest is that of data backup, the protection of the company's critical intellectual property and capital. Always claimed to be a matter of major importance in software houses (a claim not always borne out by inspection), the issue sometimes becomes a Cinderella activity in hardware companies, where the focus of most of the activity is on the tangible that can be seen and touched. Nevertheless, effective data backup is vital to the continued ability of both software integrators and hardware manufacturers to support their products; excuses by the latter that, because there are multiple copies in existence (embedded in every delivered machine), neglect here can be justified should be dismissed.

In most cases, an effective backup strategy will be described and should relate not only to the frequency with which files are changing but also to the secure storage of the media. The huge capacity and low cost of digital audio tape (DAT) and digital linear tape (DLT) cartridges and their associated drives makes these media ideal, though the recent appearance of the more robust rewriteable compact disk (CD-RW) offers an attractive alternative. Full daily backups (usually made automatically overnight) are typical, though incremental backups (i.e., of only the files that have changed since the most recent full backup) are common but require a clearer process to ensure that in the event of a restoration becoming necessary, the data are restored in the correct order to ensure the recovery of the latest version of every file.

A written procedure is essential for the secure management of data backup, but frequently no procedure exists. The absence of such should always be viewed as a serious shortcoming and feature in any list of corrective actions. The procedure must address not only the management of the media but also their labeling and storage. Labeling is vital since the media used are often part of a set that is periodically recycled. A common approach is for one cartridge to be devoted to each day of the week, thus giving a set of five that are recycled weekly. The label on each should record not only the day of the week but also the day of its latest use.

Storage must not be overlooked, either in practice or in the procedure describing it. On-site storage should always be in a fireproof safe for security against robbery and fire. Off-site storage, effected at least occasionally but always done systematically, is much preferred so that very recent copies of all critical data are held in at least two separate geographical locations.

Finally no procedure, however well thought out and carefully drafted, is of any use if it is not followed assiduously. To ensure this happy condition, some kind of log is essential. While a few contemporary backup tools achieve this automatically, confirming successful completion and warning personnel of failures via E-mail messages, there needs to be some kind of public confirmation that the backup exercise is actually being carried out. One of the simplest methods of achieving this is a handwritten log posted close to the backup device, requiring that the person physically changing the tape cartridge or CD confirms that he or she has done so by signature and date. In this way, any breakdown of the arrangements intended to address absenteeism (for whatever reason) is immediately apparent, with steps being taken to continue to protect the company's data (and perhaps, very much as a secondary matter, to chase up any individual negligence).

Disaster Recovery

The third related matter, only seldom addressed in software and hardware companies in general, is that of disaster recovery. While this will include the recovery of data from storage media, it also includes all of the activities necessary to restore the essential functions of the business in the event of certain conceivable contingencies. These would certainly include fire and theft but might also envisage other calamities. The point here is that in such an event, it will be difficult for the key individuals to react dispassionately; therefore, wisdom dictates that it is better to do such thinking in advance in the calm of normality. Once the contingency arises, the plan may be immediately recovered from safe storage, and its provisions implemented. Keeping the plan current is achieved by an annual or six-monthly simulation drill, where the contingency is imagined to have happened, and the provisions are tested for concurrency. Like all insurance, such plans are often regarded as a waste of resources if they are not needed, but they are a magnificent investment if the worst happens! All companies, regardless of size, should have such a plan and be able to provide evidence for it to reassure their customers of their willingness to remain in business to support their products.

The third and final issue in an era dominated by the PC and its networks is that of virus protection. While the digital delinquency that underlies this phenomenon is universally deplored, the prevalence and continued growth of this menace makes protection against it mandatory for every business using this technology. Automatic sentry protection should be in place on each networked workstation, with strict rules in place governing the acquisition of executable code either by floppy disk physically brought onto the site or downloaded from Internet sites of dubious integrity.

The risk of false alarms (indistinguishable from genuine alarms) can result in firms behaving like the proverbial headless chicken in the event of a perceived infection. Firms are advised to have a second antivirus product in place, so that any alarm may be corroborated in an attempt to confirm the genuine nature of the infection.

TOPICAL ISSUES

Object-Oriented Programming

The almost ubiquitous use of the Microsoft Windows® operating system throughout industry and commerce today has led to much programming being conducted using object-oriented tool sets. The extreme rapidity with which an embryonic application can be produced, compared to the process-oriented traditional languages of 20 years ago, is not entirely the unclouded blessing that it is frequently portrayed to be, for it brings its own difficulties. One stems from the racy, fertile, creative enthusiasm that assumes the documentary controls of the more pedestrian processes of yesterday are as obsolete as the older languages clearly are. While some adaptations in approach are clearly necessary and must be conceded, in principle nothing could be further from the truth. This can-do, throw-it-together culture expresses itself especially clearly in the misconceptions entertained over the use of prototypes. Assembled quickly, such illustrations of the look and feel of a working system (for that is often all that they are) are frequently seized upon by eagerly expectant customers, thrilled at the apparent pace of development but ignorant of the superficial nature of what is being seen. Adequate process QA has seldom been imposed in these circumstances, regrettably, and what is being demonstrated bears the same relationship to a thoroughly engineered system as a cardboard cutout does to the thing or person represented.

Prototypes are of enormous value provided their use is hedged with safeguards. Both developers and customers must understand and agree on the purpose of the prototype. This is often simply a temporary artifact aimed at clarifying their mutual understanding of the true functional requirements (over which remaining uncertainty must be quickly resolved), thus an "application" may quite legitimately be of extremely poor quality. This stage is followed by two necessary (and sometimes unwelcome) chores: (1) the documentation of the clarified requirements for acceptance by everyone as the basis for the next life-cycle stage and (2) the discarding of the prototype as a tool having fulfilled its usefulness and having no further place in the process. Under these arrangements, prototypes may be kept as a blessing rather than becoming a curse to the project.

Standard Software

The development of bespoke (totally original) code is not only very expensive but also very hard to produce with a low level of defects. Size for size, bespoke code is much more difficult to validate to an adequate level than standard software. We might define standard software as that in widespread use in business or industry, thus providing much opportunity for innate defects to emerge quickly and be resolved either through updates or workarounds. Operating systems are, of course, the apotheosis of standard software, their internal pathways being intrinsically exercised every time an application program executing them is run. It is the innate integrity of standard software that is thus implied by thousands of satisfied users. Both of these considerations underline the advantages to software integrators and customers of using standard software packages whenever possible. In developing requirements for new systems, all customers, whether pharmaceutical or not, should carefully weigh their musts and wants, subjecting the latter to remorseless critical analysis, in order to try to bring the final requirements within the scope of widely distributed contemporary software. Obviously, this is often not possible; far too often the issue is dismissed without adequate research, and sometimes it must be acknowledged to satisfy the personal tastes or ambitions of influential individuals. The risks of package modification with native code, or the development of bespoke code, carries a much higher level of risk which the enterprise may have to endure long after the departure of the powerful interests that insisted upon it, for perhaps partisan or personal reasons.

The Year 2000

By the time this book is in print, the full impact of the inherent quality defect known as the *millennium bug* (Figure 15.6) will be well known around the globe. While further speculation on the impact is fruitless at this late stage, it is vital to reemphasize that the phenomenon is at heart a software QA issue. It is also a critical business issue. However, these two statements have, until now, not been associated in the mind of industry and commerce at large. If

Figure 15.6. The Millennium Bug

the costs, litigation and software companies' insolvency forecast to result from the millennium bug achieve an increase in awareness of the critical importance of software quality, not just to the esoteric world of pharmaceutical validation but to all of us, the incident will not be seen in retrospect as a totally negative occurrence. Perhaps it is too much to hope that the ideals behind pharmaceutical validation to ensure the fitness for use of computer-related systems will spread to other industries. Getting it right first time has been exhaustively proved to reap greater economies in the long run than the additional costs needed to ensure it. Computer validation is good for business, and ensuring the innate quality of software through a structured assessment of the development process that created it lies at its heart.

SUMMARY

The goal of this chapter was to show that a successful validation exercise, just like any other major project, must be built on sure foundations. Since validation involves documentation and testing of various kinds, the question arises as to whether this activity on its own is sufficient to demonstrate innate fitness for use. In the case where software forms a central part of the product being validated, such as a modern analytical instrument, manufacturing machine or other integrated software/hardware system, such activity is fundamentally limited to those parts of the code executed in the course of normal operation. This will often leave the majority of possible pathways through the code unexamined and bereft of the necessary evidence.

The practical means of deriving assurance as to the innate fitness for use of software is to measure the level of innate structural integrity of the entire product. The only effective way of measuring this is to examine the software development process, from which the product is the inescapable result—a fact consistent with all that is known from statistical process control. The main principles of such an assessment were briefly outlined. In doing so, we are seeking to avoid validation exercises proceeding other than on the firm foundation furnished by an audit assessment, confirming the fundamental innate structural integrity of the product. There is a sound basis on which a pharmaceutical manufacturer's validation project can be safely built. Under this approach, such a project can be confidently expected to deliver documentary evidence demonstrating an adequate level of assurance of the product or system's fitness for use in appropriate depth and at an economic cost.

REFERENCES

1. UK GAMP Forum. 1998. *The Supplier's Guide for Validation of Automated Systems in Pharmaceutical Manufacture,* version 3. Available from the International Society for Pharmaceutical Engineering.
2. Myers, G. J. 1979. *The Art of Software Testing.* New York: Wiley.
3. ISO 8402:1994. *Quality Management and Quality Assurance—Vocabulary.* Geneva, Switzerland: International Organization for Standardization.
4. ISO 10011-1:1990. *Guidelines for Auditing Quality Systems.* Geneva, Switzerland: International Organization for Standardization.
5. TickIt. 1995. *The TickIT Guide (A Guide to Software Quality Management System Construction and Certification to ISO 9001),* Issue 3.0 London: BSI/DISC TickIT Office. Copyright BSI and quoted by permission.
6. ISO 9000-3. (1994) *Guide to the Application of ISO 9001 to the Development, Supply and Maintenance of Software.* Geneva, Switzerland: International Organization for Standardization.
7. Crosby, P. B. 1979. *Quality Is Free.* New York: McGraw-Hill.
8. *Computer Weekly,* 13 June 1996.
9. *The STARTS Guide,* 2nd ed. 1987. Manchester, UK: National Computing Center. Crown Copyright and quoted by permission.
10. Royce W. W. 1970. Managing the Development of Large Software Systems: Concepts and Techniques. In *Proceedings of WESCON.* Conference in August 1970 in San Fransisco, Calif.
11. Garston Smith, H. 1997. *Software Quality Assurance: A Guide for Developers and Auditors.* Buffalo Grove, Ill., USA: Interpharm Press, Inc.
12. ISO 10007. 1995. *Quality Management: Guidelines for Configuration Management.* Geneva, Switzerland: International Organization for Standardisation.

APPENDIX 15.1: GAMP AUDIT CHECKLIST

The following checklist is that published in Version 3 of the GAMP Guide [1] and is intended to be used when auditing suppliers of automated systems. This list is intended only as a guide, and there may be other factors which require consideration when auditing suppliers. Other similar lists have been published (e.g., Garston Smith [11]) and cover essentially the same ground. In some circumstances, these may be found easier to use, as they are often accompanied by detailed guidance as to their use, particularly in respect of the weighing of the data obtained during the audit and the conclusions that should be inferred from them. In the list given here, references to other sections of the GAMP Guide are given where relevant.

A. Company Overview

A.1. Audit details (address, audit team, supplier representatives)

A.2. Company size, structure and summary of history (number of sites, staff, organizational charts, company history)

A.3. Product/service history (main markets, how many sold, use in pharmaceutical sector)

A.4. Summary of product/service under audit (product literature)

A.5. Product/service development plans

A.6. Tour of facility (to verify housekeeping, general working environment, working conditions)

B. Organization and Quality Management

B.1. Management structure (roles, responsibilities)

B.2. Method of assuring quality in product/service (quality system, responsibilities for quality)

B.3. Use of documented Quality Management System (QMS), e.g., policy, manual, procedures, standards (see Section 8)

B.4. Maturity of QMS (relevance to product/service under audit)

B.5. Control of QMS documentation (reviews, approvals, distribution, updates) (see Section 8.2.3 and Appendix E)

B.6. Maintenance of QMS documentation (regularly updated)

B.7. QMS certified to a recognized standard (e.g., ISO 9001)

B.8. Method of checking compliance with QMS (internal audits, reviews)

B.9. Qualification and suitability of staff (see Section 8.2.6)

B.10. Independence of auditors, inspectors, testers, reviewers

B.11. Staff training (general, QMS, product/service related, new staff, changes to QMS, regulatory issues, training records) (see Section 8.2.6)

B.12. Use of subcontractors (individuals, companies) (see Section 8.2.5)

- method of selection
- subcontractor qualifications and training records
- specification of technical and quality requirements in orders placed
- method of accepting product delivered by subcontractor

B.13. Experience of validation process (with other customers, previous Supplier Audits, services provided by supplier, involvement in regulatory inspections)

B.14. Awareness of pharmaceutical regulatory requirements (knowledge of regulations, subscription to publications, attendance of relevant events, involvement in industry groups)

B.15. Continuous improvement program (use of metrics to evaluate and improve effectiveness of QMS)

C. Planning and Product/Project Management

C.1. Use of quality and project plans (per project/product, defining activities, controlling procedures, responsibilities, timescales) (see Section 8.1.2 and Appendix G)

C.2. Status of planning documentation (reviews, approvals, distribution, maintenance and update) (see Section 8.1.2)

C.3. Documentation of customer/supplier responsibilities (see Section 9)

C.4. Use of Validation Plan where supplied by pharmaceutical manufacturer (see Section 8.1.2)

C.5. Project management and monitoring (mechanism, tools, progress reports) (see Appendix G)

C.6. Accuracy of, and conformance to, planning and management procedures

C.7. Use of formal development life cycle (see Section 8.1.3)

C.8. Evidence of formal contract reviews where applicable

D. Specifications

D.1. User Requirements Specifications (Planning) (see Section 8.1.2 and Appendix A)

D.2. Functional Specifications (see Section 8.1.3)

D.3. Software Design Specifications (see Section 8.1.3)

D.4. Hardware Design Specifications (see Section 8.1.3)

D.5. Relationship between specifications (together forming a complete specification of the system which can be tested objectively)

D.6. Traceability through specifications (e.g., for a given requirement)

D.7. Status of specifications (reviews, approvals, distribution, maintenance and update) (see Section 8.2.3 and Appendix E)

D.8. Accuracy of, and conformance to, relevant procedures

D.9. Use and control of design methodologies (CASE tools) (see Section 8.1.3)

E. Implementation

E.1. Specification of standards covering use of programming languages (e.g., naming and coding conventions, commenting rules) (see Section 8.1.3 and Appendix D)

E.2. Standards for software identification and traceability (e.g., for each software item—unique name/reference, version, project/product reference, module description, list of build files, change history, traceability to design document) (see Appendix C, section 5, Item I.2)

E.3. Standards for file and directory naming

E.4. Use of command files (e.g., make) for build management, to compile and link software

E.5. Use of development tools (e.g., compilers, linkers, debuggers)

E.6. Evidence of Source Code Reviews prior to formal testing (checking design, adherence to coding standards, logic, redundant code, critical algorithms) (see Sections 8.1.3, 8.2.1 and Appendix B)

E.7. Independence and qualifications of reviewers

E.8. Source code reviews recorded, indexed, followed-up and closed off (with supporting evidence); evidence of management action where reviews not closed off satisfactorily

E.9. Listings and other documents used during Source Code Reviews retained with review reports (see Appendix B)

F. Testing

F.1. Explanation of test strategy employed at each level of development (e.g., module testing, integration testing, system acceptance testing or alpha/beta testing) (see Section 8.1.4)

F.2. Software Test Specifications (see Section 8.1.3)

F.3. Hardware Test Specifications (see Section 8.1.3)

F.4. Integration Test Specifications (see Section 8.1.3)

F.5. System Acceptance Test Specifications (see Section 8.1.3)

F.6. Structure and content of each test script (unique reference, unambiguous description of test, acceptance criteria/expected results, cross-reference to controlling specification) (see Appendix N)

F.7. Relationship between test specifications and controlling specifications (demonstrating system has been tested thoroughly)

F.8. Evidence that test specifications cover (see Section 8.1.3):

- both structural and functional testing
- all requirements
- each function of the system
- stress testing (repeat testing under different conditions)
- performance testing (e.g., adequacy of system performance)
- abnormal conditions

F.9. Status of test specifications (reviews, approvals, distribution, maintenance and update) (see Section 8.2.3)

F.10. Formal testing procedure to execute test specifications (method of recording test results, use of pass/fail, retaining raw data, reviewing test results, progressing and resolving test failures) (see Appendix N)

F.11. Status of test results and associated review records (indexed, organized, maintained, followed-up on failure) (see Appendix N)

F.12. Involvement of QA function (as witnesses and/or reviewers)

F.13. Independence and qualifications of testers and reviewers

F.14. Accuracy of, and conformance to, relevant test procedures

F.15. Control of test software, test data, simulators

F.16. Use of testing tools (documented, controlled)

G. Completion and Release

G.1. Documented responsibility for release of product, such as certificate of conformity, authorization to ship (including evidence that testing has been accepted with/without reservations) (see Section 8.1.5)

G.2. Handover of project material in accordance with quality plan/contract (e.g., release notes, hardware, copies of documentation/software)

G.3. Provision of user documentation (user manuals, administration/technical manuals, update notice with each release)

G.4. Records of releases (i.e., which customers have which version of system/software)

G.5. Warranties and guarantees

G.6. Archiving of release (software, build files, supporting tools, documentation)

G.7. Availability of source code and documentation for regulatory inspection (use of ESCROW accounts, for which releases)

G.8. Customer training (summary of courses, given by staff or third parties)

G.9. Accuracy of, and conformance to, release procedures

H. Support/Maintenance

H.1. Explanation of support services (agreements, scope, procedures, support organization, responsibilities, provision locally/internationally, use of third parties, maintenance of support agreements) (see Section 8.1.6 and Appendix Q)

H.2. Duration of guaranteed support (number of versions, minimum periods)

H.3. Provision of help desk (levels of service, hours of operation)

H.4. Fault reporting mechanism (logging, analyzing, categorizing, resolving, informing, closing, documenting, distribution, notification of other customers with/without support agreement)

H.5. Link between fault reporting mechanism and change control

H.6. Method of handling of customer complaints

H.7. Accuracy of, and conformance to, support procedures

I. Supporting Procedures and Activities

I.1. Documentation management (covering QMS and product/project documents) (see Sections 8.2.1, 8.2.3 and Appendices B and E):

- in accordance with QMS/Quality Plan
- following documentation standards
- indexed, organized
- reviews carried out prior to approval and issue
- reviews recorded, indexed, followed-up and closed off (with supporting evidence)

- evidence of management action where reviews not closed off satisfactorily
- formal approvals recorded, meaning of approvals defined
- distribution controlled
- document history maintained
- removal of superseded/obsolete documents

I.2. Software configuration management (see Section 8.2.2 and Appendices B, D, and Appendix C, section 5, Item E.2):

- system for identifying, controlling and tracking every version of each software item
- system for recording configuration of each release (which items and versions)
- identification of point at which change control is applied to each software item
- control of build tools and layered software products, including introduction of new versions

I.3. Change control covering software, hardware, documentation (see Section 8.2.4 and Appendix F):

- all change requests formally logged, indexed, assessed
- rejected requests identified as such, reasons documented, signed by those responsible, originator informed
- changes authorized, documented, tested and approved prior to implementation (except emergencies)
- emergency procedure documented, covering reviewing, testing, approving, recording
- impact of each change (on other items and on requirements for retest) assessed and documented

I.4. Security procedures (physical access, logical access to accounts/software, virus controls)

I.5. Backup and recovery procedures (secure storage and handling of media, on-site, off-site, recovery procedure exercised)

I.6. Disaster recovery procedure (tried)

I.7. Control of purchased items bought on behalf of customer (e.g., computer hardware, layered software products), including associated packaging, user documentation, warranties

I.8. Accuracy of, and conformance to, relevant procedures

16

Practical Implications of Electronic Signatures and Records

David Selby
Selby Hope International
North Yorkshire, United Kingdom

IF YOU DON'T ASK, YOU DON'T GET!

For the pharmaceutical industry supplying products to the United States, 20 August 1997 was the start of a new era. This was the day that the U.S. Food and Drug Administration (FDA) finally introduced its regulation entitled "Electronic Records; Electronic Signatures" [1]. This opened the door unequivocally to the use of electronic recording and signature technologies in the regulated pharmaceutical environment.

Such a momentous event should have lit up the pharmaceutical world, offering us the opportunity to increase the efficiencies of our development and manufacturing processes and the associated recording activities. At the same time, it offered the prospect of exchanging the mountain of paper, which has been associated with such activities, for much more compact and easily manageable electronic files.

But it didn't. Instead, the new regulation has been the source of much debate, even furious argument. More than two years later, the argument still rages. Why is this and what can we do about it?

Uniquely, I believe, the industry requested a new rule to enable the use of electronic signatures where the existing regulation demanded handwritten signatures, which the industry rightly perceived would open the door to the very efficiencies mentioned above. As the FDA drafted the rule, it realized that regulating electronic signatures without regulating the electronic records to which they were to be applied would be difficult, and in unscrupulous hands was a mechanism for preventing the FDA from doing its investigatory job. So the rule detailed in 21 CFR 11 was born.

Much dialogue took place between the industry and the FDA during the development of the rule, and the final result is better and more practical as a result. However, there are significant difficulties with the rule as it stands, particularly with its application to noncompliant electronic records generated with systems that were in place when the rule became law, the so-called "legacy systems" problems.

Much discussion has taken place about how the rule should be interpreted generally; in relation to legacy systems, discussion and lobbying continues. However, whilst not belittling the issues to be settled, I believe the industry must look beyond the current difficulties and focus much more on how it can best use the rule available to us and exploit it for our future competitiveness. The legacy system problem needs to be separated out or "ring fenced" and effort focused on the systems now being designed and developed for use in the new millennium.

At the time of writing, the FDA was about to embark on a training program for all field investigators on how to interpret and apply the rule. A few specialist investigators would be trained in more depth. Until now, there has been little evidence of their seeking to assess compliance, although at least two warning letters citing the regulation have been issued in the devices industry. Following this general training, we can expect the "honeymoon period" we have enjoyed (?) to end, and we should soon expect site inspections to include an assessment of compliance not only of new systems but also the projects we have to bring legacy systems into compliance.

Before moving on, a word on the European situation is appropriate. Most European regulators are interpreting the current EU Guide to Good Manufacturing Practice (GMP), and particularly Annex 11 on "Computerized Systems", as already permitting the use of electronic signatures and records, so there are no new issues (although this is not a universally held view). The UK Medicines Control Agency (MCA) has issued a MAIL Letter to this effect [2], and I know of no inspector's observation which has criticized *the principle* of the use of electronic signatures.

BALANCING INDUSTRY AND REGULATORY NEEDS

The industry view of the rule, as indicated above, was that the rule was necessary to enable the industry to embrace fully the new technologies which will in turn allow it to adopt new technologies, force down costs and maintain competitiveness. This situation has not changed.

The key FDA player in this story in Mr. Paul Motise. He is on record as saying that ". . . the FDA must operate on the same technological plane as the industry which it regulates". Whilst this is undoubtedly true if it is to do an effective job, it was not the only consideration of the FDA.

The FDA approached this piece of rule-making with three things in mind, in addition to enabling the use of electronic signatures (see Figure 16.1):

1. *Reliability:* The need to ensure that the records would be available for inspection for the required period of time and would not be degraded or accidentally corrupted.
2. *Trustworthiness and,* particularly, *prevention of fraud:* The content of a record is an accurate reflection of that record when it was first created, together with the history of any changes which subsequently occurred. The appended electronic signature unequivocally identifies the individual performing a function with the record (e.g., create, add, delete, review, approve, etc.).
3. *Compatibility with the role of the FDA:* The records must be available for review and copying, together with audit trail information in such a way as to allow the relationship of the record to others to be traced.

These three principles underlie all of the clauses in the ruling. Fraud prevention was obviously a key part of their thinking, as is evident from a review

Figure 16.1. Concerns of the FDA

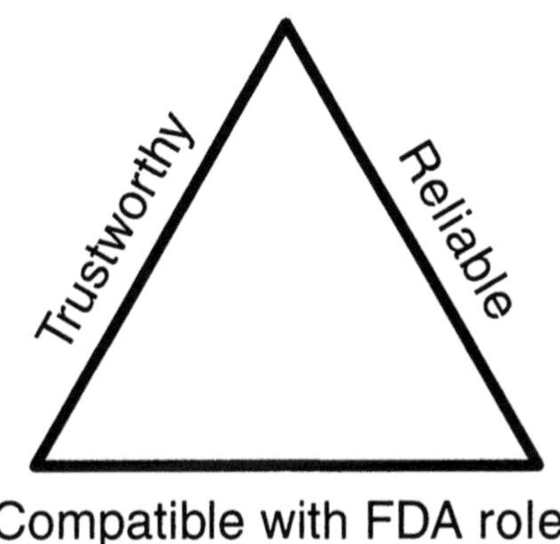

of the preamble to the rule [1]. The question they will continue to ask as they monitor the use of the new rule is:

> *Can we be sure we have eliminated all reasonable possibility of fraud whilst allowing the use of electronic signatures?*

If ever they come to the answer "No", they will legislate again (of which more later).

At the same time, the new rule has not superseded any existing regulation. The rules for making submissions and compliance in manufacturing have not changed. The new rule has simply enabled the use of technology to assist in the process.

So where does the balance between the industry and the regulators lie? Some in industry are arguing that the enabling of electronic signatures is so hemmed in with caveats that it is not worth adopting, and we should retain the status quo. Others say that the cost of bringing legacy systems into compliance will be comparable with that for fixing the Year 2000 problem.

I believe we are not in a position to turn the clock back. Our technical organizations cannot operate without the use of electronic records. A court ruling (still to be challenged) has confirmed that in the United States, the electronic record is the definitive record, even if it is subsequently printed on paper, and the paper signed by an authorized person as being a true and accurate copy of the original electronic record. (The justification is that the electronic record contains valuable audit trail information, not normally available on the printout, which if not available to the FDA may prevent it from doing its job effectively.) We must therefore make the rule work for us. It is possible there will be minor changes of interpretation, but there is unlikely to be any substantive change to the rule. Industry cannot wait. It must move forward, and it is up to the industry to identify pragmatic solutions to the issues, to embrace the new technologies and to move forward.

Heroic efforts are not necessary. Infallibility is not demanded. The major section of this chapter offers some practical ideas for interpreting the rule from this perspective. The ideas are not exclusive. Other better ones will evolve (or may already have been developed). They are simply presented here as examples of what is pragmatically possible and may be used to show an investigator that a genuine effort has been made to comply with the ruling in the current climate of immature understanding.

GENERAL CONSIDERATIONS

The regulation, despite its lengthy gestation period, does not fit seamlessly into the GxP context with which we are familiar. As indicated earlier, all of the existing rules of GxP continue to apply, including, for instance, the need to validate the systems (both quality and computer) where electronic signatures are used and electronic records are generated.

The ruling makes it plain that if we are compliant with 21 CFR 11, the FDA will consider electronic records equivalent to paper records and electronic signatures equivalent to handwritten signatures, initials and other personal marks, from the standpoint of GxP compliance.

Definitions of electronic record, electronic signature and some other terms are given in the rule and will not be repeated here. It is worth noting, though, that a definition of a handwritten signature has been included since none existed. The effect of the rule is then to make electronic signatures acceptable where the GxP regulations previously demanded a handwritten signature. In the words of Paul Motise, this ruling ". . . wipes the slate clean of handwritten signatures".

The ruling does not apply where the computer system is incidental to the creation of a paper record which may bear a handwritten signature. Examples include the following:

- Laboratory instruments, such as some older style integrators or spectrophotometers, which may generate an electronic record but do not retain it, reproducing it as a chart or report in real time or at the end of the run. Subsequent runs overwrite the existing records and cannot be stored electronically.
- Facsimiles may exist initially as paper records, be transmitted by electronic means and printed out as paper records with no retained electronic copy.
- More contentiously, if a word processor is used to generate a document but no electronic record is retained, i.e., it is used like a typewriter, it could be argued that the rule does not apply.

One other interesting aspect of the rule is the statement that a handwritten signature, even if captured electronically, is still a handwritten signature. Under these circumstances, Part C of the rule does not apply and is effectively circumvented.

This has led a number of companies to develop technologies that capture handwritten signatures electronically and append them indelibly to electronic records. Such systems comply with the rule, provided that they satisfy all of the other criteria for reliability, integrity and access by the FDA.

A reading of the preamble to the rule reinforces the impression that the drafters have been careful to avoid prescription and bureaucracy wherever possible. However, some has inevitably crept in. There is clearly sensitivity on this point, as a minimalist approach is evident.

It is necessary to tell the FDA Office of Regional Operations in Rockville, Maryland, in writing, signed by hand(!), that when employees in the organization use electronic signatures, those signatures will carry the same meaning and gravity as a traditional handwritten signature. Clearly, this letter is not required until you intend to use electronic signatures; if they are already in use, you should continue to use them. No acknowledgment will be issued.

Internally, this has implications for the organization. Most companies these days have a policy on data security, reinforcing the need to keep passwords secret. To this should be added a policy statement which says that employees must regard electronic signatures as having the same meaning and gravity as handwritten signatures. The rule has implications for managers, users and information technology (IT) professionals, and so it is necessary to develop and implement an associated communication program to make sure everyone understands not only the implications of using electronic signatures but also the regulation as it applies to electronic records. The rule effectively reinforces the concept of "system ownership", where every system that generates GxP data should have a named owner.

In GxP terms, the system owner is one who understands and takes responsibility for the integrity of data generated, manipulated and held by the system. Take the example of a Laboratory Information Management System (LIMS). In the paper world, the laboratory manager would expect to accept responsibility of the information in laboratory notebooks, because he or she continually checks the work of individuals who have been trained personally or by delegation. The manager will know how the data are generated; how the data are stored; the rules for recording, manipulating and correcting data; and the archiving procedures. The system owner has exactly the same responsibility for electronic data and records. He or she should be familiar with the following procedures which require Quality Assurance/Quality Control (QA/QC) approval:

- Giving and removing access to the system at the required levels
- the training program for managers, users and maintainers (the IT staff) of the system
- The security measures in place associated with the storage, archival and retrieval of records
- Change control and data maintenance
- Ensuring periodic audits are carried out

If further reinforcement of the concept of system ownership is necessary, it comes with the final definitions of "open" and "closed" systems (Figure 16.2).

A closed system is one where access "is controlled by persons responsible for the content of the electronic records that are on the system". This applies even if others, such as employees on a remote site, or people outside the organization, such as suppliers or service personnel, have access to the system. Clearly, the system owner is that responsible person.

Open systems are those where access is not entirely in the hands of the person responsible for the content of the electronic records. This can arise through the use of a service provider, for instance, though access over telephone lines does not necessarily mean the system is open. For instance, private networks run by large companies are clearly closed, and dial-back facilities offer the necessary degree of control for the system to be "closed", when operated together with the usual security features such as firewalls.

Figure 16.2. Open Versus Closed Systems

11.3(b)(4) Closed system means an environment in which system access is controlled by persons who are responsible for the content of electronic records that are on the system.	11.3(b)(9) Open system means an environment in which system access is not controlled by persons who are not responsible for the content of electronic records that are on the system.

Closed Systems

- Access controlled from the company.
- Can extend to clinicians, suppliers, etc.
- Encryption and digital signature standards unnecessary.
- Communication via secure network, including modem.

Open Systems

- Company delegates control.
- Can extend to clinicians, suppliers, etc.
- Use encryption and digital signature standards.
- Communication control by on-line service.

- Procedures for data ownership necessary.

Open systems require appropriate security measures to protect the integrity of the data in place, such as an encryption techniques and the use of digital signature standards. Technology is currently evolving rapidly in this area, often driven by government legislation, not only in the United States but in Europe too, so new solutions are likely to emerge quickly. Furthermore, the FDA reserves the right to access open systems at any time. Open systems clearly make life much more complicated, both technically and procedurally, and should therefore be avoided if at all possible.

All of the corporate systems discussed in this book are subject to 21 CFR 11, along with other enterprise applications not specifically discussed in this book. These other applications include Intranet, Internet, and E-mail. The classification of a system as open or closed has a great impact on the expectation for compliance.

NEW SYSTEMS—THE PRACTICAL IMPLICATIONS

Only a very few computer systems were compliant with 21 CFR 11 when it came into force. The FDA recognized this and has allowed a de facto transition period for the industry to put plans in place to remedy the situation (see "Strategies for Bringing Existing Systems into Compliance"). However, for new systems, no such transition period is being allowed, and they must be compliant from handover following sign off of the Validation Report. It is therefore important to consider the regulation's requirements when the User Requirements Specification (URS) is being written.

Whatever one's opinion of the rule, this is a much easier situation to manage than that of bringing legacy systems into compliance. However, it is not without its challenges. The interpretation of the rule for each organization needs to be worked out, and, as in other GxP areas, each organization will develop its own, sometimes unique interpretation.

This section suggests some pragmatic interpretations that may be applied to the clauses in the rule, although the detail will inevitably be modified by current policies, procedures and practices in your organization.

Electronic Records

For the FDA, the key issues surrounding electronic records are those of authenticity and integrity. Consequently, the rule sets out a series of conditions that are designed to protect these principles. Many of them are not new to us, at least in part, since they are practices already adopted by many organizations to protect their data from corruption or loss for sound commercial reasons—what might be called "Good IT Practices".

It is no surprise to find that systems generating electronic records should be validated, staff should be trained, access must be controlled and many other such points. Clearly, validation now needs to include tests to show not only that true records are generated but also that they contain relevant checks, for example, audit trail information, the identity of the record generator and links to any subsequent changes or manipulations (e.g., review, authorization etc.), the ability to make copies for regulatory purposes and access control.

This brings up issues of archiving and retrieval. As long as the system is running, this is not problematical. However, when the system is retired, provision needs to be made to continue to retrieve the records for the requisite period. System retirement is not a trivial task, and procedures need to be developed for this activity if they do not already exist.

The volumes of records for archiving can be daunting. Some firms are concentrating on archiving key GMP records relating to batch records and batch release that would be used to support a product recall. Other GMP records, however, are also important candidates for archiving, for instance, annual product reviews, QC analytical results and so on. Archive solutions are being suggested and include

- retention of the original hardware and applications;
- record storage in some "universal" format, such as the PDF (Portable Document Format); and
- WORM (write once, ready many) drives which can retain high volumes of data.

A common problem shared with many solutions is the need to retrieve records at a later date. Records should be preserved for a defined period, usually seven years, but this can vary depending on the characteristics of the drug product being supported. Technology has been rapidly developing in recent years. Ongoing maintenance is likely to be a major problem with superseded products either no longer supported or withdrawn by their supplier. All archive solutions must be able to comply with the audit trail requirements of the regulations, and procedural measures to control transfer to the long-term storage medium, restrictions on access to authorized persons, disposal and so on are inevitable.

The rule requires the use of secure, time-stamped audit trails. This provision is key for the FDA; during validation, tests should be run to ensure that each record is time and date stamped. The question has arisen "which time?" Mr. Paul Motise has been unequivocal in stating that the time must be local time, though this will cause some confusion across time zones and may impact company policies on time recording. The case for international time stamping has been eloquently argued by Ken Chapman and Paul Winter [3] in a "white paper" they have recently made available.

Stamping records whenever they are created or modified with both local and international time is quite acceptable if clumsy. Time standards need to be more accurate than those normally found on personal computers (PCs) which are notoriously inaccurate. Network time is a better standard and quite acceptable, and one may gain access to international time standards via the Internet for the ultimate in accuracy and expense. Whatever solution is adopted, action will be necessary to agree on a standard and approach for each organization.

Operational checks need to be in place to force sequential operations in the right order (Figure 16.3). This discipline is in place in many existing systems and is not seen as onerous or difficult. Operational checks are also demanded to ensure that users are authorized to carry out the activities. Again, these are widely practiced and traditionally controlled by means of hierarchical access procedures.

Checks to verify the source of data or instructions are possible and feasible in some cases, and the rule is suitably equivocal about these, by inclusion of the words *as appropriate*. This is a manifestation of the need to be practical and not take heroic measures.

The rule is quite specific on the controls that should be in place to prevent falsification of records and signatures, whether by accident or design. Policies detailing the responsibilities and accountabilities of managers, users and IT professionals to protect the integrity of records and ensure attributions are correct and maintained (i.e., who did what when, without erasing the original record in the case of changes) are necessary (e.g., the use of passwords).

One area where considerable tightening up in procedure may be necessary is in the control of system documentation. It is very common to find manuals from earlier versions of software still on office shelves, long after the applications have been updated. On-line help is quite acceptable, but features which allow it to be updated and changed by individual users without the benefit of change control should be disabled.

Documentation controls apply to system development and maintenance as well as to use. Development practices and procedures need to documented according to a Quality Management System (QMS) [4] and controlled in the same way as other GxP documents. Of particular importance is the establishment of an audit trail for documentation and management of change control and the associated records. The FDA needs to be able to establish unequivocally which version of software was running on what equipment, at which time, when every record was created. One of the effects of this rule is clearly to bring the IT Department, traditionally seen as being on the periphery of GxP, firmly within the fold!

A summary of technological and procedural controls in relation to the use of electronic records is presented in Table 16.1.

Figure 16.3. Sequential Operations

11.200(a)(1)(i) When an individual executes a series of signings during a single, continuous period of controlled system access, the first signing shall be executed using all electronic signature components; subsequent signings shall be executed using at least one electronic signature component that is only executable by, and designed to be used only by, the individual.

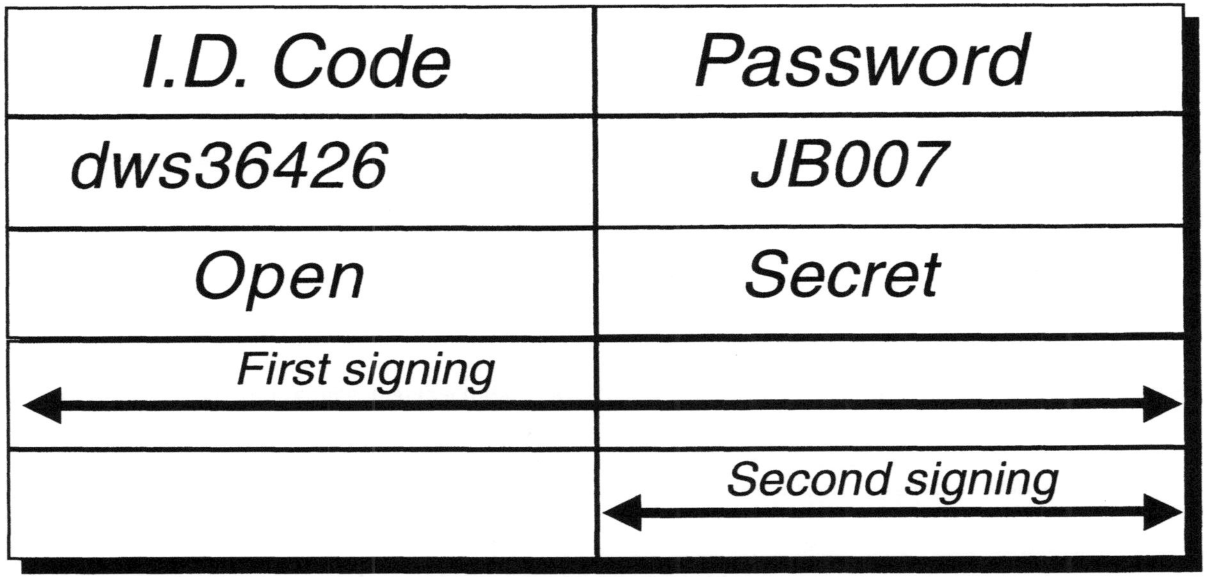

I.D. Code	Password
dws36426	JB007
Open	Secret
← First signing →	
	← Second signing →

Table 16.1. Procedural and Technological Controls for 21 CFR 11

Clause	Type of Control	Responsibility	Notes
11.10	Procedural	Pharmaceutical manufacturer	This clause specifies a number of specific controls. The pharmaceutical organization will need to demonstrate a system of self-inspection audits to demonstrate compliance with the procedures and controls listed below.
11.10(a)	Procedural	Pharmaceutical manufacturer	ER/ES (electronic record/electronic signature) systems need to be validated.
	Technological	Supplier	ER/ES system should be able to identify changes to electronic records in order to detect invalid or altered records. In practice, this means having an adequate audit trail that can be searched for information, e.g., to determine whether any changes have been made without the appropriate authorizations.
11.10(b)	Technological	Supplier	ER/ES systems should allow electronic data to be accessed in human readable form.
	Technological	Supplier	ER/ES systems need ability to export data and any supporting regulatory information (e.g., audit trails, configuration information relating to identification and status of users and equipment).
11.10(c)	Procedural	Pharmaceutical manufacturer	Pharmaceutical organizations should specify retention periods and responsibilities for ensuring data are retained securely for those periods.
	Procedural	Pharmaceutical manufacturer	Pharmaceutical organization needs a defined, proven and secure backup and recovery process for electronic data.
	Technological	Supplier	ER/ES systems should be able to maintain electronic data over periods of many years, regardless of upgrades to the software and operating environment.
11.10(d)	Procedural	Pharmaceutical manufacturer	Pharmaceutical organization needs procedures defining how access is limited to authorized individuals. Managing superuser account should be given special consideration.

Table 16.1 continued on the next page

Table 16.1 continued from the previous page

Clause	Type of Control	Responsibility	Notes
	Technological	Supplier	ER/ES systems should restrict access in accordance with preconfigured rules that can be maintained. Any changes to the rules should be recorded.
11.10(e)	Procedural	Pharmaceutical manufacturer	Pharmaceutical organization needs procedure to maintain the audit trail [see 11.10(c) above].
	Technological	Supplier	ER/ES systems should be capable of recording all electronic record create, update and delete operations. Data to be recorded must include, as a minimum, time and date, unambiguous description of event and identity of operator. This record should be secure from subsequent unauthorized alteration.
11.10(f)	Technological	Pharmaceutical manufacturer & supplier	Where operations are required in a predefined order, for example, in batch manufacture, the ER/ES system should enforce that ordering through the system's design.
11.10(g)	Procedural	Pharmaceutical manufacturer	Pharmaceutical organization needs procedures defining how the operations are to be performed and that staff have been trained in their use.
	Technological	Supplier	ER/ES systems should restrict use of system functions and features in accordance with preconfigured rules that can be maintained. Any changes to the rules should be recorded.
11.10(h)	Technological	Pharmaceutical manufacturer & supplier	Where pharmaceutical organization requires that certain devices act as sources of data or commands, the ER/ES system should enforce the requirement.
11.10(i)	Procedural	Pharmaceutical manufacturer	Pharmaceutical organization staff who use electronic record/electronic signature systems must have the education, training and experience to perform their assigned tasks.
		Supplier	Supplier requires procedure to demonstrate that persons who develop and maintain electronic record/electronic signature systems have the education, training and experience to perform their assigned tasks.

Table 16.1 continued on the next page

Table 16.1 continued from the previous page

Clause	Type of Control	Responsibility	Notes
11.10(j)	Procedural	Pharmaceutical manufacturer	Policy needed to describe the significance of electronic signatures and the consequences of falsifying them, both for the pharmaceutical organization and the individual.
11.10(k)	Procedural	Pharmaceutical manufacturer	Pharmaceutical organization needs procedures covering distribution of, access to and use of operational and maintenance documentation once the system is in operational use.
	Procedural	Pharmaceutical manufacturer	Pharmaceutical organization must ensure adequate change control procedures for operational and maintenance documentation.
	Technological	Supplier	Where systems documentation is in electronic form, an electronic audit trail should be maintained, in accordance with 11.10(e) above.
11.30	Not covered by this table		Requirements for open systems.
11.50	Technological	Supplier	ER/ES systems must ensure signed electronic records contain information associated with the signing that clearly indicates all of the following: (1) The printed name of the signer (2) The date and time when the signature was executed (3) The meaning (such as review, approval, responsibility or authorship) associated with the signature These items are subject to the same controls as other electronic records. The information can be stored within the electronic record or in logically associated records but must always be shown whenever the record is displayed/printed.
11.70	Technological	Supplier	ER/ES systems must provide a method for linking electronic signatures, where used, to their respective electronic records in a way that prevents the signature from being removed, copied or changed to falsify that or any other record.
11.100(a)	Procedural	Pharmaceutical manufacturer	Pharmaceutical organization must ensure uniqueness of electronic signature and that it is not reused or reallocated.

Table 16.1 continued on the next page

Table 16.1 continued from the previous page

Clause	Type of Control	Responsibility	Notes
	Technological	Supplier	ER/ES system should enforce uniqueness, prevent reallocation of electronic signature and prevent deletion of information relating to the electronic signature once it has been used.
11.100(b)	Procedural	Pharmaceutical manufacturer	Pharmaceutical organization needs to verify the identity of individuals being granted access to ER/ES system.
11.100(c)	Not applicable		Not a functional requirement for electronic records and signatures.
11.200(a)(1)	Technological	Supplier	ER/ES systems providing nonbiometric electronic signatures need at least two distinct components.
11.200(a)(1)	Procedural	Pharmaceutical manufacturer	Pharmaceutical organization needs to establish how it will ensure that both components of electronic signature are entered if session has not been continuous (this can be through system design or operating procedure if necessary).
	Technological	Supplier	ER/ES system should enforce that both components are entered at least at the first signing and following a break in the session.
11.200(a)(2)	Procedural	Pharmaceutical manufacturer	Pharmaceutical organization must ensure staff use only their own electronic signature, not anyone else's even on their behalf, as that would be falsification [see also 11.10(j)]
11.200(a)(3)	Procedural	Pharmaceutical manufacturer	Pharmaceutical organization needs procedure that users do not divulge their electronic signature (e.g., passwords).
	Technological	Supplier	ER/ES system should not provide any ordinary means of accessing electronic signature information.
11.200(b)	Not covered by this table		Biometrics requirements

Table 16.1 continued on the next page

Table 16.1 continued from the previous page

Clause	Type of Control	Responsibility	Notes
11.300(a)	Procedural	Pharmaceutical manufacturer	System users must be identifiable through unique combination of user identification and user password. Passwords must not be disclosed and should be regularly changed.
	Technological	Supplier	
11.300(b)	Procedural	Pharmaceutical manufacturer	Pharmaceutical organization needs procedures to cover removal of obsolete users, changing of profiles as user roles change, periodic checking of identification codes and passwords for inconsistencies with current users and periodic changing of passwords.
	Technological	Supplier	System should force passwords to be periodically changed and also enable ID/password combinations to be rendered inactive without losing the record of their historical use.
11.300(c)	Procedural	Pharmaceutical manufacturer	Pharmaceutical organization needs procedure for management of lost passwords.
11.300(d)	Procedural	Pharmaceutical manufacturer	Pharmaceutical organization needs procedure to describe how response to attempted or actual unauthorized access is managed.
	Technological	Supplier	System should provide notification of attempted unauthorized access and should take preventative measures (e.g., lock a terminal after three failed attempts, retain card).
11.300(e)	Procedural	Pharmaceutical manufacturer	Pharmaceutical organization should define how any devices or tokens that carry user/ID or password information are periodically tested and renewed.

Note: Pharmaceutical manufacturing is the organization that will use the ER/ES system in a regulated environment. The supplier of ER/ES system (could be a separate internal function of the pharmaceutical organization, such as the Information Systems department).

Electronic Signatures

There is no doubt that the FDA would prefer the use of biometric signatures, i.e., a signature based on a unique physical attribute, such as a fingerprint or a retinal scan. When the rule was first drafted, this type of technology was not nearly robust enough. That situation is changing quickly, and devices are appearing where the necessary robustness is within reach. Nevertheless, systems will continue to be developed for some time where a unique combination of an identifier and a secret password will be necessary.

So, 21 CFR 11 enables the use of electronic signatures in GxP environments provided that the applications are designed for those signatures to be used only by the authentic owner. As already indicated, it is necessary to tell the FDA that you are using them.

There is a requirement that the identity of the individual be verified. This is most easily done from an organizational identity badge (e.g., with a photograph), but this could be more problematical when access from remote sites is needed. In any case, there will need to be a record of the issue or authorization of an electronic signature, and to this could be appended a verification of the identity from an authorized person at the remote location. Similarly, procedures need to be in place to remove electronic authority from individuals when their responsibilities change (or they leave the company), an activity which is frequently neglected in my experience.

When electronic signatures are applied to electronic records, it is necessary to unequivocally name the signer in the record and on any printout. The time and date of the execution of the signing and the reason for the signature (e.g., checked by, authorized by, released by, etc.). Where the regulation demands it, there should be provision for a second person check.

The signature needs to be unequivocally linked to the electronic record to which it relates and in such a way that it cannot be removed, as the preamble to the rule says "by ordinary means". I interpret that as meaning by cutting and pasting or other such standard functions that are frequently part of an application. This unequivocal linking may present something of a technical challenge but has been eloquently achieved in some applications designed to capture and embed handwritten signatures to documents, e.g., PenOp®.

These are the only controls currently required for the use of biometric signatures. Nonbiometric signatures are quite a different matter. A range of additional controls is necessary to give the FDA the confidence it needs to ensure that the signatures were applied by their real owners.

Nonbiometric signatures must consist of two different parts, one of which is a unique identifier and the other a secret password known only to the user. The combination is thus secret. The unique identifier could be a personal identifier. It does not need to be secret. Tried and trusted technologies such as a log on, entered from the keyboard, or more effectively from a card reader or bar code are satisfactory, but these are being superseded by newer ones that are on the way.

The secrecy of the password is paramount so that the integrity of the nonbiometric signature can be guaranteed. Thus a policy must be in place making this clear and rigidly enforced. The deliberate sharing of a password should be a dimissable offense. Such a policy should be publicized within the organization as a mechanism for ensuring its importance.

Secret passwords need to be sensibly constructed and maintained. They should be memorized and changed at regular intervals. These requirements are often seen as mutually exclusive! Frequent changes mitigate against remembering the password whilst never changing or "flip-flopping", i.e., changing between two passwords at prescribed intervals risks accidental exposure. Guidelines need to be developed to manage passwords and should include the following:

- A minimum password length of six characters
- Mixed alphanumeric characters
- Changes every three months (the compromise between frequent changes and not writing it down)
- Avoiding obvious combinations like one's car registration number or the dog's name
- Incrementally changing a character, so that it is possible to work out the current password from the key (starting combination) and the date.

It was not uncommon for passwords to be legally shared between teams of staff working together, especially if they had read-only access to data. This is not compatible with 21 CFR 11; everyone needs a unique password. Since users are likely to need unique passwords for some applications, this seems somewhat onerous but should be included in the URS.

There has been much discussion about the use of electronic signatures when the computer terminal is temporarily vacated with the application open. Clearly, there is a potential for unauthorized access in this situation; in my view, however, it is greatly exaggerated.

The default situation must be an automatic lock out of the access device after a defined period of time, but very short times will pose excessive inconvenience. The security situation needs to be seen in the context of the total security system from the perimeter fence to the seat in front of a terminal in a manufacturing suite or a dedicated office. Access around sites is often controlled and restricted; frequently, for all but the most sensitive of tasks, others trained and authorized to carry out the same tasks will be present in the same area. These factors all mitigate against the need to have a very short lock-out time.

Where multiple sequential activities are to take place (e.g., using a manufacturing batch recording application or a series of repetitive tasks in a laboratory or office), the regulation requires that both parts of the nonbiometic signature be used for the first record entry but that only the secret part be used subsequently, *provided the session is uninterrupted*. This provides for additional security in the situation described above. If a terminal were left open inadvertently and another person (authorized or not) entered the secret part of his or her password combination, the application should reject it as being incompatible with the identifier entered earlier. If a session is interrupted, the application must demand that both parts of the identifier/password combination be reentered.

Procedures need to be in place to manage the occasions when staff forget their passwords or an attempt at intrusion is made. The software governing access should react to multiple attempts to gain access using an invalid password (say three) by locking out the individual and sending an alarm to a responsible person to investigate, take appropriate action and record the outcome. Often this will be the result of a lost or forgotten password, so the procedure for issuing electronic signatures will have to be reinvoked. Old identifiers should be removed when staff leave and must not be reissued for at least 10 years or there will be potential for repeating identifier/password combinations and confusing audit trails.

Finally, security devices such as strip or barcode readers need to be tested periodically. These usually, though not exclusively, fail safe. (Any Visa credit card was once found to give access to parts of a large pharmaceutical manufacturing site in the United States. History does not record what it did to the credit card!) This check is therefore best incorporated into routine internal audit procedures.

Many of these checks and procedures may already be in place as part of Good IT Practice. However, not all may be, and this rule should occasion a thorough review of IT security procedures and practices to ensure not only compliance with the regulation but also the adequacy of measures for the protection of commercial information. This rule is an example of a regulation enforcing sensible commercial protection measures.

A summary of technological and procedural controls in relation to the use of electronic signatures is presented in Table 16.1.

STRATEGIES FOR BRINGING EXISTING SYSTEMS INTO COMPLIANCE

Whilst compliance with the rule is not without its challenges for new systems, they are small by comparison with those involved in bringing existing (sometimes known as legacy) systems into compliance.

The regulation does not apply to records created before 20 August 1997. However, those same systems that were perfectly satisfactory on 19 August 1997 were in most cases producing noncompliant records on 20 August when the rule came into force. The FDA wants to see plans to ensure that all records are compliant. The complexities of achieving this have been the source of endless debate. These difficulties have in many cases prevented companies from looking beyond these difficulties to exploit the rule for their competitive advantage in the future.

The situation is closely paralleled by that in the early 1990s when regulators in both the United States and Europe started enforcing the validation of computer systems; the industry had to work its way back into compliance. The solution is similar too.

A "ring fence" has to be drawn around the noncompliant systems, and a methodology worked out for bringing them back into compliance. There are five strategies available to us:

1. *Stop the activity* (this is unlikely to be applied in many cases).
2. *Return to paper* (there are still a few activities that were computerized by an enthusiastic amateur and add complexity for little or no benefit).
3. *Develop a hybrid solution* (putting manual procedures in place as an extra layer of control to prop up the computerized system).
4. *Upgrade the computerized system.*
5. *Replace the computerized system* (migration, record retention and retrieval become serious issues).

The last three options are the most realistic, with the last the most expensive, involving specialist programming in often superseded languages for an application with a limited life. There may be some examples of applications that are widely used where the supplier may be reluctantly persuaded to upgrade the system to a compliant state for his or her customers, agreeing to share the cost mutually, provided they can agree on a set of requirements. The only feasible route to this nirvana is through a user group.

Developing a hybrid solution may first appear to be the easiest and most practical way forward. Experience has shown, however, that this solution is usually intricate, and it is easy to get tied up in knots. For corporate computer systems, there may be hundreds of new manual procedures with all their implications on higher staffing levels and training.

Only a superficial treatment of the problem of bringing existing systems into compliance can be given here. The key steps presented in Table 16.2 are as follows:

1. Agree on the objective with senior management and gain their support and approval. This is not a trivial task and may require the approval of significant resources.
2. Form a team to assess the level of compliance for every legacy GxP system against the interpretation. This is most easily done with a checklist developed from the rule and should be done with the system owner.

Table 16.2. Planning for 21 CFR 11 Compliance

Step	Activity	Deliverable
1	Agree on the objectives of the plan.	Set of objectives agreed by senior managers.
2	Form team.	Resources assigned to perform the evaluation task.
3	Communicate objectives to team and computer system owners.	Understand the implications of 21 CFR 11. Obtain commitment to resolve the noncompliance.
4	Agree on compliance requirements.	Defined interpretation of 21 CFR 11.
5	Assess the level of compliance for each system.	List compliant computer systems. List non compliant computer systems and their noncompliance.
6	Evaluate the extent of noncomplianceand agree on actions.	Prioritized list of computer systems to bring into compliance.
7	Develop plan to achieve compliance for all computer systems.	Master plan against which to progress to ward compliance can be measured.
8	Execute the plan.	Computer systems brought into compliance.

3. Communicate the objective, including the support of senior management, to everyone involved, especially the IT department. You will need their help!
4. Meanwhile, an *agreed-on* interpretation of the rule for your organization must be developed. This is (politically) the most difficult step and is best done with a small team of informed individuals led by a senior technical manager. Adequate time for debate is necessary to allow all team members to justify the decisions to others when challenged later.
5. Assess the level of compliance for each system. Compile a list of systems (which every company now has following the Y2K problem) and identify those which need to be brought into compliance. This is conveniently done with a simple record sheet that checks whether electronic records are generated and stored (in which case the rule applies) or electronic signatures are used. Assign an owner to every GxP module of every system (or the whole system if it is not too big).
6. Evaluate the strategic options for each system and agree on the actions. This is the most difficult technical step because sufficient

knowledge of the application software in order to make realistic estimates of the effort involved in updating versus replacement may take some time. The IT Department has a key role to play here. Priorties will need to be established.

7. The resulting action list from the last step will allow a master plan to be developed. It is sensible to include a prioritization step in assessing which systems should be replaced/upgraded first, a decision which should involve the system owner.
8. Execute the plan developed in a timely manner.

The master plan is likely to be requested by investigators in the near future and should be reviewed and maintained on a regular basis, as business conditions may dictate changes to the actions originally developed. Showing progress against this plan is a vital part of being able to demonstrate progress toward compliance for existing systems. Arguing that computerized systems cannot be rescued in terms of 21 CFR 11 compliance, or that there was no point in archiving data from a unvalidated system (possibly considered such because it does not comply with the regulation), is not a defensible position. As indicated earlier, the FDA is gearing up with internal training and inspection aids [5] to enforce this regulation, whether it is welcome or not.

ORGANIZATIONAL AND HUMAN FACTORS

The rule on electronic records and signatures emphasizes, more than any other piece of pharmaceutical legislation so far, the importance of the IT function in the maintenance of GxP. Traditionally, the IT Department has had followed a relatively independent line, seeing itself as outside the general influence of GMP regulations and feeling uncomfortable where it rubs shoulders with pharmaceutical regulatory activities. This is exemplified by making changes to parts of the IT infrastructure or the introduction of a new operating system or desktop environment without considering of the GxP impact or even letting users know of the changes taking place! The justification is often that the changes are transparent and will not be noticed. This independence has been reinforced by reporting relationships, where IT has often reported to Finance, which sees itself as remote from the influences of GxP, and "technological blindness", where IT professionals and QA professionals profess not to understand the other's language.

Fortunately, this remoteness is disappearing, but it still needs a friendly push. Measures that should be taken to reinforce the understanding that the IT Department is part of the GxP environment include the following:

- GxP training for IT staff
- Basic training for all staff, particularly QA/QC personnel, about the safety and security features of the IT operation, the policies and procedures in place to manage them and the importance of discipline when handling data
- The establishment of a quality unit in the IT Department (or in small operations, a person, part of whose responsibility is quality), responsible for not only the GxP aspects of the IT operation but also other quality management procedures in IT, such as data security, access and control procedures, backup, archiving and restore, disaster planning and so on.
- Inclusion of the IT Department in the internal audit program and IT staff on (other) audit teams

In most large organizations, these measures are already in place, but organizations are not yet comfortable with them. They will take time to mature, but the future is pushing these two strange bedfellows closer together, so mature they must.

EXPLOITING THE RULE

Have the regulators stolen the march on us? Is 21 CFR 11 a Trojan horse for the industry? By including electronic records in the regulation, has the baby been strangled at birth?

The rule is very flexible. The preamble is littered with examples to indicate how the FDA has avoided being prescriptive. Many of the questions asked seek to restore some of this prescription. This should be resisted. The industry has a great deal of expertise in the pragmatic interpretation of the regulations and should use it to exploit the rule.

If we abuse the freedom inherent in the rule, we may well live to regret it. Throughout the preamble, there are a number of references to the need for further legislation. Whilst the FDA has no intention of doing so at present, it will not hesitate to do so in the future if it feels this is necessary to allow it to do its job. We wish to avoid this at all costs. If we comply responsibly (and dialogue with the FDA is also encouraged in the preamble), then I believe a win-win situation will evolve.

I do not believe the rule is a Trojan horse. Whilst it is true that there are a number of things in the rule that the industry does not welcome, particularly the problems associated with existing systems, these are temporary difficulties and of little significance for the long-term future of the industry.

It is easy to forget the most important point, which is that electronic signatures are now enabled. Many of the ideas and efficiencies that we have dreamt of for some time can now be realized. Automatic clinical data collection and reduction is a reality. Electronic submissions are a reality. Electronic laboratory notebooks are a reality. True electronic batch record systems are a reality. These are just a few of the dreams. There will be many more, and enormous benefits will flow from the closer integration of the regulated parts of the value chain.

We must look beyond the legacy system problem and not dismiss it. It must be contained and managed to ensure continuing compliance. This will not be inexpensive, but the cost pales into relative insignificance when compared with the benefits to be won from the exploitation of the rule.

EDITOR'S NOTE

As this book was going to press, the GAMP Forum published a draft guide to 21 CFR 11. Copies are available through the GAMP web site—www.actwa.co.uk/gamp.

REFERENCES

1. U.S. Code of Federal Regulations, Title 21, Part 11: *Electronic Records; Electronic Signatures* (last revised 1 April 1999).
2. MCA. 1997. *Electronic Signatures* (Manufacturing and Wholesaling). MCA MAIL 104, November/December.
3. Chapman, K. G., and P. Winter. 1999. A Way Forward. *Pharmaceutical Technology* (September): 44–52.
4. UK GAMP Forum. 1998. *Supplier's Guide for Validation of Automated Systems in Pharmaceutical Manufacture,* Version 3.0. Available from the International Society for Pharmaceutical Engineering.
5. FDA. 1999. *Enforcement Policy: 21 CFR Part 11; Electronic Records; Electronic Signatures.* Compliance Policy Guide 7153.17. Rockville, Md., USA: Food and Drug Administration.

17

Concluding Remarks: Validation in the 21st Century

Guy Wingate
Glaxo Wellcome
Barnard Castle, United Kingdom

Validation practices for corporate computer systems have been presented in line with current expectations of GxP regulatory authorities and industry practice. This chapter now concludes the book by reviewing some fundamental concepts, survival hints and industry trends that will affect how we validate in the 21st century.

QUALITY ASSURANCE FOR CORPORATE COMPUTER SYSTEMS

The various GxP regulations used around the world are based on three fundamental concepts of quality assurance (QA):

1. Quality, safety and effectiveness must be designed and built into a product.
2. Quality cannot be inspected or tested into a finished product.
3. Each step of the manufacturing process must be controlled to maximize the likelihood that the finished product will be acceptable.

The implementation of these concepts requires validation. This includes validating the use of computer systems supporting the manufacture of drug products. Only pharmaceutical manufacturers, however, can conduct validation; suppliers cannot validate. Pharmaceutical manufacturers can, nevertheless, use information provided by suppliers to support their validation.

A common approach to validating corporate computer systems has emerged during the mid-1990s and is now maturing. The aim of this book has been to present this maturing approach and, with case studies, bridge the gap that has existed between theory and practice. A glossary of terms giving definitions from various regulatory and industry sources has been included as Appendix C to this book.

The benefits of a widely adopted industry framework for validating computer systems have been recently extolled by Tony Trill of the UK Medicines Control Agency (MCA). The benefits include the following [1]:

- Establishing a common management practice
- Defining customer and supplier responsibilities
- Stabilizing standards and their interpretation at a realistic level
- Linking validation to existing standards (e.g., ISO 9000)
- Reducing the costs of validation
- Shortening the time required for validation
- Ensuring appropriate documentation is produced for validation projects
- Improving the quality and reliability of delivered systems

This all makes good business sense and should eliminate the need for corrective work because the original validation was deficient. Both pharmaceutical manufacturers and their suppliers should know what is required, and where points of clarification with GxP regulatory authorities are required. An established industry approach may also provide the basis for the international acceptance of validation work by various GxP regulatory authorities. The acceptance by GxP regulatory authorities of each other's inspection findings, however, is not expected in the near future.

A VALIDATION SURVIVAL GUIDE FOR THE NEW MILLENNIUM

A positive attitude is essential to beating the validation blues! This does not mean, however, that managers and practitioners will not come across problems when validating computer systems. In such situations, it is often useful to focus on why validation is needed and reflect on the following advice:

- Do not assume validation has an intrinsic value—it has none. But your core product or service does. Can you quantify the benefits of validation for your business?
- Aim to build the principles of validation into the culture of your organization. It was never intended to be a separate function.
- Mold the principles of validation to suit your own working practices. It is often more cost-effective to use the services of an outside consultant than your own in-house staff.
- Err on the side of caution when estimating the cost of validation. During projects, do not be complacent: challenge any overspending or project slippage. Resist making changes to project requirements that are not really necessary, as these could considerably increase costs.
- Avoid validation for GxP becoming validation for great mounds of paper. Documentation for its own sake will not ensure GxP compliance. If documentation is not useful, it is not likely to be needed for validation.
- Training is critical, but it must be the right training. Most training courses do not consider the practical application of validation and are a complete waste of time.
- You will not get your validation practices perfect the first time. They will need to be refined, so expect and plan for continuous improvement.
- A proactive management style will minimize any problems experienced. The motivation and commitment of practitioners must be maintained. They must feel that they have management support in their work.

GOLDEN RULES REMAIN UNCHANGED

Many light-hearted publications have published lists of golden rules for validation practitioners. None of these lists, however, is carved in stone, and there is some flexibility in deciding what should be included in the way of guidance. The checklist given below was published in *Validating Automated Manufacturing and Laboratory Applications* [2] and essentially remains unchanged. It should help practitioners concentrate on 10 key validation issues.

1. Plan and monitor validation: adopt a proactive project management style.
2. Use competent personnel and train where necessary.

3. Document validation, including collating raw data as supporting evidence; ensure everything is reviewed and approved.
4. Implement a regime of change control covering projects and operational use of the computer system.
5. Specify procedures for validation and follow them.
6. Develop system specifications with testing in mind and test using preapproved qualification protocols.
7. Use the approval of summary validation reports to authorize the use of a computer system.
8. Operate and maintain validated computer systems in a state of control.
9. Periodically review the validation status of computer systems and initiate revalidation where necessary.
10. Archive validation evidence for retrieval.

Remember that the GxP requirements for software are the same as for manufacturing records and documentation. Similarly, computer hardware (operating platforms and supporting networks) should fulfill the GxP requirements of manufacturing equipment.

Specific guidance on the validation of different types of computer systems can be found in the case studies presented as Chapters 6 to 12 in this book. The authors of these case studies are themselves experienced practitioners, and they were encouraged to focus their papers on the key issues affecting their case studies.

GOOD BUSINESS SENSE

The introduction of computerized systems should bring benefit to a pharmaceutical manufacturing organization. Menial tasks can be removed; manual data entry reduced (with associated error rates); processes can be made faster and more consistent. All this counts for naught, however, if either the system does not operate correctly or it has insufficient validation. It has been suggested that perhaps less than 1 in 10 corporate computer systems deliver in the first instance the benefits on which a decision to implement was made. Validation can help by bringing development methodology and project management to discipline what is often a solution based on new rather than matured technology implementation.

The failure to validate to a regulator's satisfaction can have significant financial implications. Noncompliance incidents can lead to the withdrawal or delayed issue of a license to pharmaceutical manufacturers to distribute their product on the intended market. The U.S. Pharmaceutical and Research

Manufacturers of America (PhRMA) has estimated that a company will spend between U.S.$50,000 and U.S.$100,000 for every day of delay during the pharmaceutical submission process. This is a small amount compared to the typical investment of U.S.$250,000,000 to bring a new drug to market. The real financial impact of GxP noncompliance, however, is the loss in sales revenue. For top-selling drugs in production, citations for noncompliance by GxP regulatory authorities can cost a manufacturer upwards of U.S.$2,000,000 per day in lost sales revenue. If the manufacturing of multiple drug products across multiple production sites was halted by the noncompliance of a corporate computer system, the pharmaceutical manufacturer would struggle to stay in business. Compliance very visibly makes good business sense in this scenario.

The challenge is to conduct cost-effectively sufficient validation to ensure GxP compliance, and there is always debate over how much is sufficient to fulfill the regulator's expectations. Chapters 2, 4 and 13 described the basic management requirements for projects and support. Excessive validation may increase confidence in regulatory compliance, but it is expensive. Inadequate validation may be cheaper, but, in the long term, the cost of regulatory noncompliance could be devastating.

KEY ROLE FOR SUPPLIERS

To avoid insufficient and excessive standards of work, and to avoid duplicating tasks in their entirety or in part, pharmaceutical manufacturers and suppliers should work together in partnership. They must be able to work efficiently as a combined team and, as such, must be able to communicate effectively to streamline validation. The parallel mutual benefits of cooperation between customers and suppliers are outlined in Table 17.1.

The MCA acknowledges the key role suppliers have in the future of validation and are very keen that the supplier's capability is appropriately assessed (refer to Chapters 14 and 15); complementary support is given by the pharmaceutical manufacturer to ensure all validation requirements are satisfactory covered. Issues that need to managed include the following:

- Ensure validation standards are maintained when a major upgrade release of hardware or software does not maintain compliance standards established on the previous release by the user.
- Fulfill 21 CFR 11 requirements when many suppliers do not satisfy the requirements embodied within 21 CFR 11 in their product offerings.
- Conduct business impact assessments when supplier notes accompanying an upgrade are too brief to make meaningful assessment.

Table 17.1. Benefits of Cooperation Between Pharmaceutical Manufacturers and Suppliers [2]

Pharmaceutical Manufacturer	Supplier
Meet user needs	Satisfy customer
Be easier to set up	Be handed over sooner
Be in production sooner	Be paid sooner
Break down less often	Fewer warranty visits
Be easier to repair	Shorter warranty visits
Be easier to further develop	Be easier to modify and/or upgrade
Be used more effectively	Good reference sites for new customers
Cheaper overall	Cheaper overall
Preferred supplier	Repeat business

- Understand need for surveillance/monitoring of supplier capability, possibly part of a partnership relationship, and avoid repeat use of a supplier who does not deliver required compliance.

The reliance on suppliers to support the validation of applications using their products is set to increase in the coming years. Supplier assessments and selection will become much more important. In the future, pharmaceutical manufacturers will continue to be accountable to regulatory authorities for validation. Instead of being able to make up any shortfall in a supplier's capability, it is likely that in more and more cases this will not be practical, and implementations will have to be abandoned if the supplier cannot deliver what was expected. This is expensive and wasteful, so it is worth getting right.

A key topic for debate in the near future will be the acceptability of "freeware" software products distributed via the World Wide Web at no cost to the user. The problem with freeware is that it often comes with little or no development documentation and, as such, has an unknown development history. Freeware distributed by what might appear as an altruistic individual may not be what it seems. Before using freeware, a pharmaceutical manufacturer must ensure it does not have any viruses or Trojan horses embedded in it, or that "back doors" exist that hackers could open. Freeware may also be distributed as part of a supplier's marketing ploy. The idea is that users will become familiar with using certain free software packages (e.g., spreadsheets and databases) from a particular supplier and then follow up with purchases from this supplier when other software products are required. This seems set as a new trend in software marketing and includes the provision of electronic services

(E-services) such as E-mail. Some major software development companies have already announced their intention to adopt this strategy. Pharmaceutical manufacturers are faced with the problem of demonstrating the quality of the development of "freeware" and "E-services", as there is no contractual agreement between the user and supplier, and the obligation for software quality and bug fixing is perhaps not as strong. It is currently advised that freeware and no charge E-services should not be used in the pharmaceutical industry for the reasons discussed above. Pressures from evolving E-commerce, however, will mean that the pharmaceutical industry and regulators alike will need to readdress this issue and the suitable controls to enable it to be used.

MARCH OF TECHNOLOGY

Computer technology and application development methodologies are continuously evolving, as discussed in Chapter 1. There are many themes, a few of which are introduced here. Validation practitioners are encouraged to monitor the technical press to keep up to date with developments.

In the 1980s, the use of computer systems was largely limited to Programmable Logic Controllers (PLCs) and Supervisory Control and Data Acquisition (SCADA) systems. More recently, a greater variety of computer systems has been used and integrated. Figure 2.6 illustrated how computer systems can support various aspects of pharmaceutical manufacturing operations. Different selections of computer systems can provide similar collective 'patchwork' coverage of manufacturing operations. Pharmaceutical manufacturers must choose their preferred solution with a selection of systems that are not only technically capable but also validation compliant.

There is a growing trend toward decentralized process control and the interconnection (possibly with Fieldbus) of computer systems, including intelligent instrumentation, PLCs, SCADA systems, and Distributed Control Systems (DCSs) under the umbrella of a Manufacturing Execution System (MES). The batch control solutions provided by these systems are subject to GxP because they implement manufacturing control of the production process. Standards such as S88 will encourage standard batch control solutions in the future. The MES is an example of a new breed of computer system from the 1990s that are subject to GxP scrutiny because it coordinates, supervises, and manages production.

Laboratory systems such as Laboratory Information Management Systems (LIMS) have also come under increased regulatory scrutiny as the complexity of the software they deploy has become more advanced. Currently, there are often two LIMS tiers of control, data acquisition and data processing, often implemented by two coupled LIMS applications. The data acquisition LIMS is likely, however, to disappear in the longer term as analytical equipment starts to incorporate data preprocessing, and standard interfaces can connect straight to the data processing LIMS.

Business systems such as Manufacturing Resource Planning (MRP II) have become regulatory concerns because they are no longer used solely for financial information and production forecasting; they now manage production materials, plant configuration, production routings, operating procedures, batch records, and distribution details—all of which are subject to GxP. Most MRP II systems are not highly integrated, although to release many of the benefits promised by these systems, they must be integrated with other corporate systems such as a LIMS and a MES. This would appear to be the next challenge when moving from MRP II to Enterprise Resource Planning (ERP) implementations.

The implementation of multisite MES, LIMS and MRP II systems has raised the issue of how to deal with supporting information technology (IT) infrastructure (networks and desktop environment). This is a significant issue, as this area has typically not been thought of as a subject for validation. Chapter 11 suggested that qualification might be more appropriate. Validation requirements for infrastructure have yet to be settled. Practitioners are strongly recommended to watch developments, including publication of the UK GAMP Forum's proposed guidance on this topic. Meanwhile, IT infrastructure tools are already available to automate aspects of the installation and testing of corporate computer systems. The day may not be far away, for instance, where Installation Qualification (IQ) can be fully automated. The benefits of such facilities to speed up and massively reduce the effort of some validation activities cannot be ignored. The application of this technology should be proactively encouraged as long as basic quality controls for these tools exists. The temptation, as with so many new technology opportunities, is to implement it first and then consider what validation is needed. No matter how attractive a new technology is, if its functionality is GxP related, then regulators will expect it to be prospectively validated.

The use of Web browsers and Intranet facilities looks set to become very popular in the next generation of computer systems. Many suppliers of corporate computer systems are already building such facilities into their current products. Issues concerning security and data integrity, however, do exist. Although any problems will undoubtedly be overcome, it may take a few years before this technology is mature enough for exploitation in the pharmaceutical industry.

Looking further into the future, totally automated facilities loom that are truly paperless. Robotics, automated guided vehicles (AGVs), and barcode readers will be commonplace, whilst conventional operator interaction will virtually disappear. Some pharmaceutical manufacturers claim to have paperless facilities now, but in the experience of the author, they still employ a lot of paperwork (e.g., printed SOPs and work orders). Paperless facilities will bring their own set of problems, such as the use and control of electronic signatures, which is already beginning to be experienced by firms implementing

Electronic Document Management Systems (EDMSs). Once more, there is a challenge to validate appropriately and not get bogged down by the complexity of a system in its wider context.

With ever larger and more complex computer system applications, characterized by longer and more costly projects, there has been a business need to try and accelerate system implementation to speed up the return on investment. New application development methodologies being promoted to address this include Rapid Application Development (RAD). This approach endeavors to bring some structure and management controls to software, which is developed through prototyping rather than through conventional User Requirements Specification (URS), Functional Specification and design phases. Regulators, however, have not been comfortable with the RAD approach to date because, from their perspective, it seems to rush application development, losing the benefits of conventional development quality control (QC). Some pharmaceutical manufacturers are promoting modern IT development tools such as PRINCE II® (Projects in Controlled Environments) to ensure a quality management approach is adopted within RAD. PRINCE II tightly defines input, outputs and activities with software development phases. Whilst it does not directly fit the validation life cycle, it can be aligned. PRINCE II is a useful tool to help IT professionals, not used in a quality management culture, make the transition to pharmaceutical validation. Whilst regulators have been very hesitant to accept RAD, it would appear to be a way forward for corporate computer system projects. Incorporating PRINCE II within RAD could be the bridge that brings regulatory acceptance of RAD. These are the early days of this topic, and it may take many years to reach a solution acceptable to the pharmaceutical industry and regulators alike.

ORGANIZATIONAL CHANGE

The regulatory trends identified above represent significant industry deficiencies with computer system validation. The deficiencies contravene fundamental requirements in the GxP regulations. There is a continuing need to educate pharmaceutical manufacturers and their suppliers in practical validation compliance. Validation principles need to be incorporated into the culture of an organization. Senior management must give and hold to a clear vision for compliance. Those tasked to champion the compliance cause must believe in its intrinsic value to the business. Without sustained management backing computer validation, skills will remain in the domain of consultants instead of being instilled as another competency by their permanent employees.

Independent Validation Departments should be established to demonstrate impartiality in regard to pressures from system development and operation functions. Typically, the Validation Department might be expected to reside under the QA umbrella. One problem to date many pharmaceutical manufacturers have had, however, is ensuring that QA staff have sufficient knowledge of computer technology and applications to ensure appropriate validation to meet the business need. QA Departments need to develop their capability in this respect, which has been identified as a key issue by many regulators.

Outsourcing tends to be encouraged by many pharmaceutical manufacturers to reduce the number of full-time employees in a company. The Validation Department will be included in this, too, by using individual contractors or managed supplier project capabilities. There is nothing wrong with outsourcing, but the pharmaceutical manufacturer must maintain a critical mass internal capability to effectively manage adopted validation standards. The cost of getting the balance wrong is burnout and resignation of key staff. Computer validation experts are already at a market premium, and this situation is unlikely to change. Organizations must manage to retain and develop key staff for computer systems validation.

Pharmaceutical manufacturers tend to have an ebb and flow in regard to centralized and decentralized organizations. This is often reflected in the harmonization or disparity in validation practices adopted between sites or geographic regions of a company. The cyclic nature of organizational changes must be managed to minimize the impact on consistent validation standards and practices. Centralized Validation Departments must not lose touch with the hands-on experience of the operating site. Decentralized Validation Departments must ensure that a suitable support network is established, with focal points for maintaining a common vision and approach. Both extremes of organizational structure are hard for maximizing validation efficiencies; it gets easier in the middle ground.

Organization structures must not hinder exploitation of new technology. In particular, the Validation Department in many pharmaceutical manufacturing organizations has a reputation for slowing down or even preventing the implementation of new technology. Validation must not be seen as a constraint and should not be managed as such. Pragmatic solutions should be found to enable the implementation of new technology whilst ensuring compliance, perhaps by refining validation approaches. Validation Departments are not known for their inspiration, but there is room for the development of novel approaches! Perhaps in the next few years we will see the emergence of new ways of validating corporate computer systems, which tend to have their own characteristics and needs compared to process control systems and laboratory analytical equipment.

REGULATORY INSPECTION TRENDS

The validation of computer systems has been a topical issue for over a decade and still accounts for up to 10 percent of the citations for GxP noncompliance by regulatory inspections [3]. Chapters 3 and 5 discussed the current state of play with regulatory compliance. The importance of computer systems validation is set to continue, with regulatory interest in the role of new technology and the integration of systems within the supply chain.

The MCA said back in 1996 that they expected an increase in the number of reported problems with computer systems validation [4]. This prediction has been fulfilled. Chapter 5 discussed some recent U.S. Food and Drug Administration (FDA) and MCA computer systems inspections from 1999 reported by members of the UK GAMP Forum. The theme of noncompliance findings within these inspections is very similar to those reported by the MCA and the FDA in the early and mid-1990s. The five categories with the most observations that had been reported in 1996 by the FDA, in descending order, covered testing and qualification, change control, validation approach, specification and design and QA [5].

In 1999, the MCA noted the following top concerns they share with the FDA [6]:

- Validation approach
- Security practice
- Backup and restoration
- Change control
- Audit trails

Other MCA concerns include archiving and the decommissioning of systems, training, level of QA awareness and appreciation for computer systems technology and applications, standards of documentation and Supplier Audits. These concerns were born out with regard to corporate computer systems by specific GxP inspection failures on MRP II systems, warehouse control systems, MES and Enterprise Asset Management (EAM) systems.

One-third of FDA Warning Letters cite batch records [7]; with the move to electronic batch records in corporate computer systems, the impact on computer validation inspection findings cannot be far away. Indeed, it would seem that the FDA has anticipated this with an intent to enforce 21 CFR 11 concerning the use of electronic records and electronic signatures. The FDA began training their investigators in the summer of 1999 on what they should expect in the way of computer systems validation in this regard. The prediction made by the Australian Therapeutic Goods Administration (TGA) inspector Paul Hargreaves a couple of years ago that half of all GxP inspections would include a dedicated computer system validation audit [8] would appear not to be too far off the mark in the early years of the 21st century.

Reflecting on technology developments outlined earlier and potentially higher regulatory expectations and concerns, it is expected that the pharmaceutical industry will request a wide-ranging debate on how much validation is enough. The cost of compliance can be enormous for corporate computer systems. There needs to be a sanity check to verify that industry practice has not drifted to a position where the benefit does not justify the cost.

ELUSIVE HARMONIZATION

There has been a need for greater consistency in the use of computer validation terminology. Some initiatives have tried to address this by proposing a standard lexicon, but limited input from various industry representations has ultimately led to their failure. Recently, a new initiative was established by the U.S. Parenteral Drug Association (PDA). This initiative is soon to come to fruition with the publication of an industry position paper collating a glossary of computer validation terminology. The initiative has taken input from the PDA, the FDA and the GAMP Forum. The success of this initiative, however, will not be known for many years.

One way in which GxP regulators can assist in this process is to harmonize their GxP requirements. The European Union's (EU) GxP Directive, which has already been adopted by countries outside the European Union, has shown that harmonization can work. Mutual Recognition Agreements (MRAs) between EU national regulatory authorities and the FDA offer a similar opportunity for harmonization across the Atlantic.

Regulatory inspectorates/agencies under the banner of the Pharmaceutical Inspection Convention Scheme (PIC/S) are also trying to harmonize their inspection practices in terms of inspecting computer systems validation. The project started in 1997 and is still ongoing. Participants include the well-known regulatory authorities—the FDA, the MCA and the TGA. Nothing is published yet, but work has progressed to the point where the only outstanding issue to be agreed upon is the appropriate regulatory expectation concerning the use of electronic records and signatures. The practical issues of 21 CFR 11 raise much debate, as described in Chapter 16, and the PIC/S regulators will have to try and reconcile the lesser requirements of some regulatory authorities compared to the FDA.

Chapter 4 introduced the aim to produce a "Body of Knowledge" collating summaries of all major industry guides for computer validation as a single reference source explaining how industry guidance fitted together. Version 3 of the GAMP Guide (GAMP 3) has already started this process by including works within its content from other industry bodies, notably Arbeitsgemeinschaft fur Pharmazeutische Verfahren Stechnik (APV), Gesellschaft Meβ- und Automatisierungstechnik/Normenarbeitsgemeinschaft für meβ-und Regelungstechnik (GMA/NAMUR) and PhRMA. The UK GAMP Forum is now planning a new revised edition of the GAMP Guide. GAMP 4 is planned for publication in 2001. It will include new guide procedures for validation reporting, archiving and decommissioning, the use of electronic records and signatures, IT infrastructure and postal audits. Additional good practice examples for different types of systems are also planned for inclusion; some of these are expected to have been provided by other industry groups. By bringing the main guidance together, the UK GAMP Forum hopes to acknowledge potential discrepancies between different guides whilst emphasizing a broad consensus.

THE FINAL ANALYSIS

In the final analysis, pharmaceutical manufacturers have no choice but to validate their computer systems; otherwise their license to market a drug will be revoked or not issued in the first place. The cost of validation need not be excessively high if the exercise is focused, its success being dependent on three main factors:

1. The extent and scope of validation needs to be easily determined using simple and well-understood rules. The GAMP Guide's validation strategies for different categories of software is a start but needs to be developed further.
2. Validation life cycles and associated deliverable documentation needs to be scaleable to take due account of the size, complexity and criticality of a computer system.
3. Validation needs to examine the balance of work conducted by pharmaceutical manufacturers and their suppliers. More leverage needs to be made with the development work done by suppliers, so that less supplementary work is required by pharmaceutical manufacturers.

Validation should be commensurate with the potential impact of the computer system on GxP. If there is no potential impact on GxP, then validation is not required. If there is a potential impact, then the whole system should be validated, with particular attention on the GxP aspects of the system's functionality.

Abridged validation life cycles must be facilitated to avoid needless overkill on small and/or simple projects. Validation already has a reputation for great mounds of paper; its time to change perceptions and work smarter.

Pharmaceutical manufacturers are not IT experts; they are users of the technology. Suppliers should be selected who are capable and competent. Greater emphasis and use of the supplier's own Quality Management Systems (QMSs) and documentation needs to be made. The original intent of the GAMP Guide was very much along these lines. Future editions of the GAMP Guide are likely to reemphasize this theme and build upon it further.

A flexible approach to validation must be maintained so that the most effective validation practice is adopted for the characteristics of the system in question, within the business and technical constraints that may exist. Constraints must not be accepted without a challenge. Constraints may be mitigated or managed by considering alternative manufacturing processes, manufacturing equipment, computer technology and even the replacement of a computer system by a manual system. It is always worth considering the consequences of taking no action. New systems are not necessarily better systems. Unless a new system or upgrade is widely used, the number and severity of any bugs is unknown and will not be discovered until it is used. Meanwhile, operational experience with an existing or widely used system should have uncovered most of the operational problems. If measures can be or have been introduced to control or contain these known problems to an acceptable level, then it may be safer not to implement the new system or upgrade.

The validation life-cycle approach has proven itself as an effective method of ensuring the quality, safety and efficacy of manufactured drug products. It should be no more than common sense good practice, and it should deliver tangible business benefits. Pharmaceutical manufacturers must not, however, confuse current industry practice with good practice. Standards are improving, and drug manufacturers must get used to the increasing expectations of regulators. It is also important to realize that this does not necessarily mean more paperwork and bureaucracy. Validation is often misunderstood in this respect. GxP regulatory authorities are not enforcing validation for its own sake but to protect the end users of drug products. Validation cannot be ignored; it is here to stay.

REFERENCES

1. Trill, A. J. 1996. *MCA View on the Presented Initiatives.* Presentation at the ISPE/PDA Joint Conference on Computer-Related System Validation, 2–3 May in Basel, Switzerland.

2. Wingate, G. A. S. 1997. *Validating Automated Manufacturing and Laboratory Applications: Putting Principles into Practice.* Buffalo Grove, Ill., USA: Interpharm Press, Inc.

3. *The Gold Sheet.* 1996. Vol. 30, No. 7. Chevy Chase, Md., USA: F-D-C Reports Inc.

4. Trill, A. J. 1996. *Regulations and Computer Systems Validation in Context.* Presentation at a Management Forum Seminar: Computer Systems Validation: A Practical Approach, 6–7 November in London.

5. Dillinger, A., and H. Hambloch. 1996. *R/3 Validation at Pharmaceutical Companies.* Process Industry Workshop at SAPPHIRE '96, 12–14 June in Vienna, Austria.

6. Trill, A. J. 1999. *A Current MCA View of Computerized Systems Used in Good Practice Applications.* Presentation at Achieving Cost-Effective Computer Systems Validation for cGMP in Pharmaceuticals—Business Intelligence, 27–28 April in London.

7. *The Gold Sheet.* 1999. Vol. 33, No. 7. Chevy Chase, Md., USA: F-D-C Reports Inc.

8. *Inspection of Computer Systems.* Presentation at Pharmaceutical Inspection Cooperation Scheme Seminar, 18–20 September 1996 in Sydney, Australia.

Appendix A

Contributors' Biographies

GUY WINGATE, EDITOR
Computer Compliance Manager, Glaxo Wellcome, United Kingdom

Dr. Wingate is responsible for the computer validation of business systems, process control systems and laboratory systems within Glaxo Wellcome's UK secondary manufacturing business. Until recently, Dr. Wingate was validation manager at ICI Eutech. His personal validation experience covers more than 100 projects dealing with process control systems, laboratory systems, MRP II and document management systems. He is a Chartered Engineer and a member of the ISPE, the PDA (Parenteral Drug Association), and the IEE. Dr. Wingate has published papers widely on the subjects of dependability and validation of computer systems in journals and books and regularly chairs and speaks at validation conferences in the UK and Europe. He is author/editor of *Validating Automated Manufacturing and Laboratory Applications: Putting Principles into Practice,* published by Interpharm Press. Dr. Wingate was the founding chairman of the Supplier Forum for providers of computer systems, software and associated services to the pharmaceutical industry and now chairs the UK GAMP Forum's Industry Board.

Address:
Glaxo Wellcome Operations
Harmire Road
Barnard Castle
County Durham, England
Tel: +44-1833-693106
E-mail: gw69325@glaxowellcome.co.uk

ROGER DEAN
System Support Manager, Pfizer Limited, United Kingdom

Mr. Dean is currently the system support manager for various computer systems in the Manufacturing Division at Pfizer Limited. He is a chemist by profession, and after 27 years in quality operations, moved into the information technology (IT) field where he has been since 1994. In this role, he is responsible for coordinating desktop operations and managing various projects, including the implementation of the Manufacturing Division's Electronic Document Management System (EDMS). A large part of his involvement in the implementation of the EDMS has been in ensuring that it is a validated, GMP–compliant system and that it remains in a validated state, which includes planning the validation and preparing and running the protocols.

Address:
Pfizer Limited
Ramsgate Road
Sandwich, Kent, CT13 9NJ, England
Tel: +44-1304-646770
E-mail: deanr@pfizer.com

ANTON DILLINGER
Development Manager, SAP, Germany

Mr. Dillinger has studied economics and computer science at Karlsruhe and has more than 20 years' experience in software development and quality management. Mr. Dillinger was responsible for developing software for quality management. In 1993, he became responsible for the Quality Management System (QMS) within logistics development. In 1996, he became the overall responsible person for quality management of the entire development and developed SAP's qualification strategy. Since 1998, Mr. Dillinger has been responsible for the internal qualification progam in SAP Development.

Address:
SAP AG
Neurottstrasse 16
69190 Walldorf, Germany
Tel: 06227-34-43-14
Fax: 06227-34-43-74
E-mail: anton.dillinger@sap-ag.de

PAUL N. D'ERAMO
Executive Director, Worldwide Policy and Compliance Management, Johnson & Johnson, United States

Mr. D'Eramo is responsible for establishing worldwide quality and compliance policies; he proactively represents Johnson & Johnson with industry trade associations and regulatory authorities on the development of new regulations and guidelines. Prior to joining Johnson & Johnson, Mr. D'Eramo worked for the U.S. Food and Drug Administration (FDA) for 20 years, most recently as the manager of Technical Operations for the Mid-Atlantic Region. He is a member of the Good Automated Manufacturing Practice (GAMP) Forum Steering Committee and is on the International Society for Pharmaceutical Engineering (ISPE) Board of Directors and chairs the Technical Documents Steering Committee. He was awarded the ISPE Professional Achievement Award and Vice President Al Gore's "Hammer Award" in 1997.

Address:
Johnson & Johnson
Quality & Compliance Services
410 George Street
New Brunswick, NJ 08901-2021
Tel: +1-732-524-1140
Fax: +1-732-524-6838
E-mail: pderamo@corus.jnj.com

CHRISTOPHER EVANS
Lead Consultant for Computer Validation, Eutech, England

Mr. Evans is a lead consultant in computer systems and medical device validation. He has 27 years' experience with Supervisory Control and Data Acquisition Systems (SCADA) and Distributed Control Systems (DCSs) within the chemical and pharmaceutical industries. During the last 7 years, Mr. Evans has had specific responsibilities for ensuring the GMP compliance of these systems to U.S. Food and Drug Administration (FDA) and UK Medicines

Control Agency (MCA) requirements. He is the current chairman of the UK Supplier Forum, which is a body supported by the UK government to encourage a wider understanding of computer system validation requirements in suppliers to the pharmaceutical industry. Mr. Evans regularly makes presentations on the subject of computer validation and was a contributor to the *Validating Automated manufacturing and Laboratory Applications: Putting Principles into Practice,* published by Interpharm Press. Since completing Chapter 6, Mr. Evans has joined Glaxo Wellcome as the operations manager for computer validation for international product supply and is based at their Barnard Castle, UK, facility.*

Address:
Eutech Engineering
Belasis Hall Industrial Estate
Billingham, Cleveland, TS234YS, England
Tel: 01642-372000
E-mail: Chris.Evans@Eutech.com

HOWARD GARSTON SMITH
Software QA Auditor, Pfizer Limited, England

Mr. Garston Smith held positions at Upjohn Limited and Glaxo Laboratories before joining Pfizer in 1982. Here he gained experience in BASIC (Beginner's All-Purpose Symbolic Instruction Code) and FORTRAN (Formula Translator) programming, writing code associated with laboratory automation and information management systems. He has since been appointed software quality assurance auditor for the European Region of Corporate Information Technology with Pfizer. A renowned lecturer and member of the UK GAMP Forum, he is a chartered chemist and a qualified person under the European Union Directives' Medicines Regulations.

Address:
Pfizer Limited
Ramsgate Road
Sandwich, Kent, CT13 9NJ, England
Tel: +44-1304-646770
E-mail: garsth@pfizer.com

*Christopher Evans can be reached at Glaxo Wellcome as follows: Glaxo Wellcome, Harmire Road, Barnard Castle, County Durham, DL12 8DT. His telephone number is 01833 692955; his E-mail address is CE58727@glaxowellcome.co.uk.

HEINRICH HAMBLOCH
GITP, Germany

Dr. Hambloch is director of Good Information Technology Practices in Krefeld, Germany, and consultant for the validation of computer systems. Previously, he worked for Digital Equipment GmbH in Frankfurt in the European Application Center for Technology for the Chemical and Pharmaceutical Industry. Dr. Hambloch has lectured at and chaired numerous seminars in the field of IT validation. He is chairman of the specialist group "Information Technology" of the International Association for Pharmaceutical Technology (APV, Arbeitsgemeinschaft fur Pharmazeutische Verfahrenstechnik).

Address:
GITP
Immenhofweg 23
D-47803 Kredfeld, Germany
Tel: +49-2151-538210
Fax: +49-2151-538211
E-mail: dr_heinrich_hambloch@csi.com

NORMAN HARRIS
20CC, England

Mr. Harris's career has been concerned with the impact of computers on the design and operation of process plants and projects. He has been employed in senior positions with ICI, Davy, BNFL, ComputerVision and Intergraph. He has worked in Malaysia, Japan, Holland, Norway and the United States. From 1960 to 1972 he worked for ICI. Through the rest of the 1970s, his work was with the international contractor Davy. This experience covered all aspects of the application of computers to the design and management of major engineering projects in many countries. In the early 1980s, he joined BNFL as head of the Engineering Computer Applications Department. Subsequently, he was marketing manager for the construction industries for ComputerVision Europe, spending nearly two years as both country and project manager of the Offshore Support Group based in Norway. He was the process industry coordinator at Intergraph's European headquarters. Currently, he provides consultancy on business-driven strategies for information management to vendors and blue-chip companies. Mr. Harris is a Fellow of the Institute of Management and of the Institution of Mechanical Engineers (IMechE) and is chairman of the Information Management Committee for the Process Division of IMechE.

Address:
20CC Ltd
P.O. Box 674
Croydon, CR9 5DA, England
Tel: +44 (0)20 8680 3511
Fax: +44 (0)20 8686 9691
E-mail: 20CC@compuserve.com

KEES DE JONG
ITQA Manager, Solvay Pharmaceuticals, The Netherlands

Mr. de Jong is currently ITQA manager and is responsible for the quality system of the IT department of Solvay Pharmaceuticals in The Netherlands. He started his career in IT in 1978 as system specialist for Honeywell Bull, and joined Solvay in 1985 as a system analyst for the R&D (research and development) automation center, especially within the Good Laboratory Practice (GLP) environment. He was involved in the translation of GxP guidelines to the IT world, in system development as well as system management. He is also member of Solvay's Computer Validation Committee.

Address:
Solvay Pharmaceuticals
P.O. Box 900
1380 DA Weesp, The Netherlands
Tel: +31-294-477515
Fax: +31-294-480727
E-mail: kees.dejong@solvay.com

ROBERT KAMPFMANN
Quality Manager, SAP, Germany

Mr. Kampfmann graduated in computer science in 1986 and started working in software development at Software AG, Darmstadt, Germany. After five years, he built up a quality assurance (QA) team and was responsible for quality management of the entire product line covering middleware and data warehouse solutions. Mr. Kampfmann joined SAP AG in 1997 and was responsible for managing Supplier Audits. Today he is quality manager of SAP's Business Information Warehouse solution.

Address:
SAP AG
Neurottstrasse 16
69190 Walldorf, Germany
Tel: 06227-34-43-14
E-mail: robert.kaufmann@sap-ag.de

ERNA KOELMAN
Global GLP/GCP QA Director, Solvay Pharmaceuticals, The Netherlands

Ms. Koelman is currently global GLP/GCP (Good Clinical Practice) QA director within Solvay Pharmaceuticals. Previously, she was head of the GLP Quality Assurance Unit of Solvay Pharmaceuticals. She has been the leader of the project team that set up the local computer validation policy of Solvay Pharmaceuticals B.V. and also leader of the team responsible for implementing this policy in the company over a period of two years, before it was handed over to a permanent Computer Validation Committee. She has been involved in a number of validation projects of different types of systems, ranging from simple equipment to complex custom-built or highly customized applications.

Address:
Solvay Pharmaceuticals
P.O. Box 900
1380 DA Weesp, The Netherlands
Tel: +31-294-479500
Fax: +31-294-415256
E-mail: erna.koelman@solvay.com

RICHARD MITCHELL
Validation Manager, Pfizer Limited, United Kingdom

Mr. Mitchell is the validation manager for the bulk pharmaceutical manufacturing plants in the Manufacturing Division of Pfizer Limited, Sandwich, UK. He is involved in all aspects of the validation of bulk pharmaceutical chemical manufacture. A microbiologist by training, he joined Pfizer Limited in 1987 as a fermentation technologist in the fermentation development laboratory. He managed the laboratory for four years before moving into validation in 1993. He became manager of the group in 1996. Mr. Mitchell has gained practical experience in the validation of information management systems, including EDMSs. Since completing Chapter 10, Mr. Mitchell has joined Glaxo Wellcome as the site validation manager for International Active Supply at their Dartford, UK, facility.**

**Richard Mitchell can be reached at Glaxo Wellcome as follows: Glaxo Wellcome, Temple Hill, Dartford, Kent, DA1 5AH. His telephone number is 01322 39 1760; his E-mail address is RJM58511@glaxowellcome.co.uk.

Address:
Pfizer Limited
Ramsgate Road
Sandwich, Kent, CT13 9NJ, England
Tel: +44-1304-646770

GERT MOELGAARD
Director, Manufacturing Automation, Novo Nordisk Engineering, Denmark

Mr. Moelgaard is a member of International Society for Pharmaceutical Engineering (ISPE) International Board of Directors and the UK GAMP Forum Steering Committee. He is director of automation engineering at Novo Nordisk Engineering A/S, being responsible for the automation of pharmaceutical and biotech production facilities as well as computer integrated manufacturing. After joining Novo Nordisk in 1982, he has been engaged in several major automation projects and Manufacturing Execution System (MES) projects within pharmaceutical production. He has been involved in validation activities since 1987 and with quality management methods on computer systems. He has been involved in planning and presentations at several international conferences on computer validation.

Address:
Novo Nordisk Engineering A/S
Krogshoejvey 55
2880 Bagsvaerd, Denmark
Tel: +45-4444-7777
Fax: +45-4444-3777
E-mail: gtm@novo.dk

KEES PIKET
QA Manager, Solvay Pharmaceuticals, The Netherlands

Mr. Piket is currently QA manager of good automation practices and responsible for the validation of all computerized systems used in GxP–related areas at Solvay Pharmaceuticals. He is also chairman of the Computer Validation Committee, which is responsible for the computer validation policy within Solvay Pharmaceuticals. Since joining Solvay Pharmaceuticals over 10 years ago, he worked on several assignments within manufacturing and quality assurance. He has experience with the implementation and validation of process control systems, EDMSs, Laboratory Information Management Systems (LIMSs), Enterprise Resource Planning (ERP) systems and analytical systems.

Address:
Solvay Pharmaceuticals B.V.
P.O. Box 900
1380 DA Weesp, The Netherlands
Tel: +31(0)294-479244
Fax: +31(0)294-418931
E-mail: kees.piket@solvay.com

BRIAN RAVENSCROFT
Engineering Quality Manager, AstraZeneca, England

Mr. Ravenscroft has worked in pharmaceutical R&D for more than 25 years. During that period, he has carried out a number of management roles in projects and maintenance operations. He is currently responsible for promoting and maintaining quality standards in the engineering environment through continuous improvement in order to ensure current regulatory requirements are met and appropriate internal standards complied with. He has recently been leading a project to develop a QMS and address the change management and information system needs necessary to implement the resultant systems and processes.

Address:
AstraZeneca
Charnwood R&D
Bakewell Rd.
Loughborough, Leicestershire, LE11 5RH, England
Tel: +44 0 1509 64 4388
Fax: +44 0 1509 64 5579
Email: brian.ravenscroft@charnwood.gb.astra.com

CHRIS REID
Integrity Solutions Limited, Cleveland, England

Mr. Reid formed Integrity Solutions Limited in 1999 in order to provide validation and quality management consultancy to the pharmaceutical industry. Mr. Reid was formerly manager of process and control systems validation for Eutech, where he was responsible for developing validation products and managing a team of validation consultants and practitioners. Mr. Reid has provided validation consultancy to senior management of a number of leading UK and European pharmaceutical companies. Prior to entering the field of validation, Mr. Reid had 8 years of software development and product management experience. Mr. Reid has extensive experience in the development and application of QMSs based on life-cycle development, operation and maintenance of computer systems within the chemical and pharmaceutical industries.

Address:
Integrity Solutions Limited
P.O. Box 71
Middlesbrough
Cleveland, TS7 0XY, England
Tel: +44-1642-320233
Fax: +44-1642-320233
Email: creid@integrity-solutions.freeserve.co.uk

TONY RICHARDS
Engineering Operations Manager, AstraZeneca R&D, Charnwood, England

Mr. Richards joined AstraZeneca, a pharmaceutical R&D facility, in 1994. At this time the Engineering Department was engaged in a major change program driven by the Engineering Quality Project. Major facets of the change program included a commitment to customer service through the introduction of multidiscipline teams, assessment centers, a teamwork training program, reliability-centered maintenance (RCM), a Maintenance Management System, electronic maintenance documentation and outsourcing maintenance. Previously, Mr. Richards worked in the manufacturing and nuclear industry.

Address:
AstraZeneca
R&D Charnwood
Engineering Dept.
Bakewell Rd.
Loughborough, Leicestershire, LE11 5RH, England
Tel: +44 1509 644420
Fax: +44 (0) 1509 645579
E-mail: tony.richards@charnwood.gb.astra.com

OWEN SALVAGE
Information Systems Business Manager, CSR, Australia

Mr. Salvage is currently the information systems business manager with CSR of Australia and is leading the company's ERP implementation with management and technical responsibility for software development and hardware infrastructure. Having spent 10 years with ICI and Zeneca, installing pharmaceutical process control and information systems, Mr. Salvage has in recent years provided project management and advice on matters relating to computer systems validation in general and particularly ERP systems validation. Mr. Salvage is a Chartered Engineer and a member of the IEEE (Institute for Electrical and Electronic Engineers) and is a member of the Australasian Production and Inventory Control Society (APICS).

Address:
CSR
9 Help Street
Chatswood
New South Wales 2067, Australia
Tel: ++612-93725495
Fax: ++612-93725476
E-mail: osalvage@csr.com.au

DAVID SELBY
Selby-Hope International, England

Until recently, Mr. Selby was general site manager for Glaxo Wellcome at the Barnard Castle, UK, facility. He has spent more than 30 years in the pharmaceutical industry, including responsibilities in R&D, international GMP (Good Manufacturing Practice) compliance, auditing, production, QA and IT. Mr. Selby was the founding chairman of the UK GAMP Forum, which has produced the internationally accepted *Supplier's Guide for Validation of Automated Systems in Pharmaceutical Manufacture* (GAMP Guide). He has published a number of papers on validation and the application of GMP to computerized and automated control systems.

Address:
Selby-Hope International
30 East Road
Melsonby
Richmond, North Yorkshire, D10 5NF, England
Tel: +44-1325-71-88-44
Fax: +44-1325-71-88-45
E-mail: selby.hope@btinternet.com

NICOLA SIGNORILE
Site IT Manager, Hoechst Marion Roussel (Gruppo Lepetit), Italy

Mr. Signorile is HMR's site IT manager in southern Italy, which manufactures secondary pharmaceuticals and is subject to FDA inspections. Mr. Signorile spent 3 years as a consultant dealing with information systems and ERP/

MRP II (Manufacturing Resource Planning) before joining HMR's IT function 6 years ago. Previously, he spent 4 years developing control software on data network communication systems for NATO (North Atlantic Treaty Organization) and as a network systems integrator for a commercial aerospace company.

Address:
Gruppo Lepetit S.p.A.
03012 Anagni (FR)
Localita Valcanello, Italy
Tel: +39-775-760309
Fax: +39-775-760224
E-mail: NicolaRosario.Signorile@HMRAG.com

MELANIE SNELHAM
Lead Consultant in Laboratory Validation, Eutech, England

Ms. Snelham is the lead consultant for laboratory validation within Eutech. Previous experience includes validation within GLP and GMP arenas. Before joining Eutech, she was head of quality assurance for Ciba Pharmaceuticals. During her career at Ciba and at Medeva Pharmaceuticals, she gained experience in the validation of LIMS, Toxicology, MRP II and other computerized systems and led regulatory inspections from the UK, U.S. and Japanese authorities.

Address:
Eutech
Brunner House
P.O. Box 43
Northwich, Cheshire, CW8 4FN, England
Tel: 01606-705248
Fax: 01606-704321
E-mail: melanie.snelham@eutech.com

DAVID STOKES
Account Manager, Motherwell Information Systems, England

Mr. Stokes has worked in the automation/IT industry for over 20 years and has had wide ranging responsibility across a number of disciplines. After completing a traditional apprenticeship, he worked with Leeds and Northrup as a control systems design engineer with additional responsibility for training, installation and commissioning. Over 15 years with the company, he played a lead business and technical role in implementing a number of pharmaceutical projects and was an active member of the SP88 committee on batch control. This has lead to a greater involvement in the IT side of the industry, leading to his current position as account manager with Motherwell Information Systems. He is also a member of the steering committee for the pharmaceutical industry's Supplier Forum.

Address:
Motherwell Information Systems
Chapter House
St. Catherines Court
Hylton Riverside
Sunderland, SR5 3XJ, England
Tel: +44 (0)191 516 3000
Fax: +44 (0)191 516 3020
E-mail: dstokes@mother.co.uk

TORSTEN WICHMANN
Quality Director of Development, SAP AG, Germany

Mr. Wichmann has about 5 years' experience focusing on quality issues in the environment of software development and computerized systems. At SAP headquarters in Walldorf, Germany, Mr. Wichmann was in charge of the administration of the QMS for development and services as well as for the coordination of its continuous revision and improvement before he was appointed quality manager for the PI development group. Since July 1998, Mr. Wichmann has headed the central quality management of SAP development.

Address:
SAP AG
Neurottstrasse 16
69190 Walldorf, Germany
Tel: 06227-34-43-14
E-mail: torsten.wichmann@sap-ag.de

PETER WILKS
Quality & Compliance Manager, Glaxo Wellcome, United Kingdom

Peter Wilks is the quality and compliance manager for Glaxo Wellcome in the worldwide information systems and technology (WISAT) UK operations group. He previously had a corporate computer validation role Rhone-Poulenc Rorer. Mr. Wilks has over 20 years' computing experience, with the last 6 years exclusively in the pharmaceutical industry. He is currently leading a special interest group within the UK GAMP Forum on infrastructure validation, with the objective of providing further guidance for pharmaceutical companies. Mr. Wilks is also a member of the Institute of Quality Assurance and the British Computer Society.

Address:
WISAT
Glaxo Wellcome
Stockley Park West
Uxbridge, Middlesex, UB11 1BT, England
Tel: +44 (0) 1438 76 2242
Fax: +44 (0) 1438 76 2303
E-mail: pw71713@glaxowellcome.co.uk

Appendix B

Abbreviations and Acronyms

10Base2 (10 Ms)	Ethernet running on RG-58 coaxial cable at a maximum distance of 185 m per segment
10Base5 (10 Ms)	Ethernet running on thicknet coaxial at a maximum distance of 500 m per segment
10BaseFL (10 Ms)	Ethernet running on fiber optic 62.5 μm cabling at a maximum distance of 2 km per segment
10BaseT (10 Ms)	Ethernet running on UTP at a maximum distance of 100 m per segment
ACL	access control list
ACS	Application Configuration Specification
AGV	automated guided vehicle
ALGOL	Algorithmic-Oriented Language
APICS	Australasian Production and Inventory Control Society
APV	Arbeitsgemeinschaft fur Pharmazeutische Verfahrenstechnik (Germany)
AQAP	Association of Quality Assurance Professionals

ASCII	American Standard Code for Information Interchange
AUI	Application User Interface
BASIC	Beginner's All-Purpose Symbolic Instruction Code
BMS	Building Management System
BNC	Bayonet Neil Concelman
BPC	bulk pharmaceutical chemical
BPR	Business Process Reengineering
bps	bits per second
BS	British Standard
CAD	Computer-Aided Design
CAE	Computer-Aided Engineering
CASE	Computer-aided software engineering
CD	compact disk
CD-ROM	Compact disk, read-only memory
CD-RW	rewritable compact disk
CFR	U.S. Code of Federal Regulation
CIM	Computer Integrated Manufacturing
COBOL	Common Business-Oriented Language
COM	component object model
COTS	commercial off-the-shelf
CPU	Central Processing Unit
CRC	cross-redundancy check
CRP	conference room pilot
CSV	Computer System Validation
DAT	digital audio tape
DBA	database administrator
DBMS	Database Management System
D-COM	distributed component object model
DCS	Distributed Control System
DDE	dynamic data exchange
DECnet	Digital Equipment Corporation network

DLL	device descriptor language
DLT	digital linear tape
DOS	Disk Operating System
DQ	Design Qualification
EAM	Enterprise Asset Management
EBR	electronic batch record
EBRS	Electronic Batch Recording System
EDI	Electronic Data Interchange
EDMS	Electronic Document Management System
EIA/TIA-568	Commercial Telecommunications Standard
EMI	electromagnetic interference
EOLC	Environmental/Operation Life Cycle
EPROM	electronically programmable read-only memory
ERP	Enterprise Resource Planning
ESD	electrostatic discharge
EU	European Union
FAT	Factory Acceptance Testing
FDA	U.S. Food and Drug Administration
FDDI	Fiber Distributed Data Interface
FDS	Functional Design Specification
FMEA	Failure Modes and Effects Analysis
FORTRAN	Formula Translator
FTP	File Transfer Protocol
GAMP	Good Automated Manufacturing Practice
GCP	Good Clinical Practice
GDP	Good Distribution Practice
GLP	Good Laboratory Practice
GMA	Gesellschaft Meβ- und Automatisierungstechnik (Germany)
GMP	Good Manufacturing Practice
GPP	Good Programming Practice

GUI	Graphical User Interface
GxP	GCP/GDP/GLP/GMP
HDS	Hardware Design Specification
HMI	human machine interface
HPLC	high performance liquid chromatography
HQ	headquarters
HTML	Hyper Text markup Language
HVAC	heating, ventilation and air-conditioning
IBM	International Business Machines
ICT	information and communications technology
IEEE	Institute for Electrical and Electronic Engineers
IIP	Investors in People
I/O	input/output
IPC	in-process control
IPX	Internet Packet Exchange
IQ	Installation Qualification
IS	information systems
ISO	International Organization for Standardization
IT	information technology
KPI	key performance indicator
LAN	local area network
LIMS	Laboratory Information Management System
MAU	media attachment unit
MCA	UK Medicines Control Agency
MCC	motor control center
MES	Manufacturing Execution System
MMI	man machine interface (see also HMI)
MMS	Maintenance Management System
Mps	Megabits per second
MPS	master production schedule

MRA	Mutual Recognition Agreement
MRP	Materials Management System
MRP II	Manufacturing Resource Planning
NAMUR	Normenarbeitsgemeinschaft für Meβ- und Regelungstechnik (Germany)
NOS	network operating system
O&M	operation and maintenance
OCS	Open Control System
OLE	Object Linking and Embedding
OMM	object management mechanism
OQ	Operational Qualification
OS	operating system
OSI	Open System Interconnection
PC	personal computer
PDA	Parenteral Drug Association
PDA	Personal Data Assistant
PDF	Portable Document Format
PhRMA	U.S. Pharmaceutical and Research Manufacturers of America
PIC/S	Pharmaceutical Inspection Convention Scheme
PID	proportional, integral and derivative
PLA	Performance Level Agreement
PLC	Programmable Logic Controller
PMA	Pharmaceutical Manufacturers Association
PQ	Performance Qualification
PRINCE II®	Projects in Controlled Environments
PRM	Process Route Map
QA	Quality Assurance
QC	quality control
QMS	Quality Management System
QS	Quality System

QTS	Quality Tracking System
RAD	Rapid Application Development
RAID	redundant array of inexpensive disks
RAM	random access memory
RCM	reliability-centered maintenance
RDB	relational database
RDBMS	Relational Database Management System
RFI	radio-frequency interference
RTM	Requirements Traceability Matrix
SAP	Systems, Applications and Products in Data Processing (German company)
SAT	Site Acceptance Testing
SCADA	Supervisory Control and Data Acquisition
SCR	Source Code Review
SDLC	System Development Life Cycle
SDS	Software Design Specification
SGML	Standard Generalized Markup Language
SHE	Safety, Health and Environment
SLA	Service Level Agreement
SMDS	Software Module Design Specification
SMTP	Simple Mail Transfer Protocol
SOP	Standard Operating Procedure
SQL	Structured Query Language
STEP	Standard for Exchange of Product Model Data (also known as ISO 10303)
TCP/IP	Transmission Control Protocol/Internet Protocol
TGA	Therapeutic Goods Administration (Australia)
UAT	user acceptance testing
UK	United Kingdom
UPS	uninterruptible power supply
URS	User Requirements Specification

U.S.	United States (of America)
UTP	unshielded twisted pair
UV	ultraviolet
VMP	Validation Master Plan
VMS	Virtual Memory System
WAN	Wide Area Network
WIP	Work in Progress
WORM	write once, read many
WYSIWYG	what you see is what you get
XML	Extensible Markup Language
Y2k	Year 2000

Appendix C
Definitions

Acceptance Testing

Testing, usually conducted by or on behalf of a user, to decide whether or not to accept a system or software from its supplier.

Active Paper on Glass

Paper-based information transferred into an electronic document with properties such as calculations, data verification and quality limits automatically conducted within each electronic document.

Bespoke Code

Software produced for a customer, specifically to order, to meet a set of user requirements. Bespoke code includes so-called standard software, where the version of the software to be used has not been market tested over a period of time by other customers.

Biometric

A method of verifying an individual's identity based on measurement of the individual's physical feature(s) or repeatable action(s), where those features and/or actions are both unique to that individual and measurable.

Black Box Testing
Functional testing that ignores the internal mechanism or structure of a system or component and focuses on the outputs generated in response to selected inputs and execution conditions.

Bridge
OSI data link layer device. Used to connect multiple local area networks (LANs) together.

Business Continuity Plan
See Contingency Plan.

Change Control
A formal system by which qualified representatives of appropriate disciplines review proposed or actual changes that might affect a validated status. The intent is to determine the need for action that would ensure and document that the system is maintained in a validated state.

Client-Server
A term used in a broad sense to describe the relationship between the receiver and the provider of a service (e.g., a networked system where front-end applications, as the client, make service requests upon another networked system).

Closed System
An environment in which system access is controlled by persons who are responsible for the content of electronic records that are on the system.

Communications Software
Software that runs under the control of operating system software. It is responsible for the provision of data communication to other hardware platforms.

Computer System
A group of hardware components and associated software designed and assembled to perform a specific function or group of functions.

Computerized System
A term used to cover a broad range of systems, including automated manufacturing equipment, control systems, automated laboratory systems, Manufacturing Execution Systems (MESs) and computers running laboratory or

manufacturing database systems. The automated system consists of the hardware, software and network components, together with the controlled functions and associated documentation.

Concentrator

See Hub.

Configuration Management

The process of identifying and defining the configuration items in a system, controlling the release and change of these items throughout the system life cycle, recording and reporting the status of configuration items and change requests and verifying the completeness and correctness of configuration items.

Contingency Plan

Definition of the measures to be taken to allow an operation or process to continue as associated computer systems become unavailable. Sometimes known as a Business Continuity Plan or Disaster Recovery Plan.

Corrective Maintenance

Retrospective maintenance to repair or remedy equipment, including computer systems.

Customized System

A standard system which has been uniquely modified to meet the user's needs.

Dead Code

Program logic that cannot execute because the program path does not permit the logic to be reached. Newly developed programs should be reviewed for the presence of dead code. Dead code must be removed prior to compilation and submission for production implementation. In instances where program logic becomes dead code as a result of program modifications, the associated dead code should be removed from the program before recompilation and submission for production implementation. Commented source code is not dead code because it is ignored by the compiler and does not become program logic. Code rendered inaccessible by configuration (e.g., switches, parameters, calls, etc.) is not dead code because this code is intended to be available for use depending on the need of a particular implementation. Similarly, code residing within a standard library, which is not accessed by the calling program, is not considered dead code because this code is intended to be available for use depending on the need of a particular implementation. Code that has been included for the purposes of testing or for later diagnosis during support work, and which can be configured on or off, is not regarded as dead code. If the code is configurable for use in many different projects, each with a different configuration of options, the unused options should not be removed. However, the source code and configuration review and testing processes must demonstrate that the correct options have been correctly de-selected and do not function.

Design Qualification

Formal and systematic verification that the requirements defined during specification are completely covered by the succeeding specification, design and implementation.

Desktop

The end user's workstation and local software environment. Normally provides a Graphical User Interface (GUI) front-end menu that provides users with easy access to required applications.

Disaster Recovery Plan

See Contingency Plan.

Documentation

Any written or pictorial information describing, defining, specifying, reporting or certifying activities, requirements, procedures or results.

Electronic Document

A document created, stored and maintained electronically, usually as a file.

Electronic Record

Any combination of text, graphics, data, audio, pictorial or other information representation in digital form that is created, modified, maintained, archived, retrieved or distributed by a computer system.

Electronic Signature

Based on cryptographic methods of originator authentication, signatures are computed by using a set of rules and parameters such that the identity of the signer and the integrity of the data can be verified. An entry in the form of magnetic impulses or computer data compilation of any symbol or series of symbols executed, adapted or authorized by a person is equivalent to the person's handwritten signature.

Environmental Control Equipment

All equipment used to maintain the environment for hardware within the manufacturer's specified operating limits (e.g., air-conditioners for temperature and humidity).

Environment Power Equipment

All power sources, earthing requirements and power conditioners required by computer hardware. This includes protection and emergency devices such as surge suppressors and uninterruptible power supplies (UPS).

Escrow

An agreement between a customer and a supplier to guarantee the availability of the source code in case of force majeure, e.g., bankruptcy.

Ethernet Baseband

Network specification as defined by the IEEE 802.3 series of standards.

Functional Testing

See Black Box Testing.

Handover

Completion of a project to a point where documentation and responsibility for operation and maintenance is passed from the project team to the support organization.

Hardware Platform

All computing hardware deployed in running business application programs. The definition covers servers, central processing units (CPUs), memory, internal disk drives and peripheral controllers.

Hub

An electronic device that serves as the center of a star topology.

Hybrid System

Automated system that requires additional procedural and physical controls to maintain its integrity and the integrity of its operations.

Information System

Computer system used to implement business processes, provide data management or support administration activities.

Installation Qualification
Establishing that the hardware and software making up a computer system are those specified and installed in accordance with any governing practices, and that supporting documentation including operating procedures are available.

Legacy System
Existing system whose validation does not necessarily meet current compliance requirements.

Local Area Network
Network spread over a local vicinity (up to site-wide application).

Multitasking
Multiple jobs performed via time sharing on a single processor.

Network
A data communication system that links two or more computers and peripheral devices. It comprises the cabling, the network hardware and communications software.

Network Cabling
All physical cables and ancillary nonelectronic hardware involved in carrying network data, up to and including terminators. It includes physical switches, cable patch panels, cable and all items used in the routing of cables.

Network Hardware
All electronic network equipment, including network interface cards, hubs, routers, switches, bridges and other stand-alone network equipment.

Nondisclosure Agreement
Confidentiality agreement not to share information (business or technical) with third-party organizations.

Object Brokering
Management of software objects (specification and testing) for reuse.

Open System
An environment in which system access is not controlled by the persons who are responsible for the content of electronic records that are on the system.

Operational Qualification

Establishing confidence that process equipment and subsystems (including computerized systems) are capable of consistently operating within established limits and tolerances.

Operating System

Software that enables the general operation of the hardware platform. It controls system resources such as central processing unit (CPU) time, disk access or memory and allocates it to other software running on the hardware.

Passive Paper on Glass

Existing Standard Operating Procedures (SOPs) and operational documents such as master production and control documents are directly transferred into electronic documents.

Paperless System

Highly automated system processing information electronically without the need for paper printouts.

Performance Qualification

Establishing confidence that the process (including the computerized system) is effective and reproducible.

Peripheral Equipment

Hardware deployed to extend the capability of the hardware platform, including printers, modems, keyboards, tape drives, screens and scanners.

Platform

The hardware and software that must be present and functioning for an application program to perform as intended. A platform includes, but is not limited to, the operating system or executive software, communication software, microprocessor, network, input/output hardware, any generic software libraries, database management, user interface software and the like.

Port

An interface on a hub (what the devices connect to).

Process Control System

Computer system providing real-time control and/or monitoring of a manufacturing or packaging process.

Process Route Map
Formal definition of manufacturing and packaging process flows.

Prospective Validation
The validation of new or recently installed systems following a life cycle concept.

Preventive Maintenance
Prospective maintenance activities aimed at anticipating breakdowns of equipment, including computer systems.

Quality
The totality of features and characteristics of a product or service that bears on its ability to satisfy given needs.

Quality Assurance
A planned and systematic pattern of all actions necessary to provide adequate confidence that the item or product conforms to established technical requirements.

Quality Control
The regulatory process through which the industry measures actual quality performance, compares it with standards and acts on the difference.

Quality Management System
A formal system of policies and procedures to manage the implementation of quality within a scope of operations.

Retrospective Validation
Establishing documented evidence that a system does what it purports to do based on an analysis of historical information.

Risk Assessment
A comprehensive evaluation of risks and associated impacts.

Router
An OSI network layer device used to connect multiple local or remote local area networks (LANs) together.

Service Level Agreement
A formal agreement, possibly contract, defining the services to be provided by a supplier to a customer.

Source Code Review
An examination of software listing for quality attributes and functional correctness.

Structural Testing
See White Box Testing.

Supplier Audit
A formal examination of a supplier's operational practices and product integrity in order to assess whether they are fit for purpose.

Support Organization
Department or group with responsibility for providing assistance to the organization with prime accountability for a service.

System Specification
A description of how the system will meet its functional requirements.

Switch
OSI data link layer device used to connect multiple local area networks (LANs) together.

Third-Party Software
Software used in conjunction or to enhance the functionality of an application but not produced by the organization supplying the application.

Token Ring
Token passing local area network (LAN) technology developed by IBM (International Business Machines).

Topology
The physical arrangement of devices and media within a network structure.

Transceiver
Electronic device used in a network that is capable of transmitting and receiving data.

User Requirements (Specification)
Describes what the equipment or system is supposed to do and, as such, is normally written by the pharmaceutical manufacturer. This links to Performance Qualification (PQ) which tests the user requirements.

Validation
Establishing documented evidence which provides a high degree of assurance that a specific process will consistently produce a product meeting its predetermined specifications and quality attributes.

Validation Plan
A plan created by the user to define validation activities, responsibilities, procedures and milestones for delivery.

Version Control
Sequential reference allocated to progressive revisions of documentation, software, equipment (including hardware), data and so on.

Warning Letter
Formal letter from the U.S. Food and Drug Administration (FDA) to a pharmaceutical manufacturer warning of significant noncompliance.

Wide Area Network
Network spread outside the local vicinity (linking multiple sites).

White Box Testing
Structural testing that takes into account the internal mechanism of a system or component.

Index

Milton Keynes UK
Ingram Content Group UK Ltd.
UKHW052026071024
449327UK00027B/2440